Series in Laboratory Medicine

Leo P. Cawley, M.D., Series Editor

Exfoliative Cytopathology

Exfoliative Cytopathology

Second Edition

Zuher M. Naib, M.D.

Professor of Pathology, Professor
of Obstetrics and Gynecology,
and Director of the Division
of Cytopathology, Emory University
School of Medicine, Atlanta

Little, Brown and Company   Boston

To the screening cytotechnologist,
whose patient work is saving
innumerable lives

Preface

The purpose of this book, as with the first edition, is to present in the simplest way possible the wealth of specific morphological details that can be seen in the cells exfoliated from different sites. Many suggestions for improvements on the first edition have been incorporated. A number of illustrations were changed, and some new ones were added. Because of the increased importance of cytology in the detection and diagnosis of oral, ocular, joint, and skin lesions, discussions of these subjects were augmented. Four new chapters (Chapters 15–18) have been added. The references at the end of each chapter, although expanded and updated, are again my own selection from the ever-increasing current literature.

I thank the staff of the Cytopathology Division at Grady Memorial Hospital in Atlanta for their help in the preparation of the manuscript. I am especially grateful to Susan Elliott, C.T. (ASCP), my chief technologist, who helped me review the manuscript, and to Melanie DeFrain and Marie Spinks, my secretaries, for their patient assistance. The editors of Little, Brown and Company continue to deserve much credit for their professional assistance. Finally, I owe thanks to my wife, Nelly, for her help, patience, and understanding.

Z. M. N.

Contents

Exfoliative Cytopathology

Introduction

Exfoliative cytopathology is the study of the normal and the disease-altered desquamated cells from various sites (Fig. 1). Because of the need for continuous renewal of the body tissues, cellular exfoliation is an unceasing process. The rate of desquamation varies with each tissue, its function, and metabolic capacities. Some of these desquamated cells accumulate in natural cavities and in recesses. The majority are lost from the surface or through the gastrointestinal tract. There are two types of cellular exfoliation from which samples are studied.

Natural Spontaneous Exfoliation

The physiologically desquamated cells will often show, besides the pathologic changes and the normal changes of natural aging, the results of their separation from confinement in the organized structures. The samples of cells to be studied are usually suspended in fluid and removed by bulb or syringe aspiration. Good examples are the accumulation of exfoliated vagino-cervical cells in the vaginal pool secretion in the posterior vaginal fornix and of mesothelial cells in the effusions of the pleural and abdominal cavities.

Artificial Exfoliation (Surface Biopsy)

Artificial exfoliation occurs when the surface of the mucosa is scraped and viable cells are traumatically exfoliated before their natural time of shedding. According to the type of information desired, the scraping will be energetic, as in the case of cervical carcinoma where the deep basal cells are needed for examination, or very gentle, as in the case of hormonal studies where the cells desired must come from the surface layer of the vaginal mucosa. The artificially exfoliated cells often appear in sheets and are smaller and less mature than the spontaneously desquamated ones.

Cytopathology in Relation to Histopathology

ADVANTAGES

Cytopathologic methods provide a rapid, inexpensive, simple means of diagnosis that can, on occasion, supplement or replace a frozen section or a biopsy.

1

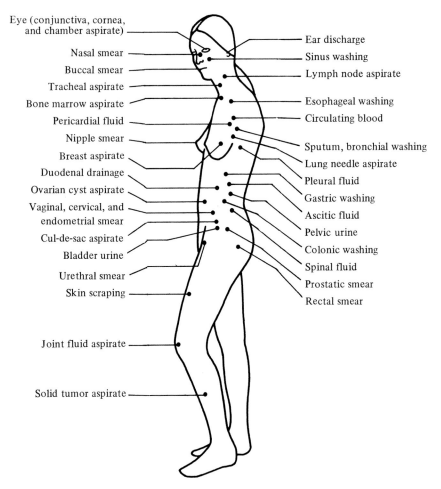

Figure 1
The most common sources of cytologic specimens.

No injury to tissue is produced, and this allows frequent repetition of cellular sampling, which is important in the evaluation of the progression or posttreatment regression of a lesion.

The smears cover a wider surface than that involved in a biopsy.

The cells can be obtained from areas inaccessible to a biopsy, for example, from the bottom of a crypt, the renal pelvis, or the bottom of a diverticulum.

The intimate cellular structures are often more clearly seen in an isolated cell of a smear because of the minimum shrinkage and distortion in such cells. Furthermore, the entire cell, not merely a section from it, is studied in a three-dimensional view by focusing the microscope up and down.

The smears permit a better evaluation of the nature of the inflammations or infections. Fungi and parasites are usually easier to see in a smear. Special stains can always be used, as in tissue sections.

Changes due to irradiation and other forms of therapy are often easier to evaluate.

LIMITATIONS

The interpretation of the morphologic cellular changes is based mainly on individual observation and often cannot be forced into rigid criteria. The diagnostic features are often different from the histopathologic ones.

The cytologic diagnosis is not final; it must be confirmed by histology.

The location of the lesion cannot be pinpointed by cytology. A squamous cancer cell found in a sample of sputum, for example, may have originated anywhere from the buccal mucosa, pharynx, larynx, or bronchi.

The screening of a smear can be time consuming. Often the nature of the lesion is not as obvious as in a histologic section.

The interrelation and arrangement of the cells cannot be established. Neighboring cells in a smear often originate from different parts of an organ.

The relation of the cells to the supporting stroma cannot be determined by cytology, which is important in the diagnosis of an invasive carcinoma as compared with an in situ one.

The size of the lesion cannot be approximated by cytology, since the number of exfoliated cells often has little relation to the size of the lesion.

The type of lesion—e.g., in situ as compared with early invasion, adenocarcinoma or sarcoma—is more difficult to determine by a smear.

The sample of the cells studied may originate from an unwanted site (liver cells in ascitic fluid aspiration, bronchial carcinoma cells in a gastric washing, or rectal cells in the vaginal smear).

The exfoliated cells may not represent the true nature of the entire lesion. Poorly differentiated cancer cells, for example, are often the only cells exfoliating from a neoplasm with mixed components.

Technique

No amount of skill and experience will enable the cytopathologist to render an accurate interpretation from a poorly obtained or fixed cellular sample. He does a disservice to the patient if he accepts such a smear for interpretation. Furthermore, a direct and close relationship and understanding should always exist between the clinician, the histopathologist, and the cytopathologist if the full benefit of this method is to be obtained. Cytopathology should not compete with histopathology but rather should complement it. Any abnormal cell seen in the smear should be explained by study of histologic sections.

The majority of the recommended methods for obtaining satisfactory specimens are described at the beginning of the different chapters, but, based on the most common mistakes in technique seen in our laboratory, the following recommendations are emphasized:

Do not use a lubricant on the instruments or gloves. If moisture is necessary, use warm saline solution.

Do not use talcum powder on the gloves. Eliminate all dust before touching the patient, instruments, or slides.

Do not permit the cells to dry while trying to spread them perfectly on the slide. A poorly distributed thick smear is only a slight handicap for the diagnosis and is preferable to a smear that has been well spread but dried and then fixed.

Do not perform the biopsy before taking the smear. The presence of excess blood will render interpretation of the smear difficult by masking or diluting the epithelial cells.

Do not allow the patient to douche or use an intravaginal medication or contraceptive substance for at least 24 hours before the examination. If she has done so, postpone the cellular sampling unless a visible lesion that can be scraped is present.

A history of irradiation or of previous biopsy or conization, with an indication of the time lapse, should always be given because of the eventual diagnostic pitfalls presented by the regenerating cells.

The smears should be taken before swabbing the cervix with acetic acid as used during colposcopy. A serious effort should always be made to obtain a sample of the endocervical canal mucus plug.

SLIDE EXAMINATION

The examination of a slide should be systematic from one end to the other, in a series of overlapping horizontal sweeps, preferably with the use of a mechanical stage holder on the microscope. One should not attempt to follow the cellular streaks trapped in the mucus. Too many important clues needed for the diagnosis can be missed in this way.

To begin with, the observer should, with a low-power magnification (about $\times 50$), get a general impression of the quality of the smear, the possible presence of tissue structures, and the location of the best cells to be examined. Under higher magnification (about $\times 150$) the actual screening of the smear can begin by moving the slide horizontally from one end to the other. The oil immersion examination of a cell is permitted only when an answer to a specific inquiry concerning a definite cellular structure is demanded; for example, does the nuclear membrane touch the cytoplasmic border? The continuous use of high magnification for screening will produce, besides eyestrain, an increased number of false-positive interpretations. Nothing looks more malignant than a lymphocyte seen through an oil immersion objective ($\times 1250$).

The significant cells found in the slides should be marked with a small ink dot, preferably with green semitransparent ink, so that the dot will not completely obscure any ink-covered cells. To call the attention of the pathologist to an especially diagnostic structure, a red line drawn with a wax pencil can be added on the cover slip. If more characteristic cells are found later, the screener should make an effort to remove from the slide, if need be, the earlier markings placed on the first but less diagnostically significant cells encountered. It is as bad for a screener to place too many dots on a slide as not enough.

Poor staining may be detected by the examination of polymorphonuclear leukocytes, the nuclei of which should normally be dark blue, and the cytoplasm translucent blue. Greenish-brown staining of the cytoplasm often indicates the eosin stain has deteriorated and needs to be changed.

References and Supplementary Reading

Andrews, G. S. *Exfoliative Cytology*. Springfield, Ill.: Thomas, 1971.

Bamforth, J. *Cytological Diagnosis in Medical Practice*. Boston: Little, Brown, 1966.

Bosvhaun, H. W. *Praktische Zytologic*. Berlin: Walter de Gruyter, 1960.

Carvalho, G. *Vaginal Cytology*. New York: Vantage Press, 1968.

Frost, J. K. Clinical Cytology. In E. R. Novak and G. E. S. Jones (Eds.), *Novak's Textbook of Gynecology* (6th ed.). Baltimore: Williams & Wilkins, 1961.

Frost, J. K. *Concepts to General Cytopathology* (2d ed.). Baltimore: Johns Hopkins Press, 1965.

Frost, J. K. Gynecologic and Obstetric Cytopathology. In E. R. Novak and J. D. Woodruff (Eds.), *Novak's Gynecologic and Obstetric Pathology* (6th ed.). Philadelphia: Saunders, 1967.

Frost, J. K. *The Cell in Health and Disease*. Baltimore: Williams & Wilkins, 1969.

Graham, R. M. *The Cytologic Diagnosis of Cancer* (3d ed.). Philadelphia: Saunders, 1972.

Grunze, H. *Klinische Zytologie der Thorax-Krankheiten*. Stuttgart: Ferdinand Euke, 1955.

Hopman, C. B. *Clinical Cytology and Cytologic Research*. Miami: Miami Post Publishing Co., 1960.

Kaston, C. S., and Bamforth, S. B. *Atlas of in Situ Cytology*. Boston: Little, Brown, 1962.

Koss, L. G. *Diagnostic Cytology and Its Histopathologic Bases* (2d ed.). Philadelphia: Lippincott, 1968.

Moore, K. L. *The Sex Chromatin*. Philadelphia: Saunders, 1966.

Mouriquand, J. *Cytodiagnostic mammaire*. Basel: F. Hoffman-LaRoche et Cie, 1962.

Nieburgs, H. E. *Diagnostic Cell Pathology in Tissue and Smears*. New York: Grune & Stratton, 1967.

Papanicolaou, G. N. *Atlas of Exfoliative Cytology*. Cambridge: Harvard University Press, 1954.

Patten, S. F., Jr. *Diagnostic Cytology of the Uterine Cervix*. Basel: Karger, 1969.

Prolla, J. C., and Kirsner, J. B. *Handbook and Atlas of Gastrointestinal Exfoliative Cytology*. Chicago: University of Chicago Press, 1972.

Pundel, J. P. *Les frottis vaginaux endocriniens*. Paris: Masson et Cie, 1952.

Pundel, J. P. *Diagnostic cytologique du cancer génital chez la femme*. Paris: Masson et Cie, 1954.

Pundel, J. P. *Précis de colpocytologie hormonale*. Paris: Masson et Cie, 1966.

Reagan, J., and Ng, A. *The Cells of Uterine Adenocarcinoma*. Baltimore: Williams and Wilkins, 1965.

Riotton, G., and Christopherson, W. M. *Cytology of the Female Genital Tract*. Geneva: World Health Organization, 1973.

Schade, R. O. K. *Gastric Cytology*. London: Arnold, 1960.

Smolka, H., and Soost, H. J. *An Outline and Atlas of Gynaecological Cyto-diagnosis* (2d ed.). Baltimore: Williams & Wilkins, 1965.

Takahashi, M. *Color Atlas of Cancer Cytology*. Philadelphia: Lippincott, 1971.

Taylor, R. C. W. *Practical Cytology*. New York: Academic, 1967.

Tweeddale, D., and Dubilier, L. *Cytopathology of Female Genital Tract Neoplasms*. Chicago: Year Book, 1972.

Wachtel, E. *Exfoliative Cytology in Gynaecological Practice*. London: Butterworth, 1964.

Tissue and Basic Cellular Structures 1

General Cellular Morphology

Cells are the essential units of the structure of all living organisms. Their structure varies with their function, but a large number of their components are common to all. The nucleus, needed for reproduction, is surrounded by a cytoplasm supplying most of the life-sustaining functions. A cell can survive but not multiply without a nucleus (red blood cell), but it cannot survive without cytoplasm (stripped nucleus). The cells contain multiple structures, some living (organelles), others without life (inclusions), each having different specific functions to perform. If we are to examine these different components with the light microscope, the cells must be stained. As illustrated in the diagram of a ciliated cell (Fig. 2), the most important components are as follows.

CELL MEMBRANE (PLASMALEMMA)

The cytoplasmic membrane is a semipermeable membrane made of a complex union of protein and lipid molecules that allows a selective two-way traffic of substances needed by the cell for growth and multiplication or rejected as metabolic waste. It also holds adjacent cells together and probably plays a role in intercellular communication.

CYTOPLASMIC MATRIX

The consistency of the cytoplasmic matrix (cell sap or hyaloplasm) varies from liquid to a firm jellylike viscosity. It fills the space between the various organelles. Usually finely granular, it is composed mainly of water that contains (in solution or suspension) inorganic and organic ions of varying molecular sizes, anabolic and catabolic compounds, nuclear proteins, RNA (riboneucleic acid), glycogen (contained in large vacuoles), lipids (in fat droplets), and different enzymes and catalysts needed for the breakdown and absorption of its nourishment (e.g., oxidizing and reduction phenomena). The centrosomes containing the dark-staining centrioles are important for cell division.

ORGANELLES

Of the different organelles contained in the cytoplasm, the most common are:

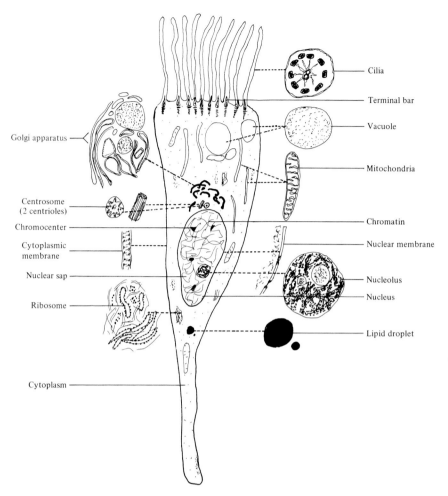

Figure 2
A ciliated columnar cell. Some of its components are enlarged to show the structural details seen only with the electron microscope.

MITOCHONDRIA.   Small rodlike, ovoid structures produce some of the enzymes needed for cellular metabolism, and they are the site of the energy-yielding catabolic process. In a live cell they move actively. They contain most of the extranuclear DNA (desoxyribonucleic acid) and are capable of division. Their injury may result in a degenerative cloudy swelling of the cells and an eventual necrosis.

GOLGI APPARATUS.   The function is not well known, but it has a possible action in the production and concentration of cellular secretion and protein enzymes. When the cells are silver stained, the Golgi apparatus appears as a complex array of narrow parallel channels.

CENTROSOMES. These are important in cell division (they direct the chromosomal movement) and production of the basal bodies from which the cilia emerge.

RIBOSOMES. Ribosomes are needed in the synthesis of cellular protein.

CELL INCLUSIONS
Cell inclusions differ from organelles in their lack of organization and are usually composed of a single substance. Some are phagocytized and some are produced by the cells. They can be contained in vacuoles (lipid droplets and secretory granules) or are free (melanin pigment and glycogen granules).

NUCLEUS
The shape and size of the nucleus vary with the type and function of the cell and the stage of its mitosis. During interphase, the following can be readily seen with a light microscope:

NUCLEAR MEMBRANE. The nucleus has a double osmiophilic, parallel membrane containing many multiangular pores that open to allow the interchange of substances absorbed or excreted by the nucleus. Usually thin and regular, it will appear thick and irregular when chromatin clumps adhere to it.

CHROMATIN. Composed mainly of basophilic nuclear protein and DNA (desoxyribonucleic acid), the chromatin can be uniform or irregular, fine or coarse, filamentous or granular, according to the type and activity of the nucleus. The chromatin is contained in a jelly-like nuclear sap (karyolymph) that stains poorly.

CHROMOCENTERS (KARYOSOMES). The size and number of chromocenters vary. They are the result of regional condensation of the chromatin and differ from the nucleoli in that they are primarily composed of DNA rather than RNA.

NUCLEOLI. Although usually spherical and central, the nucleoli may vary, according to the type of cell, in size, shape, number and position. They are composed mainly of ribonucleic acid (RNA) and give an acidophilic reaction with Papanicolaou stain.

Life of a Cell

The life of a cell can be summarized as follows:

1. Birth (mitotic division from a basal or reserve cell);
2. Maturation and differentiation (for instance, increase in size, formation of cilia, formation of secretory vacuoles, or pigmentation);
3. Functioning period, when the full-grown cell secretes, filters, absorbs, etc.;

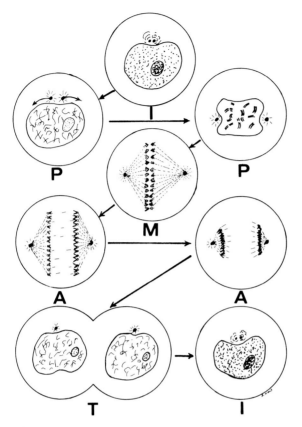

Figure 3
Various stages in cellular mitosis: (I) Interphase. (P) Prophase. (M) Metaphase. (A) Anaphase. (T) Telophase.

4. Degeneration, slow or rapid decrease of its functions, followed by its
   death and desquamation.

Cellular Division

The growth, regeneration, and healing of wounds of the tissues depend on
the capacity of all cells to divide and form new cells where needed. These
processes, described in 1875 by Eduart Strasburger in plant cells and about
the same time by Walter Flemming in animal cells, are called mitosis in
somatic cells and meiosis in sex cells.

MITOSIS
This cellular division, as illustrated in Figures 3 and 4, is a complex process
in which the chromatin arranges itself into a long chain that breaks up into
46 pairs of discrete threads called chromosomes. The nucleolus and nuclear
membrane begin to disintegrate, and asters may be seen at the horizontal

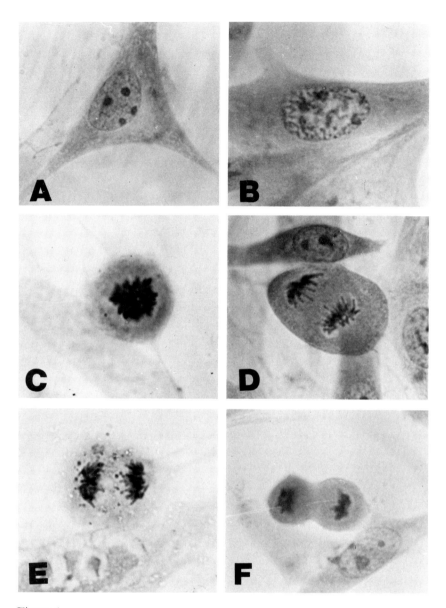

Figure 4
Mitosis of a cell in a culture. (A) Interphase. Note the fine, regular chromatin pattern and prominent nucleoli. (B) Early prophase. The nucleoli and nuclear membrane are less distinct and the chromatin is coarsely granular. (C) Metaphase. An equatorial plate is formed. (D) Early anaphase. The divided chromosomes are moving toward the opposite poles. (E) Late anaphase. The chromosomes are fusing. (F) Telophase. The cytoplasm is beginning to divide.

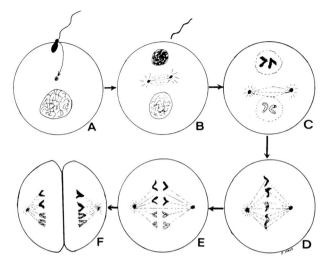

Figure 5
Various stages in fertilization and second meiotic division. (A) The head of a sperm penetrates the egg, loses its tail, and yields a centrosome. (B) the centrosome splits and forms a spindle. (C) In each male and female nucleus, 23 pairs of chromosomes are formed. (D) The two nuclei fuse, with the disappearance of their nuclear membrane, and the restored diploid (46) chromosomes line up. (E) The mixed chromosomes (of the male and female cells) divide. (F) Two separate daughter cells are formed.

poles of the nucleus (prophase). Each chromosome divides lengthwise into two chromatids and they align themselves into an equatorial plate dividing the mitotic spindle, while the nuclear membrane disintegrates and the asters and spindle become distinct (metaphase). These divided chromosomes continue to separate into two groups and move toward opposite poles (anaphase). Each new group contains the same number of chromosomes as the original cell. The new chromosomes mass at each end of the spindle, and nuclear membranes form again around each of them. The cytoplasm begins to divide (telophase) to form two new separate daughter cells, each with its new nucleus, nucleolus, and nuclear membrane in a resting phase (interphase).

MEIOSIS AND FERTILIZATION
As illustrated in Figure 5, fertilization produces a mixture of the chromosomes from the male and the female germ cells. Before their union, these cells must have only half the number of chromosomes (haploid state) to preserve, when combined, the normal total number of 46 chromosomes per cell (diploid state). This division and reduction to half the number of chromosomes occurs during gametogenesis and is called meiosis. The chromosomes divide four times while the cells divide only twice.

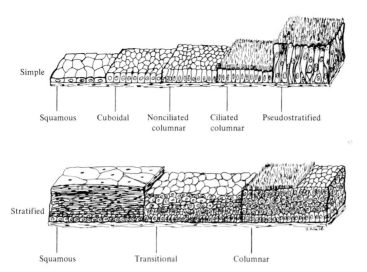

Figure 6
Various types of epithelium.

Epithelial Tissues

The majority of the exfoliated cells to be considered in this work arise chiefly from the epithelium, which covers the entire external surface of the human body and the inner surface of the hollow organs. The different epithelial tissues (Fig. 6) that are significant for cytodiagnosis are divided into:

1. Simple epithelium, where every cell is directly attached to the basement membrane, sometimes by a very narrow and elongated cytoplasmic process (cells from the pseudostratified epithelium, for example);
2. Stratified epithelium, where only the basal cells are in direct contact with the basement membrane and the remaining cells overlap each other in a variable number of layers.

SIMPLE EPITHELIUM

THE SIMPLE SQUAMOUS EPITHELIUM is found, for instance, in the mesothelial tissue lining the pleural, pericardial, and abdominal cavities and as a part of the serosa on the surface of the organs they contain. The life span and the amount of exfoliation of its component cells vary enormously and depend upon multiple external factors. The average life span of these cells is 18 days.

THE SIMPLE CUBOIDAL EPITHELIUM is found, for instance, in the endometrial glands. Massive removal of the endometrial cells occurs periodically following first an active stage of proliferation, then a secretory phase. The average life span of these cells is 20 days.

THE SIMPLE COLUMNAR EPITHELIUM (ciliated or nonciliated) is seen, for instance, in the endocervix and the gastrointestinal tract. In some parts of the intestines, the birth of the cells occurs deep in the glandular crypts. These cells, while maturing, move upward toward the surface of the villi, where they desquamate into the intestinal lumen. The life span of a cell in the duodenum is only 2 days; in the fundus or stomach, approximately 5 days; in the cardiac portion of the stomach, 9 days; in the colon, 10 days; and in the endocervix, 2 weeks. It is estimated that, during the normal life of a man, the duodenal epithelium renews itself more than 12,000 times.

THE PSEUDOSTRATIFIED COLUMNAR EPITHELIUM (ciliated or mucus-secreting goblet cells) is seen, for instance, lining the trachea and the bronchial tree. These cells, after their mitosis from the small cuboidal basal cells, progressively enlarge and grow upward until they reach the surface, always keeping a direct contact with the basement membrane by an increasingly delicate cytoplasmic process. When this contact is lost, the cells exfoliate. The average life span of this type of cell is 15 days in the nasopharynx and 30 days in the bronchial tree. It is estimated that more than 15,000,000 cells are spontaneously exfoliated per day from the normal bronchial tree of an adult. Most of these exfoliated cells are swallowed and digested in the gastrointestinal tract. The active life of a phagocytic "dust" cell, frequently seen in the pulmonary material, averages 8 days.

STRATIFIED EPITHELIUM

THE STRATIFIED SQUAMOUS EPITHELIUM lines the oral cavity, esophagus, vagina, exocervix, and skin. The cells originate by mitosis from the deep basal cells and, during their maturation and keratinization, progressively move toward the surface by changing first into parabasal cells, then into intermediate and superficial cells. After reaching the surface, partially because of the constant traumatic surface friction to which they are subjected, they exfoliate readily. The speed of the migration of a cell toward the surface varies with the location of the epithelium, its vascularity, growth, and space availability, and different hormonal effects. The life span of a cell in the vagina is approximately 8 days; in the buccal cavity, 9 days; and in the esophagus, 14 days.

THE STRATIFIED TRANSITIONAL EPITHELIUM varies greatly in appearance and thickness. This epithelium covers the renal pelvis, ureter, and bladder. In composition it is similar to the squamous epithelium, except that there is no tendency toward keratinization. The cells originate by mitosis from the basal cells. During a progressive maturation, they migrate to the surface where they exfoliate in the urine. They have an average life span of 60 days in the bladder and 50 days in the urethra.

THE STRATIFIED COLUMNAR EPITHELIUM (ciliated or nonciliated) covers very small areas of the body and is relatively rare. It can be found in some regions of the urethra, pharynx, epiglottis, fornix of the conjunctiva, and in the fetal tissues. The exfoliated cells often cannot be differentiated from other columnar cells.

References and Supplementary Reading

Bertalanffy, F.   Aspects of cell formation and exfoliation related to cytodiagnosis. *Acta Cytol.* 7:362, 1963.

Bloom, W., and Fawcett, D. W.   *A Textbook of Histology* (9th ed.). Philadelphia: Saunders, 1968.

Bratchet, J.   The living cell. *Sci. Am.* 205:50, 1961.

Bratchet, J., and Mirsky, A. E.   *The Cell.* New York: Academic, 1961.

Fawcett, D. W.   *An Atlas of Fine Structures: The Cell, Its Organelles and Inclusions.* Philadelphia: Saunders, 1966.

Fawcett, D. W., and Porter, K. R.   A study of the fine structures of ciliated epithelia. *J. Morphol.* 94:221, 1954.

Forquhar, M. G., and Palade, C. E.   Functional complexes in various epithelia. *J. Cell Biol.* 17:375, 1963.

Freeman, J.   Fine structures of the goblet cell mucous secretory process. *Anat. Rec.* 144:341, 1962.

Gross, P. R. (Ed.).   Second conference on the mechanisms of cell division. *Ann. N.Y. Acad. Sci.* 90:345, 1960.

Ham, A. E.   *Histology* (5th ed.). Philadelphia: Lippincott, 1965.

Ito, S.   The surface coat of enteric microvilli. *J. Cell Biol.* 27:475, 1965.

LeBlond, C. P., Puchtler, H., and Clermont, Y.   Structures corresponding to terminal bar and terminal web in many types of cells. *Nature* 186:784, 1960.

LeBlond, C. P., and Walker, B. E.   Renewal of cell populations. *Physiol. Rev.* 36:255, 1956.

Levine, L. (Ed.).   *The Cell in Mitosis.* New York: Academic, 1963.

Lipkin, M., Sherlock, P., and Bell, B.   Cell proliferative kinetics in the gastrointestinal tract of man. *Gastroenterology* 45:721, 1963.

Rowe, C. V.   Some aspects of structure and functions in cytoplasm. *J. Pediatr.* 60:601, 1962.

Swift, H.   Cytochemical studies on nuclear fine structure. *Exp. Cell Res.* [Suppl.] 9:54, 1963.

Cytology of the Normal Female Genital Tract

2

Anatomy and Histology

The different components of the external and internal female genital tract are illustrated in Figure 7.

The vagina and the portion of the cervix which protrudes into the upper part (portio vaginalis) are normally covered by a smooth, white, nonkeratinized stratified squamous epithelium. This mucosa terminates toward the anatomic external os of the cervix where it is replaced by the pink endocervical simple columnar type of epithelium. The transition between the squamous and columnar epithelium (squamocolumnar junction) is, in 60% of the adult patients, microscopically ill defined with the presence of an irregular intermediate (transitional or transformation) zone, while grossly, especially with a Schiller's test (staining the intracellular glycogen brown with iodine), this junction seems to be well defined and abrupt. Although the squamocolumnar junction is usually situated on the external endocervical os, its exact location varies. It moves up and down the endocervical canal according to such factors as patient's age, hormonal status, pregnancy, history of inflammation, size of the uterus, or congenital abnormalities. This squamous-columnar junction can be very high in the canal (entropion), low on the external os (ectropion), or formed by patches of columnar epithelium in the middle of the normal squamous mucosa that lines the portio vaginalis (ectopia). The last two conditions may be mistaken grossly for cervical ulceration or erosion. Colposcopy may be helpful in determining their nature.

THE ENDOCERVICAL GLANDS, formed by the branching folds (plicae palmatae) of the columnar mucosa, are mainly found in the portio supravaginalis of the cervical canal. This mucosa, 2 to 3 mm thick, is made of monolayered, tall columnar cells with oval, eccentric nuclei situated toward the base of the cells. Some are ciliated (immediate lining of the canal), whereas others have large mucin-containing secretory vacuoles (lining of the glandular acini). This mucosa remains thin and does not participate extensively in the changes of the menstrual cycle, except possibly for a slight increase of mucus secretion toward the time of ovulation. During pregnancy, the endocervical cells enlarge and proliferate, and the amount of

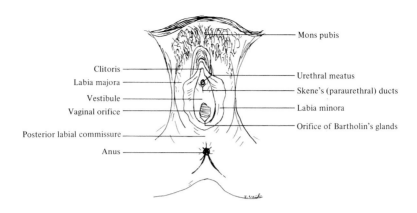

Clitoris —
Labia majora —
Vestibule —
Vaginal orifice —
Posterior labial commissure —
Anus —

— Mons pubis
— Urethral meatus
— Skene's (paraurethral) ducts
— Labia minora
— Orifice of Bartholin's glands

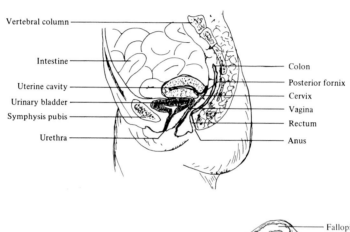

Vertebral column —
Intestine —
Uterine cavity —
Urinary bladder —
Symphysis pubis —
Urethra —

— Colon
— Posterior fornix
— Cervix
— Vagina
— Rectum
— Anus

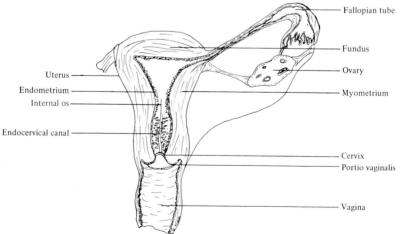

Uterus —
Endometrium —
Internal os —
Endocervical canal —

— Fallopian tube
— Fundus
— Ovary
— Myometrium
— Cervix
— Portio vaginalis
— Vagina

mucus secretion increases. When the openings of the endocervical glands become occluded, as in chronic cervicitis, they may form small cystic structures (nabothian cysts).

THE UTERINE CAVITY, normally measuring 3 to 5 cm in length and a few millimeters in diameter, is lined by a special glandular mucosa made of uterine glands (Fig. 8), which is 5 mm in thickness at the height of its proliferation in a changing stroma. The junction between the endocervical and endometrial mucosa occurs in the internal os, but its exact location varies with such factors as the age of the patient, hormonal status, and number of pregnancies. A transitional zone also exists.

Beginning with puberty, the endometrial glands undergo monthly cyclic changes as a result of the effects of various ovarian hormones. At the end of each cycle there is partial destruction and exfoliation of these glands, followed by regeneration. This destruction and the secondary hemorrhage appear as the menstrual flow, which normally lasts 3 to 5 days. During the regenerative period, the proliferative endometrial glands are lined by several layers of crowded young endometrial cells with very scanty cytoplasm. This stage is the result of the increased estrogenic hormone production by the ovary. Following ovulation, the ovarian corpus luteum produces progesterone and changes the aspect of the endometrial glands from straight and tubular to tortuous, and the lining changes from multilayered cells to monolayered mature cells with large cytoplasmic vacuoles (secretory endometrium).

THE FALLOPIAN TUBES, measuring 8 to 10 cm in length and 5 mm in diameter, are lined by an extensively folded mucosa, especially toward the ampullae. This mucosa is composed of a monolayered epithelium made of nonciliated secretory cells; slender nonciliated, nonsecretory cells (peg cells); and ciliated ones. The tubal epithelium responds, although not as extensively as the endometrium, to the cyclic or pregnancy-related hormonal changes.

THE TWO OVARIES, measuring 4 × 2 × 1 cm each, are situated on each side of the uterus in the pelvic cavity and suspended by several peritoneal folds and ligaments. Ovoid in shape, their white to pink surface is smooth during the prepubertal years, but it becomes irregular and scarred as the graafian follicles develop and rupture.

Figure 7
(A) The external female genital tract. (B) The sagittal section showing the relationship of the female genital tract to the surrounding organs. (C) The coronal section showing the relationship of the ovary to the rest of the genital tract.

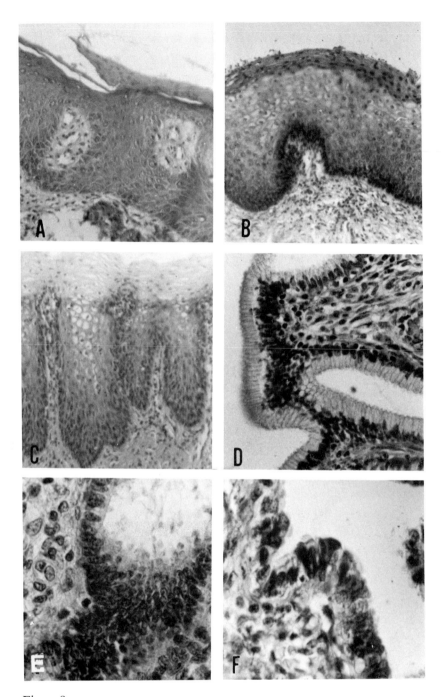

Figure 8
Histology of various mucosa covering the female genital tract. (A) Vulval. (B) Vaginal. (C) Cervical. (D) Endocervical. (E) Endometrial. (F) Tubes.

Gynecologic Smear Preparation

ROUTINE VAGINOCERVICAL SMEAR

Material Required

1. A diagonally cut tongue depressor or an Ayre spatula, preferably made of wood. (The cells have a tendency to slide from the surface of a smooth plastic or metal spatula.);
2. A glass pipette with a rubber bulb;
3. A glass slide with one end frosted, on which, before smearing, the name and identification number of the patient are written with a lead pencil;
4. A bottle of fixative or a spray fixative (see Chap. 16). (The specimen of only one patient should be placed in each bottle to prevent possible cellular cross contamination or a mixup of the patients.)

Technique (Fig. 9)

The slides should be identified and matched with the patient's requisition, which should include the history of the patient, her age, and the dates of her last menstrual period (LMP) and her previous menstrual period (PMP).

All talcum powder should be removed from gloves before touching the patient, instruments, or slides.

The speculum should be introduced without lubricant. It can be dipped in warm saline solution for lubrication. The vaginal mucosa must be carefully examined for any lesion before obstructing it with the speculum.

The cervical surface should not be wiped; wiping it would remove the cell-rich adherent endocervical mucus.

Using the spatula or the glass pipette with a rubber bulb, remove several drops of the secretion from the endocervix and place them on the slide without smearing them.

Place the small end of the spatula in the external os of the endocervical canal as deeply as possible and rotate it 360 degrees, energetically scraping the entire surface of the external os and part of the internal os. The margins, and not the bottom, of a grossly ulcerated area should be carefully and energetically scraped.

The collected material from this scraping is then mixed with the endocervical mucus droplets, and the entire specimen is smeared with a lateral motion rather than with a circular one. The slide is immediately dropped into the bottle of fixative or sprayed before it has a chance to air dry.

Precautions

The smear should be allowed to remain for at least 15 minutes in the fixative before staining.

The smear should not be left longer than one week in the fixative; otherwise cellular distortion may occur.

In the presence of an atrophic mucosa, the slide and spatula should first be moistened with saline solution.

If the smear is accidentally allowed to air dry, it can be rehydrated by placing it in tap water for a few minutes before fixation.

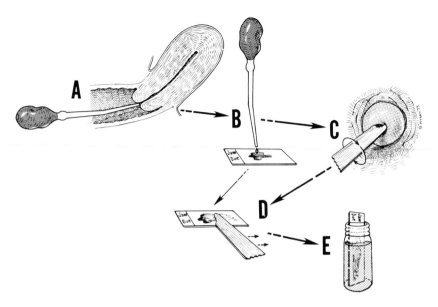

Figure 9
Technique for collecting a cervical smear. (A) Several drops of secretion are aspirated from the endocervix. (B) The specimen of secretion is placed on a labeled slide and not smeared. (C) The cervical os is scraped. (D) The scraped material is mixed with the droplet of endocervical mucus and smeared. (E) The slide is immediately fixed.

In postmenopausal women a few droplets of vaginal pool secretion may be smeared on a second slide for the detection of endometrial lesions.

SMEARS FOR HORMONAL EVALUATION
With a cotton-tipped applicator or a wooden spatula, the lateral wall of the upper third of the vagina should be very gently scraped under direct vision to obtain only the surface cells. Special precautions should be taken not to take the smear from an ulcerated or grossly inflamed area. This cellular material should be gently smeared on a slide and immediately dropped into the fixative.

ENDOMETRIAL ASPIRATION
This procedure is indicated whenever an endometrial lesion is suspected, and the aspiration or washing should be taken only after the routine vagino-cervical smear.

Endometrial aspiration or washing is contraindicated in cases of severe vaginocervical infection or suspected pregnancy.

After taking the routine cervical smear, the portio vaginalis and the endocervical canal are cleaned with a cotton-tipped applicator. One end of a thin plastic tube with five or six holes attached to a 10-cc syringe is then introduced into the uterine cavity without dilatation of the endocervix.

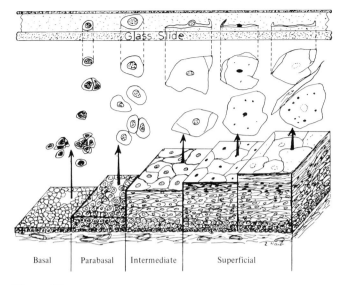

Figure 10
The different histologic layers of a vaginal stratified squamous epithelium.

Basal   Parabasal   Intermediate   Superficial

Repeated suction is applied while rotating and withdrawing the tube. The content of this cannula can be immediately smeared on a glass slide, or, preferably, the cannula and its content are washed by vigorously shaking it in a saline solution to be filtered later through a membrane filter.

The main danger involved in introducing fluid into the endometrial cavity is the possibility of spreading the uterine contents to the abdominal cavity. This has been minimized by the use of a double plastic tube, which allows a saline shower of the endometrial cavity under negative pressure (Dowling and Gravlee, 1964). The blood-stained fluid obtained usually contains fragments of endometrial and endocervical tissue that are visible to the naked eye. This fluid can be centrifuged, after which part of the deposit can be smeared and the rest can be sectioned in a cell block.

Histology of Stratified Squamous Epithelium

As illustrated in Figures 10 and 12, the section of the nonkeratinized vaginocervical stratified squamous epithelium may show four distinct zones. The deepest one, situated just over the basement membrane, is made of a layer of small cuboidal cells with rounded ends (basal layer or stratum cylindricum). This in turn is covered by several layers of larger basophilic polyhedral cells (parabasal layer or stratum spinosum). This in turn is covered with several layers of even larger cells with round vesicular nuclei, glycogen-containing cytoplasm, and intercellular bridges (intermediate layer). Toward the surface, the epithelium is composed of layers of flattened cells with pyknotic nuclei and thin, abundant, eosinophilic, kerati-

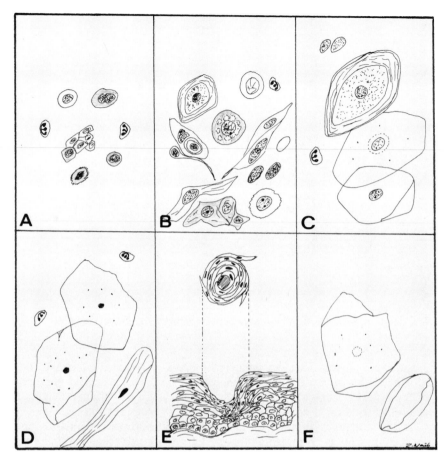

Figure 11
Various normal squamous epithelial cells found in a vaginal smear. (A) Basal cells. (B) Parabasal cells. Note the polymorphism and the occasional stripped nuclei. (C) Intermediate cells, the top cells representing navicular cells. (D) Superficial cells. (E) Pearl formation and its histologic explanation. (F) Anucleated squamous cells.

nized cytoplasm (superficial or cornified layer). In rare instances, part of the surface of the lower vagina is covered with a thin layer of anucleated, heavily keratinized, thin cells. This may occur in other areas as the result of repeated irritation (leukoplakia).

Cytology of the Stratified Squamous Epithelium

The features of the exfoliated squamous cells often differ from the ones in histologic section because of their three-dimensional image and the presence of other alterations resulting from various degrees of natural cellular aging, induced degenerative changes, and their separation from the confinement of the surrounding organized structure.

DEEP BASAL CELLS–GERMINAL CELLS (FIG. 11A)

Originating from the single-layered lining of the basal membrane, the deep basal cells are always traumatically exfoliated.

Rarely seen in a normal vaginal smear, they are more commonly found in the scraped smear of an atrophic or deeply ulcerated mucosa.

They are the smallest epithelial cells found in the vaginal smear (about the size of a leukocyte).

They are uniform in size and shape (round or oval).

They usually exfoliate in small sheets or tight cellular clusters (90% of the time). Rarely recognized as basal cells when found singly, they are usually mistaken for small histiocytes.

Their cytoplasm is scanty but thick and homogeneous with smooth borders. They stain dark to deep blue with Papanicolaou stain.

Their nuclei are round, centrally located, well preserved, and hyperchromatic with vesicular, basophilic, coarse chromatin clumping.

No prominent nucleoli are ever found.

Occasional mitoses can be seen, especially when the cells have exfoliated in sheets from an ulcerated (healing) area.

PARABASAL CELLS (FIGS. 11B, 12C)

Originating from the deep layer of the stratified squamous epithelium, they are either traumatically exfoliated, especially in cases of erosion or ulceration of the mucosa, or physiologically exfoliated from an atrophic epithelium (prepuberty, lactation, postirradiation or postmenopausal mucosa, or in estrogen deficiency).

They are fairly uniform in shape (spherical or round), except when they have exfoliated from an ulceration, where variable bizarre shapes with pseudopodlike projections (repair cells) may be seen (Fig. 52B).

Their size varies from 15 to 25 μ in diameter.

They exfoliate singly (in atrophy) or in sheets (ulceration or postirradiation).

The thick cytoplasm stains deep blue or green, except when air dried before fixation, which may produce a variable artificial eosinophilia. The outer part (exoplasm) is often more transparent than the inner part (endoplasm). They have a well-defined smooth border, and the amount of cytoplasm may vary from scanty to adequate according to the degree of their maturation.

Occasional vacuoles, cytoplasmic granules, or glycogen can be found in their endoplasm.

Their round-to-oval nuclei are usually well preserved, large (8–12 μ), hyperchromatic, and centrally located.

Their nuclear membranes are usually even, regular, and delicate.

The chromatin is granular and uniformly distributed.

Occasional minute-to-prominent nucleoli (1 to 2), depending on the activity of the cells, can be found.

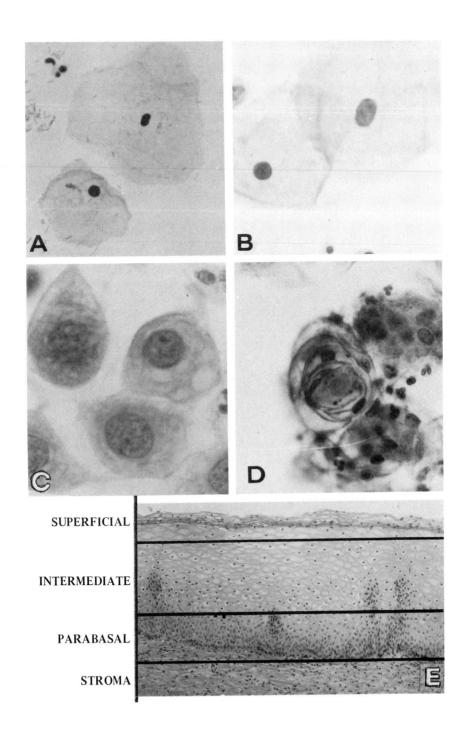

SUPERFICIAL

INTERMEDIATE

PARABASAL

STROMA

Occasional basal cells may be found with dense basophilic or eosinophilic cytoplasm and central, round, pyknotic nuclei, occasionally with prominent perinuclear halos. They are parabasal cells that have quickly degenerated or matured as a reaction to excessive irritation or excessive estrogenic hormonal effect.

In the traumatic exfoliation of an atrophic mucosa, the nuclei and cytoplasm of these cells may become distorted and elongated, imitating the appearance of atypical squamous spindle cells.

The background of the smear will often contain large numbers of cellular cytoplasmic and nuclear debris, red blood cells, and leukocytes, and it may resemble the background of smears taken from invasive ulcerated cervical squamous cell carcinoma (tumor diathesis).

The cytoplasm of some parabasal cells may contain small clusters of granular inclusions surrounded by small halos, diagnostic of inclusion vaginitis (TRIC, *Bedsonia,* or chlamydia infection).

In a long-standing atrophic mucosa, some of these parabasal cells, during degeneration, fuse together to form multinucleated giant cells, which are differentiated from histiocytic foreign-body giant cells by the irregularity of size and shape of the basal nuclei.

All or part of keratinized squamous cells may be found ingested in the cytoplasm of foreign-body giant cells. This may represent a peculiar sensitization of the patient to her own exfoliated keratinized vaginal cells.

The parabasal cells disappear very quickly in an atrophic smear when estrogen is given but persist if they are atypical or have originated from an ulcerated or irradiated mucosa.

INTERMEDIATE CELLS (FIGS. 11C, 12B)
Originating from the middle layer of the stratified squamous epithelium (precornified), they form the most common epithelial cells seen in the smear at the postovulatory time, during pregnancy, menopause, or as the result of the action of progesterone or adrenocortical hormones.

Their size varies from 30 to 60 μ, depending on the degree of their maturation.

Their shape is polyhedral with little variation, often with a folding or curling tendency of their cytoplasmic edges, especially evident in navicular

---

Figure 12
(A) Two superficial squamous cells. Note the small pyknotic nuclei and the abundant, thin, transparent cytoplasm. ($\times$ 350) (B) Two intermediate squamous cells. Note the vesicular nuclei and the abundant, thin, folded cytoplasm. ($\times$ 350) (C) Parabasal squamous cells in a smear from a postmenopausal woman. Note the large, round, vesicular nuclei and thick vacuolated cytoplasm. ($\times$ 750) (D) Squamous epithelial benign pearl. Note the multiple, vesicular, central nuclei, surrounded by layers of molded superficial cells with pyknotic nuclei. ($\times$ 450) (E) Section of benign squamous epithelium of the vaginal portio of the cervix. Note the superficial and intermediate layers of vacuolated cells with contained glycogen and the multilayered parabasal and basal cells with large nuclei. ($\times$ 120)

cells of pregnancy or postmenopause (see p. 67). They usually exfoliate singly (physiologic exfoliation) but are occasionally seen in large sheets (traumatic desquamation).

Their cytoplasm is thin, semitransparent, abundant, and pink-to-blue stained. Frequent vacuolization and glycogen deposits or granules are present in the endoplasm.

Their cytoplasmic borders can be distinct or poorly defined.

Their round-to-oval nuclei are central, well preserved, large, and vesicular with delicate, uniformly distributed chromatin clumps.

Female sex chromatin masses (Barr bodies) are often easier to see in these cells, making them the cells of choice to examine for sex chromosome determination.

Their nucleoli, if present, are minute, round, and not remarkable.

The abundance of the cytoplasm and the smaller size of the nuclei differentiate these intermediate cells from parabasal cells. The presence of a discernible chromatin pattern in their nuclei distinguishes them from the superficial cells. The yellow or brown granules found occasionally in their endoplasm are often glycogen deposits, as seen during pregnancy. These glycogen granules are sometimes lost during fixation or staining, leaving a large, empty perinuclear halo. Because of the fragility of the cytoplasm, cytolysis (destruction of the cytoplasm) by the action of Döderlein's bacillus or *Trichomonas* is common, resulting in stripped nuclei that are sometimes difficult to differentiate from the stripped nuclei originating from endocervical cells, especially in postmenopausal women. These stripped nuclei may be confused with poorly differentiated carcinoma cells, except for the regularity of their chromatin and nuclear membranes.

SUPERFICIAL CELLS (FIGS. 11D, 12A)
Originating from the upper layer of the stratified squamous epithelium, they are usually physiologically exfoliated.

They are the most common epithelial cells seen in the vaginal smear at the preovulatory time of the cycle, after estrogenic therapy, or in case of a functioning ovarian tumor.

They are the largest epithelial cells found in the vaginal smear, their size varying from 40 to 60 μ in diameter.

They are usually single, they have a polyhedral shape with little variation, and their cytoplasmic borders are sharp, smooth, and distinct.

Their thin cytoplasm is homogeneous, without vacuolization. It occasionally contains several brown-to-black, small, round, granular deposits that are found mainly toward the midzone, the result of cellular degeneration and aging. They may also contain yellowish-brown staining glycogenic granules, as are found in the intermediate cells.

Their cytoplasm stains from yellow to orange to pale blue, according to the variation of their affinity for acid dyes and the amount of keratin.

Their single, round-to-oval nuclei are central, pyknotic, dark, and regular in size (5 μ). They have uniform borders with no discernible chromatin pattern (Table 1).

If the nuclei are larger than a red blood cell, the superficial cells are

Table 1. Differential Characteristics of Normal Squamous Genital Cells

Criteria	Basal	Parabasal	Intermediate	Superficial
Size	8–10 μ	15–20 μ	30–60 μ	40–60 μ
Shape				
Polygonal	0%	5%	85%	75%
Oval	5%	40%	10%	20%
Round	95%	55%	5%	5%
Occurrence	Sheets 90%	Single 60% Sheets 40%	Single 80%	Single 90%
Amount of cytoplasm	Scanty	Adequate	Abundant	Abundant
Cytoplasmic stain	Deep blue	Blue	Pink or blue	Orange
Cytoplasmic vacuolization	None	Occasional	Occasional	None
Nuclei-cytoplasm ratio	8:10	5:10	2:10	1:10
Nuclear size	7–9 μ	8–13 μ	10–12 μ	5–7 μ
Nuclear shape	Round	Round to oval	Round to oval	Round
Chromatin pattern	Coarse	Granular	Finely granular	Pyknotic
Multinucleation	Rare	Few	Few	Rare
Nucleoli	None	Occasional and prominent	Small	None

probably abnormal. The degenerative black-brown granules sometimes seen in the cytoplasm are a lipid-containing condensation of cytoplasmic material. One should not be able to see any nuclear chromatin pattern in the nucleus; if he can, the cell is not a superficial one but an intermediate one.

SQUAMOUS PEARLS (FIGS. 11E, 12D)
Squamous pearls are the result of the exfoliation of a compressed agglomeration of squamous cells originating from the bottom of an epithelial pit or crevasse.

The component cells can be either superficial, intermediate, or parabasal in type, according to the depth of the pit.

There is usually a single or multiple, central, hyperkeratotic or degenerated parabasal or intermediate cell around which the other cells are linked in concentric rings.

Large numbers of such pearls in a smear usually indicate the presence of leukoplakia, superficial hyperkeratosis, or squamous metaplasia. Squamous pearls can also be found in combination with adenocarcinomatous cells, indicating the presence of an adenoacanthomatous neoplasm.

ANUCLEATED CELLS (FIG. 11F)
Anucleated cells are rare in a normal smear, except when contaminated by perineal cells or the hands of the doctor. They may also originate from the lower third of the patient's vagina.

Their size and shape vary, with a definite tendency to marginal folding.

Some stain yellow or orange with a glassy appearance indicating the presence of excessive keratinization. This is often diagnostic of a benign leukoplakia (hyperkeratosis) or a reaction to a chronic irritation, as seen in uterine prolapse (procedentia) (see Chap. 5).

Other cells have semitransparent or pink cytoplasm with the presence of an empty shadow where the nuclei used to be. They are usually the result of degenerative keratolysis of a superficial or an intermediate squamous cell.

During pregnancy, the presence of poorly stained, pale yellow, anucleated cells may indicate a ruptured membrane. These cells are probably fetal skin cells desquamated into the leaking amniotic fluid.

ENDOCERVICAL RESERVE CELLS (FIG. 15B)
These are young, endocervical, parabasal cells with a multipotential differentiation. They may be seen in sheets or clusters of single cells. Their cytoplasm is adequate to scanty, cyanophilic, and finely vacuolated. Their round-to-oval nuclei are centrally located, with fine, uniformly distributed chromatin. Small, round chromocenters are often multiple. Isolated and small, these cells may be difficult to differentiate from histiocytes and endometrial stromal cells.

Cytology of the Columnar Epithelium

CILIATED ENDOCERVICAL CELLS (FIGS. 13A, 14A)
Rarely seen intact, the fragile, ciliated endocervical cells are the result of direct traumatic exfoliation and are found more commonly in cervical

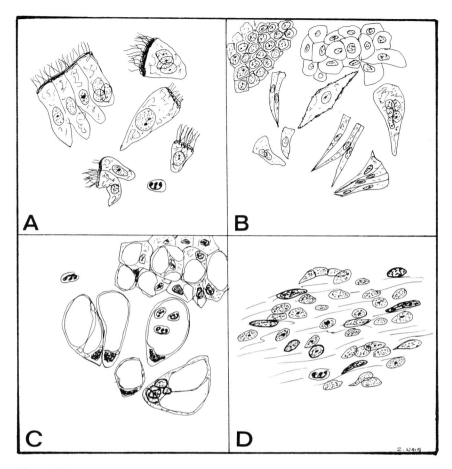

Figure 13
Morphology of the various endocervical cells. (A) Ciliated. (B) Nonciliated.
(C) Secretory. (D) Stripped nuclei.

scrapings or endocervical aspirations than in the vaginal pool smear. They are almost never seen in a postmenopausal atrophic smear.

Exfoliating usually in tight clusters, in small sheets, or in palisade formations, they can present a mosaic or honeycomb appearance when their apical ends are in focus.

Their size varies (10–25 μ), but their shape is fairly constant (cylindrical or pyramidal).

On one extremity of the cell, a straight, often hazy, granular terminal bar (dark red stained) may be present, made of basal bodies arising from the repeated reduplication of the centrioles. This terminal bar can persist even after the cilia have been lost through degeneration.

When the cell is well preserved, delicate pink cilia are attached to this lavender or red terminal bar. Their length varies according to the original position of the exfoliated cell in relation to the os of the endocervical canal. They are shorter toward the portio vaginalis.

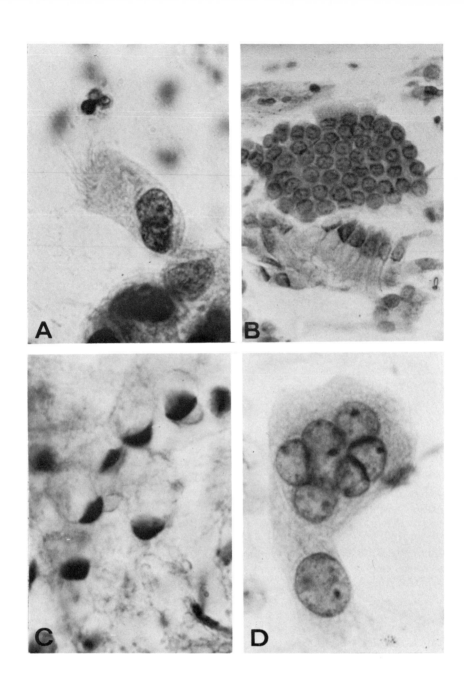

The cytoplasm of ciliated endocervical cells is elongated, adequate, with a semitransparent, lacy appearance, but darker stained than the pale mucus-producing endocervical cells. No large secretory vacuoles are present.

Their cytoplasmic borders are thin but distinct, in contrast to those found in other types of endocervical cells.

The nucleus is centrally placed or close to the apical cellular end in contrast to the position of the nuclei in nonciliated cells. The nuclei are round or oval in shape and vary moderately in size.

The chromatin is finely granular with a uniform distribution. The nuclear borders are even and smooth, and they often merge with the cytoplasmic membrane on both sides.

When multiple, the nuclei overlap, giving the impression of false hyperchromatism.

A centrally placed, small red nucleolus can be found on rare occasions (5%).

NONCILIATED ENDOCERVICAL CELLS (FIGS. 13B, 14B)
Occurring singly or in palisade formation, these tall, columnar cells are fairly uniform in size (20–30 μ) and shape (elongated).

Their adequate cytoplasm is narrow (less than 5 μ in diameter), and their borders are sharp, smooth, and delicate.

Their cytoplasm is semitransparent and finely vacuolated, and stains poorly and unevenly (pale blue). In some, fine acidophilic granules can be seen.

Their nuclei are round, elongated, or triangular in shape, with moderate size variation. Their lateral borders often merge with the cytoplasmic membranes. Occasionally a nipple-like protrusion can be seen. They are usually centrally located.

The finely granular chromatin stains dark blue and may give the impression of nuclear pyknosis.

The nucleoli are occasionally prominent. Multinucleation may occur, and the nuclei may compress each other, imitating herpesvirus-infection changes (see p. 98), from which they differ by the retention of their normal chromatin patterns and nucleoli (Figs. 14D, 15A). The reason for the multinucleation is unknown. It has no clinical significance.

HYPERSECRETORY ENDOCERVICAL CELLS (FIGS. 13C, 14C)
These cells increase in number during chronic irritation, in pregnancy, in the presence of glandular endocervical polyps, or during the use of various contraceptive pills.

Figure 14
Various aspects of endocervical cells in a vaginal smear. (A) Ciliated endocervical cell. Note the binucleation and slender cilia. ($\times$ 750) (B) Group of nonciliated endocervical cells in palisade and mosaic formation. ($\times$ 450) (C) Hypersecretory endocervical cells. Note the overdistention of their cytoplasm and the flattening of their dark nuclei by the single mucus-containing vacuoles. ($\times$ 750) (D) Giant multinucleated endocervical cell. Note the overlapping and slight molding of the nuclei and the prominence of the nucleoli. ($\times$ 750)

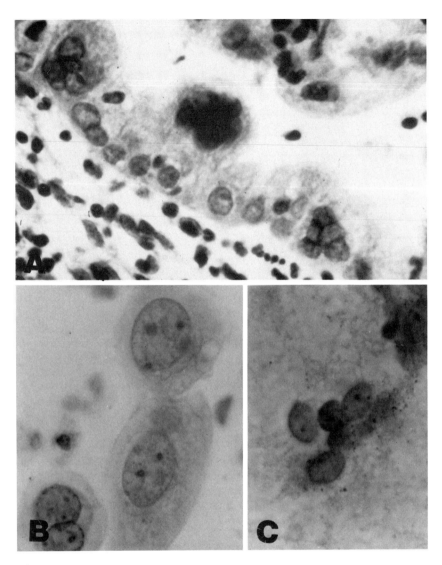

Figure 15
Endocervix. (A) Section showing the origin of the multinucleated cells. (B) Endocervical reserve parabasal cells. (C) Degenerate "naked" endocervical cells.

They vary greatly in size (15–60 μ) and exfoliate singly or in clusters. Their shape varies from round to triangular.

Their cytoplasm is usually distended by a single or multiple large secretory vacuole, and it often contains numerous large, healthy polymorphonuclear cells. The borders of their cytoplasm are often indistinct, thin, and very delicate. They often stain irregularly. Because of the fragility of the cytoplasm, it is common to find in the smear numerous stripped nuclei with only a wisp of transparent cytoplasm still attached in strands of thick cervical mucus.

The nuclei are often enlarged, oval- to crescent-shaped, and eccentrically situated toward the narrow end of the cell. The size of the nucleus may vary from 9 to 20 μ in diameter. The heavy nuclear membrane is often fuzzy.

The chromatin is coarsely clumped with a tendency to condense toward the marginal zone. The nuclei may, in cases of hypersecretion, appear almost completely pyknotic with an extremely crescentlike shape.

The nucleoli may be prominent (60%), spherical in shape, and variable in number (1 to 4).

Multinucleation is common, especially in cases of hormonal hyperplasia and chronic or acute cervicitis.

On occasion, it is difficult to differentiate a small, young, single endocervical cell from a histiocyte, both having a similar lacy cytoplasm and round vesicular nuclei. The triangular shape and the irregularity of the staining of the cytoplasm of the endocervical cells can help in differentiation. The endocervical cells are distinguished from the endometrial cells by their larger and more variable size and their abundance of cytoplasm. During pregnancy or the postpartum period, groups of endocervical cells, as a result of acute or chronic irritation, can become considerably larger, with monstrous nuclei (greater than 20 μ), having very coarse chromatin clumps and hyperchromatism. These cells can be confused with anaplastic malignant cells, except for the persisting regularity of their smooth nuclear membrane and the abundance of their cytoplasm.

ENDOCERVICAL STRIPPED NUCLEI (FIGS. 13D, 15C)
So-called bare nuclei, often having a wisp of cytoplasm still attached, are commonly seen in smears from postmenopausal or pregnant women or from women with an endocervical ectropion or ectopy.

These nuclei are uniformly round or oval in shape but may vary moderately in size (6–14 μ).

Their nuclear membrane is regular and sharp with little sign of degeneration.

The chromatin pattern is uniform and finely granular with occasional clumping, and it resembles the pattern of the normal nucleus of the intact endocervical cell. Some condensation of the chromatin material toward the nuclear rim may be seen.

On occasion, single, small reddish nucleoli can be seen centrally placed.

These stripped nuclei, especially when air dried (which causes an apparent enlargement), should not be confused with poorly differentiated squamous carcinoma cells. Although both vary in size, the shape of the endocer-

Table 2. Differential Characteristics of Normal Columnar Genital Cells

Criteria	Reserve	Endocervical	Endometrial	Stromal
Size	8–20 μ	10–25 μ	8–10 μ	8–12 μ
Shape	Oval	Cylindrical	Round	Irregular
Occurrence	Sheets 60%	Sheets 50%	Acini 40%	Single 95%
Cilia	None	Occasional	Occasional	None
Amount of cytoplasm	Adequate	Abundant	Scanty	Scanty
Cytoplasmic stain	Blue	Blue	Blue	Blue
Cytoplasmic vacuoles	Fine	Large	Rare	Fine
Nuclei-cytoplasm ratio	5:10	3:10	8:10	8:10
Nuclear size	7–12 μ	9–20 μ	8–10 μ	7–9 μ
Nuclear shape	Round to oval 95%	Round to oval 95%	Round 90%	Bean-shaped 15% Oval 85%
Chromatin pattern	Moderately granular	Fine	Granular	Coarsely granular
Multinucleation	Rare	Common	Rare	Rare
Nucleoli	Small	Prominent	Small	Rare

vical nuclei (contrary to what is found in carcinoma) is regular and round or oval with a smooth nuclear membrane, and its chromatin is finely granular, bland, and uniformly distributed.

GLANDULAR ENDOMETRIAL CELLS (FIGS. 16, 17, 18; TABLE 2)
Physiologically the glandular endometrial cells may be found normally in a routine vaginal smear only during the menstrual flow and the following few days. At any other time of the cycle, they may indicate an endometrial lesion.

The endometrial cells can be obtained at any time by an endometrial aspiration, lavage, or scraping.

They can be found in tight clusters, singly, or in acinar formation.

Their shape and size vary according to the phase of the menstrual cycle and the method used to obtain them, but in the same smear their size (8–10 μ) and shape (round to oval) show little variation.

The cytoplasm is scanty, delicate, transparent, green-to-pink stained, with an indistinct border. Often only fragments of cytoplasm are left by the time they are smeared.

On occasion, delicate cilia can be seen, especially in an aspirated endometrial smear. Their presence seems to be related to increased estrogenic stimulation. The greatest number is found during midcycle or during endometrial hyperplasia.

The nuclei are usually single, dark blue, round, hyperchromatic, and uniform in size (8–10 μ) and shape (round).

The nuclei of adjoining cells often overlap, giving the impression of an artificial hyperchromatism.

The chromatin is coarsely clumped but evenly distributed.

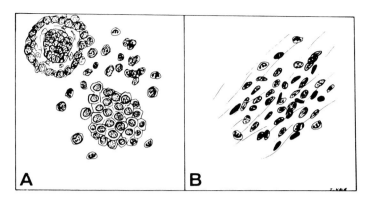

Figure 16
(A) Endometrial cells. (B) Stromal cells.

The nuclear borders are even and heavy.

No prominent nucleoli can be found.

Endometrial cells differ from endocervical cells by the regularity of the size of their nuclei, the scantiness of their cytoplasm, the coarser chromatin clumping, and their exfoliation in very tight clusters. They closely resemble clusters of small histiocytes and can be differentiated from them by the regularity of the size and shape of the endometrial cells compared to the variable size and shape, with occasional beanlike appearance of the nuclei, of the histiocytes.

During the proliferative phase, the exfoliated endometrial cells will show marked molding, giving the appearance of an irregular structure, which sometimes may be confused with the cells of a poorly differentiated small-cell carcinoma. The very tight cellular molding, the regular hyperchromatism, and the uniformity of the size are of help in the diagnosis of their benignancy.

ENDOMETRIAL STROMAL CELLS (FIGS. 16B, 18A, 19)

They may originate from the spongiosa layer (deep) or the compact layer (superficial) of the endometrium. They are usually seen in clusters or in groups of single cells. Occasionally, stromal cells are trapped in mucus and are seen aligned in a single file. The superficial stromal cells vary in size, and they have an adequate vacuolated cyanophilic cytoplasm and an eccentric, single nucleus with a prominent granular chromatin pattern.

The deep (young), smaller stromal cells are more uniform in size (8–12 μ), with scantier cytoplasm, and they resemble large lymphocytes. These cells are also sometimes mistaken for small in situ cancer cells. The presence of endometrial glandular cells in the smear and the demonstration of their more regular nuclear chromatin help in their recognition.

ACINI (FIGS. 18C, 20)

Acini are exfoliated whole glands, often lying on their sides. They can be endometrial or endocervical in type. The greater amount of cytoplasm pres-

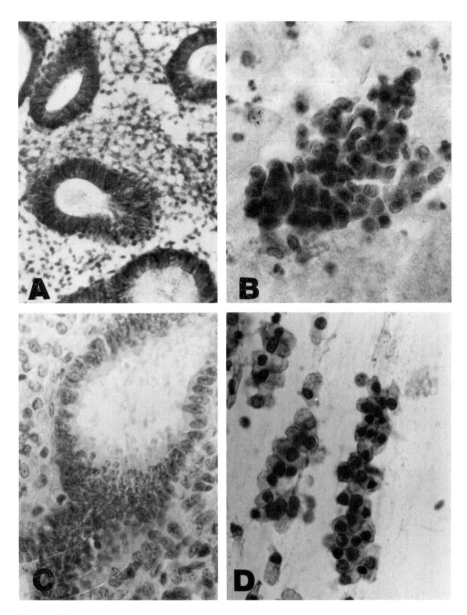

Figure 17
(A) Proliferative endometrium. (× 250) (B) Cytology of proliferative endometrium. (× 450) (C) Secretory endometrium. (× 450) (D) Cytology of secretory endometrium (endometrial washing). (× 450)

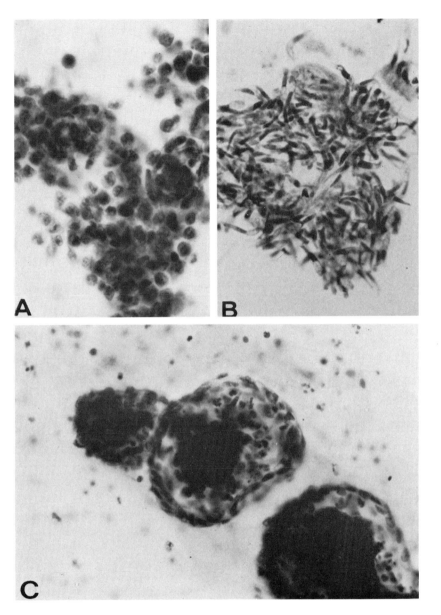

Figure 18
(A) Clusters of endometrial stromal cells and histiocytes as seen in a vaginal smear taken during menstruation. (× 450) (B) Clusters of elongated, secretory, endometrial cells obtained by endometrial lavage on day 20 of the menstrual cycle. (× 450) (C) Endometrial glandular acini. (× 450)

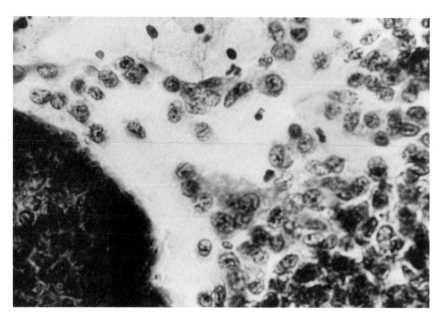

Figure 19
Endometrial stromal and glandular cells during normal menses. (× 450)

ent in the individual cells of the endocervical acini differentiates them from the endometrial ones.

They are more commonly seen in the vaginal smears during the menstrual flow, in endocervical or endometrial aspiration, or in the presence of an endocervical or endometrial benign lesion (polyp, hyperplasia).

Three distinct layers should be recognized to differentiate them from the pseudoacini, in which only one or two layers can be found. This is best done by moving the plane of focus of the microscope up and down.

The definite diagnosis of acini can be of importance when seen, for example, in effusions. Their presence in such cases indicates a metastatic adenocarcinoma in spite of the benign general appearance of the individual component cells.

Acini can be differentiated from a polypoid structure by the presence of an acellular cavity in the middle layer, instead of a solid cellular core as found in the polyp, and also by the location of the nuclei toward, rather than away from, this central core.

FALLOPIAN TUBE CELLS (FIG. 21)
Rarely seen, they are hard to recognize in a routine vaginal smear and are usually confused with endocervical cells. They are classified into ciliated cells, secretory cells, and the more abundant peg cells.

Uniformly elongated and pyramidal in shape, the nonciliated *peg cells* vary in size (15–30 μ).

Their cytoplasm is adequate to abundant, basophilic, and denser than the endocervical or endometrial cytoplasm.

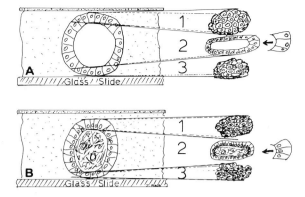

Figure 20
The three distinct layers that must be recognized for the diagnosis of (A) glandular acinus and (B) polypoid structures. They are differentiated by the presence in the polypoid structure of a central, solid, cellular core and by the location of the nuclei toward, rather than away from, the center.

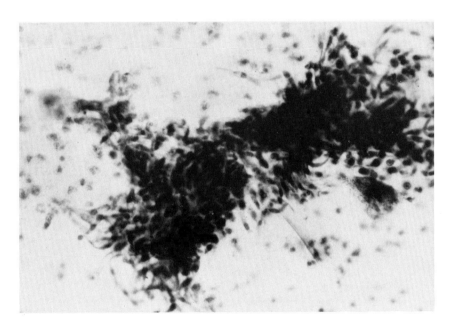

Figure 21
Fallopian tube cells obtained by direct scraping. ($\times$ 450)

The *ciliated cells* have short, heavy, red cilia on one extremity of the cell on top of a heavy terminal bar.

Their nuclei are round or oval with a moderate size variation, and their nuclear borders are sharp and even. The chromatin is finely granular and uniformly distributed.

Occasional multinucleation occurs with the nuclei overlapping as well as molding against each other.

The nucleoli, when present, are minute, red, and spheroidal, and their number varies from 1 to 4.

The size of the cells may vary during the various phases of the menstrual cycle; they reach their maximum size about the twenty-second day.

The *secretory cells* are found mainly in the luteal phase. They vary in size (15–20 μ) according to the amount of secretion contained in their cytoplasm. They are elongated, and their foamy cytoplasm stains green to pink. Their single, oval nucleus is usually eccentric. Several round, small nucleoli are present. Occasional mitoses may be seen.

Cytology of Nonepithelial Cells

MESODERMAL CELLS (FIG. 22)
Physiologically exfoliated mesodermal cells are rare in smears, except for histiocytes and the cellular contaminants from the blood elements. In some instances, well-preserved, elongated, smooth-muscle cells can be seen in cases of traumatic scraping of an ulcerated vaginocervical lesion, a fibroma, polyp, or a scraping taken during abortion. The smooth-muscle cells have elongated nuclei that are uniform with pointed ends (Fig. 22A). On rare occasions clusters of anucleated, pink, elongated smooth-muscle cells (vermiform bodies) may be seen in the vaginal smears of patients with a necrotic fibroma (Fig. 22B).

RED BLOOD CELLS (ERYTHROCYTES) (FIGS. 23A, B)
The fresh red blood cells are biconcave, round, anucleated structures, usually stained pink to orange to green. They are often seen piled up together (rouleau formation). They have little clinical significance. The uniformity of their size (6–7 μ) can be useful in the estimation of the size of the epithelial cells by comparison. Any abnormality of their size or shape—such as sickle cells, target cells, or megalocytes—should be noted and reported, since it may indicate serious blood disease.

The older red blood cells may indicate a pathologic bleeding. They are recognized by their darker brown stain, their granular aspect, and their tendency to cluster into tight irregular clumps. Some of them are found phagocytized by the histiocytes. To differentiate hemosiderin granules from melanin pigment, a special iron stain of the smear may be done.

The presence of fresh blood cells in a smear is often the result of a traumatic scraping, especially if the cervix is congested, as in pregnancy; it has no significance.

During the normal menstrual flow, a mixture of fresh and old blood in large amounts is usually found. Such a smear should be repeated toward

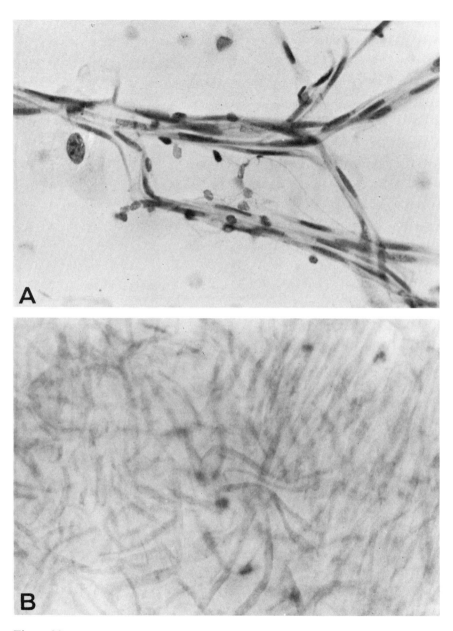

Figure 22
(A) Well-preserved, elongated, smooth-muscle cells in a traumatically scraped smear. (× 450) (B) Poorly preserved, elongated, smooth-muscle cells (vermiform bodies) in the vaginal smear of a patient with a necrotic uterine fibroma. (× 450)

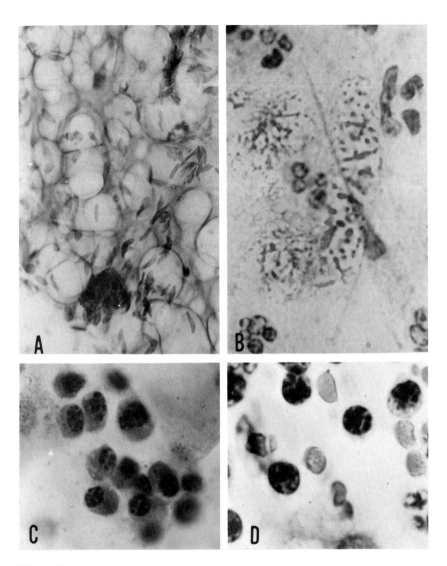

Figure 23
(A) Red blood cells in the vaginal smear of a patient with sickle-cell anemia.
(× 450) (B) Hemolysed red blood cells and neutrophils. (× 450) (C) Cluster
of plasmocytes. (× 450) (D) Scattered lymphocytes. (× 450)

midcycle. The presence of this profuse bleeding can either mask or dilute the diagnostically significant cells.

POLYMORPHONUCLEAR LEUKOCYTES (NEUTROPHILS, EOSINOPHILS, AND BASOPHILS) (FIG. 23B)

In small amounts, these cells may be found in a normal vaginal smear. They originate mainly from the endocervix. Smears from women who have had a total hysterectomy are usually void of any neutrophils. Polymorphonuclear leukocytes are abundant in acute, subacute, or chronic genital inflammatory disease.

When these cells are found in tight clusters in a postmenopausal woman, the clusters should be examined more carefully under high magnification to determine whether there are hidden neoplastic epithelial cells diagnostic of an endometrial adenocarcinoma.

The size of these cells is fairly constant ($10~\mu$), being one and one-half times the size of red blood cells. They also can serve as a measuring unit for estimating the size of other cells.

They have a poorly defined cytoplasm, which is usually transparent but may contain eosinophilic or basophilic granules, ingested debris, or microorganisms.

Their nuclei are lobulated, and each lobule is connected by a fine thread to the next lobule.

The leukocytes can become a diagnostic problem when they are very young or have degenerated, since they may assume all sorts of shapes. When spindly, with a very elongated, stretched pyknotic nucleus, they can, for example, resemble the myoblastic cells seen on occasion in abortion smears. If these inflammatory cells are obscuring the majority of the epithelial cells, the smear should be reported as unsatisfactory and a repeat smear, after clearing the inflammation, recommended.

LYMPHOCYTES (FIG. 23D)

The size of the mature lymphocytes is regular and about the same as that of the red blood cells. Their basophilic nuclei are slightly wrinkled and surrounded by a narrow band of basophilic cytoplasm.

They can be seen mixed with leukocytes in chronic cervicitis or grouped together in large clusters after the rupture, by scraping, of an inflammatory lymphoid follicle, as seen in certain cases of chronic follicular cervicitis.

These lymphocytes should be examined carefully to rule out the possibility of a lymphomatous or leukemic neoplasm. Prominence of their nucleoli and marked variation in their size are suggestive of these diseases and should be reported.

Occasionally, large immature lymphocytes may become confused with cells originating from an in situ or a poorly differentiated squamous cell carcinoma. Nothing looks more malignant than a large lymphocyte examined under oil immersion magnification.

PLASMOCYTES (FIG. 23C)

These inflammatory cells are usually seen in chronic cervicovaginitis and are often found mixed with lymphocytes.

Single, oval, with slight variation in their size (8–12 μ), they usually have an adequate blue-gray cytoplasm.

Their nuclei are eccentric, round, or oval in shape, with the chromatin arranged in large clumps in cartwheel fashion.

These cells degenerate very quickly and can be confused with endometrial stromal cells or small-cell carcinoma.

They have very little phagocytic power and have no prominent nucleoli except in cases of plasmocytic leukemia.

HISTIOCYTES (MACROPHAGES) (FIGS. 24, 25)

A good percentage of the false-positive cytologic diagnoses are due to the wrong interpretation of the nature of these cells. Depending on their activity, the size, shape, and other features of these cells will vary extensively, imitating almost any cell. The cells most difficult to differentiate from small histiocytes in a normal vaginal smear are the endometrial stromal cells (see p. 37).

Occurrence

Histiocytes may originate from the loose reticuloendothelial system, endometrium, or endocervix and are found most commonly:

1. Two to three days before, during, and after the normal menstrual flow;
2. During the early and late stages of pregnancy, in abortion, and in the postpartum period;
3. In an inflammatory erosion or ulceration of the cervix and endocervix;
4. In the presence of a foreign body (e.g., sutures, intrauterine device);
5. After irradiation, biopsy, or cryosurgery.

In postmenopausal women, the presence of histiocytes may indicate an endocervical, endometrial, or ovarian lesion, benign or malignant. After irradiation, the presence of histiocytes is a good prognostic sign. Postoperatively, after a biopsy or cone, their presence indicates normal healing of the tissues.

Types of Histiocytes

SMALL (10–15 μ). Small histiocytes are seen mainly during the menstrual flow. They can be confused with stromal or loose glandular endometrial cells. They are often found in clusters (Figs. 25A, B, and C).

LARGE (15–25 μ). These cells are mainly seen in the presence of cervical erosion or ulceration. Their nuclei will often have a prominent nucleolus and a coarse chromatin pattern (Fig. 25D).

MULTINUCLEATED GIANT CELLS (15–120 μ). These cells may be seen in postmenopausal smears, during an abortion, after irradiation, or in the presence of a foreign body (intrauterine device) (Fig. 25E).

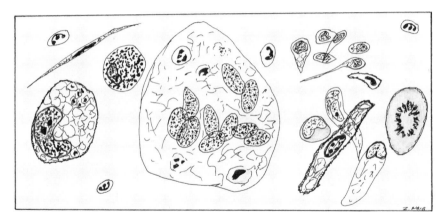

Figure 24
Various shapes and sizes of histiocytes that may be encountered in a vaginal smear.

Cellular Morphology

The most important criterion for differentiation from the epithelial cells is that histiocytes are shed only singly or in clusters of single cells, never in sheets or with cytoplasmic or nuclear molding. The presence of reciprocal molding between two cells indicates that they are probably not histiocytes.

Histiocytes vary enormously in size and shape according to their function and activity. Although usually round or oval, they can be seen as elongated, polygonal, spindle, or triangular.

The amount of cytoplasm varies from one cell to another and can be scanty or abundant. It is generally semitransparent, lacy, or foamy and may contain multiple small or large vacuoles, with or without phagocytized debris. Usually basophilic, the cytoplasm can also stain a bright pink, orange, or brown. Its borders are usually indistinct and delicate.

The nuclei are eccentric but can vary slightly in location from one cell to another. Often they touch the cytoplasmic membrane in one spot, without distending it, thus differing from a malignant cell, in which the nucleus often touches the cytoplasmic border at more than one spot and often molds against it. The nuclear shape varies from the characteristic bean or kidney form to round, triangular, or irregular. The nuclear size varies $(5-15~\mu)$ depending on its activity. Mitosis is so frequently seen that, whenever a single cell in a smear is found in mitosis, it is probably a histiocyte rather than a cancer cell.

The nuclear borders are variable, usually sharp and smooth, but they can be wrinkled and poorly defined.

The chromatin often is regular and finely granular (70%), but it can also be seen in heavy irregular clumping, which again can be confused with the abnormal chromatin of malignant cells. Generally hypochromatic, an increased basophilia can occur as the result of acute irritation or degeneration.

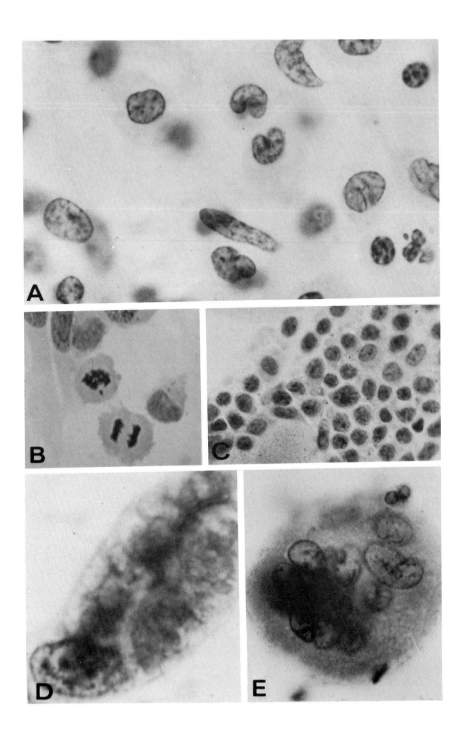

The nucleoli, indistinct most of the time (85%), can, on occasion, become prominent. This is a pitfall in the differential diagnosis between an adenocarcinoma cell and such a histiocyte.

In a multinucleated, foreign-body giant cell (3–100 nuclei), all the nuclei have the same size and shape and chromatin texture (sister image). They are often arranged in a central area. They overlap but do not mold, and this differentiates them from the other multinucleated giant cells seen in the vaginal smear (trophoblastic, endocervical, or herpesvirus infected cells).

NONEPITHELIAL CONTAMINANTS (FIG. 26)

Sperm (Fig. 26A)
The head of the sperm, stained unevenly blue-to-gray, with or without its tail, can be seen persisting in a vaginal smear even 4 or 5 days after sexual intercourse. In large numbers, they can obscure the smear and become a handicap in diagnosis. They should not be confused with the round bodies of a monilial infection or the protozoa in *Trichomonas* infestation. Their recognition can be important in the determination of a suspected rape. To prevent unnecessary complications, their presence should not be reported except in response to a specific inquiry from the physician.

PINWORM (*ENTEROBIUS VERMICULARIS*)
On occasion, this parasite may be seen in the routine vaginal smear. It is easily recognized. The orangeophilic oval eggs, containing a mature larva, measure approximately 60 by 30 μ in diameter. They are recognized by their outer refractile shell with occasional central indentation. They need to be differentiated from vegetable contaminants. No specific changes in the cellular components of the smear are found.

Pollen (Fig. 26C, D)
Size and appearance vary according to the different types of pollen. Pollen is found more frequently during spring and late summer. Because of the transparent, glassy capsule usually surrounding the cell, these structures are easy to recognize. On occasion, the deep granular, orangeophilic cytoplasm and large, uniform nuclei of pollen may be confused by an inexperienced observer with a degenerate, malignant, well-differentiated squamous carcinoma cell.

Figure 25
Histiocytes. (A) Number of small histiocytes in the vaginal smear taken before the menses. Note the marked variation of their shape with occasional bean-shaped nuclei. ($\times$ 750) (B) Two histiocytes seen in a vaginal smear in different stages of mitosis. ($\times$ 750) (C) Loose clusters of small histiocytes, which are often difficult to differentiate from endometrial stromal cells. ($\times$ 450) (D) High-power magnification view of a histiocyte. Note the vacuolated cytoplasm, the indentation of the nuclear membrane, and the prominence and number of nucleoli. ($\times$ 1500) (E) Large foreign-body giant cells. The multiple nuclei overlap without molding. ($\times$ 750)

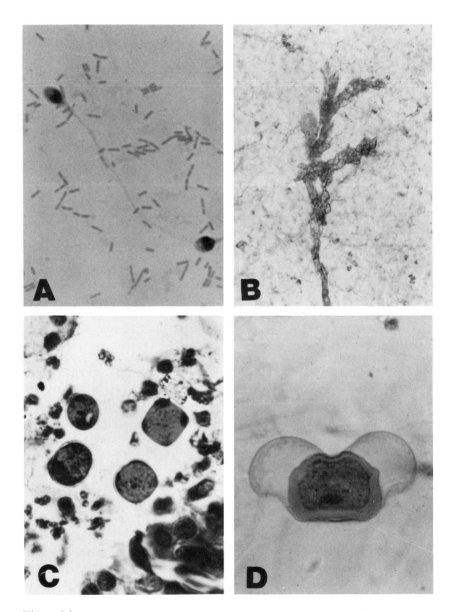

Figure 26
(A) Sperm and Döderlein bacilli. (× 750) (B) Sulfa crystals after treatment for an acute vaginitis. (× 450) (C, D) Pollen contaminants (× 450)

Lubricant

Deep-purple-stained, irregular masses of lubricant that have been used unnecessarily on the speculum may sometimes be seen masking a good portion of the smear. This should not be confused with physiologic, smeared endocervical mucus, which stains light pink to red.

Talcum

Starch talcum granules, which are deep-purple-stained or transparent with characteristic Maltese-cross-like features when viewed with a polarized light, may be seen covering the smear and hiding the cells. These can be mistaken for the nuclei of malignant cells. They are easy to recognize when examined under polarized light. The clinician should be reminded to eliminate all dust before touching his patient, instruments, or slides.

Yeast

Large numbers of round or filamentous bodies may be found in a vaginal smear, usually in groups of tight colonies, which stain dark blue to purple. Yeast can mask the diagnostically significant epithelial cells.

References and Supplementary Reading

Bertalanffy, F. D.   Aspect of cell formation and exfoliation related to cytodiagnosis. *Acta Cytol.* 7:362, 1963.

Boschaun, H. W.   Definition of a superficial cell. *Acta Cytol.* 2:52, 1958.

Boschaun, H. W.   Cytomorphology of normal endometrium. *Acta Cytol.* 2: 505, 1958.

Brux, J. de, and Bret, J.   Cells originating in ulcerated submucous myomas of uterus; so-called "vermiform bodies." *Am. J. Clin. Pathol.* 32:422, 1959.

Colmenares, R. F., and Naib, Z. M.   Significance of lymphocytic pools in the routine vaginal smear. *Obstet. Gynecol.* 26:909, 1965.

Dazo, E., Whitehead, N., and Solomon, C.   Histogenesis of microvilli and cilia in the endometrial cells. Morphologic and cytochemical study. *Acta Cytol.* 14:586, 1970.

Dougherty, C. M.   Relationship of normal cells to tissue space in the uterine cervix. *Am. J. Obstet. Gynecol.* 84:648, 1962.

Dowling, E. A., and Gravlee, L. C.   Endometrial cancer diagnosis: A new technique using a jet washer. *Ala. J. Med. Sci.* 1:412, 1964.

Eisenstein, R., and Battifora, H.   Lymph follicles in cervical smears. *Acta Cytol.* 9:344, 1965.

Fluhmann, C. F.   The squamocolumnar transitional zone of the cervix uteri. *Obstet. Gynecol.* 14:133, 1959.

Fluhmann, C. F.   *The Cervix Uteri and Its Diseases.* Philadelphia: Saunders, 1961.

Fornari, M.   Cellular changes in the glandular epithelium of patients using IUCD. *Acta Cytol.* 18:341, 1974.

Graham, C.   Cyclic changes in the squamocolumnar junction of the mouse cervix uteri. *Anat. Rec.* 155:251, 1966.

Graham, R. M.   The small histiocyte. Its morphology and significance. *Acta Cytol.* 5:77, 1961.

Hopman, B. C., Wargo, J. D., and Werch, S. C.   Cytology of vernix caseosa cells. *Obstet. Gynecol.* 10:656, 1962.

Krantz, K. E.   The gross and microscopic anatomy of the human vagina. *Ann. N.Y. Acad. Sci.* 83:89, 1959.

Lianes, A., et al.   Scanning electron microscopy of normal exfoliated squamous cervical cells. *Acta Cytol.* 17:507, 1973.

Liu, W., et al.   Normal exfoliation of endometrial cells in premenopausal women. *Acta Cytol.* 7:211, 1963.

McLennan, C. E., and Rydell, A. H.   Extent of endometrial cells shedding during normal menstruation. *Obstet. Gynecol.* 26:605, 1965.

Nasiell, M.   Histiocytes and histologic reactions in vaginal cytology. *Cancer* 14:1223, 1961.

Oudkiewicz, J.   Quantitative and qualitative changes of epithelial cells of fallopian tubes in women according to the phase of menstrual cycle: A cytologic study. *Acta Cytol.* 14:531, 1970.

Rubio, C. A., Sigurdson, A., and Zajicek, J.   Viability tests in exfoliated cells from vaginal and oral epithelia. *Acta Cytol.* 17:32, 1973.

Schuller, E. F.   The epithelia of the uterine endocervix. *Acta Cytol.* 3:333, 1959.

Schuller, E. F.   Ciliated epithelia of the human uterine mucosa. *Obstet. Gynecol.* 31:215, 1968.

Spjut, H. J., Kaufman, R. H., and Carrig, S. S.   Psammoma bodies in the cervico-vaginal smear. *Acta Cytol.* 8:352, 1964.

Wells, L. J.   Embryology and anatomy of the vagina. *Ann. N.Y. Acad. Sci.* 83:80, 1959.

Endocrine and Pregnancy Cytology 3

Physiology of the Endocrine Cycle in Women

In infancy and childhood, the ovaries are virtually inactive. Only a very small amount of estrogen, and no progesterone, is produced. At puberty, in response to stimuli from the central nervous system and the hypothalamus, the anterior pituitary gland in turn stimulates the ovaries via the bloodstream by means of the secretion of various gonadotropic hormones. The ovaries not only produce the ovum needed for reproduction but also secrete hormones that control the growth and proliferation of the endometrium, the maturation of the vaginal epithelium, and the regulation and growth of other organs.

As illustrated in Figure 27, under the influence of the follicle-stimulating hormone (FSH) released from the anterior pituitary gland, a primordial follicle of the ovary begins to enlarge and to produce increasing amounts of estrogen; these in turn stimulate the proliferation of the endometrium and the maturation of the vaginal mucosa (follicular phase of the cycle).

As the ovum matures and estrogen production increases, another pituitary hormone, the luteinizing hormone (LH) is secreted, which acts on the mature follicle, causing its rupture and the release of the ovum (ovulation). This occurs near the fourteenth day of the normal cycle, when estrogen production is at a maximum. The luteinizing hormone continues to be produced by the pituitary gland, causing the development of the corpus luteum in the ruptured follicle, and a third pituitary hormone, the luteotropic hormone (LTH), then maintains the structure of the newly formed corpus luteum and stimulates its secretion of progesterone. Increasing amounts of progesterone are then produced by the ovary (corpus luteum), transforming the endometrium from the maximum proliferative phase, reached under the influence of estrogen, to the secretory phase in preparation for the possible implantation of the fertilized ovum. The high level of progesterone acts on the pituitary gland and hypothalamus and inhibits the secretion of luteinizing hormone and luteotropic hormone; this in turn causes involution of the corpus luteum. If there is no implantation, progesterone then diminishes to a level insufficient to maintain the integrity of the endometrium, which begins to break down and slough off, appearing as the menstrual flow, which lasts for four or five days. Without progesterone-estrogen

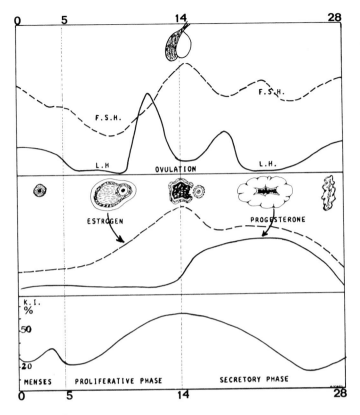

Figure 27
Curves of the level of different hormones and the karyopyknotic index during a normal menstrual cycle. FSH = follicle-stimulating hormone; LH = luteinizing hormone; and KI = karyopyknotic index.

inhibition, the cerebral nervous system once again starts stimulating the hypothalamus and pituitary gland to release follicle-stimulating hormone for the next cycle.

If the ovum is fertilized and implants in the prepared secretory endometrium, the corpus luteum continues to produce progesterone, and the gonadotropic hormones produced by the growing placenta almost completely supplant the function of the corpus luteum by the third month of pregnancy.

At menopause, in a majority of women, the function of the ovaries progressively decreases. Ovulation becomes irregular, then ceases, with a corresponding decrease of estrogen and progesterone production.

CYTOLOGIC HORMONAL EVALUATION
Hormones may influence, more or less, the morphology and staining characteristics of the endocervical, endometrial, vaginal, urethral, and bladder mucosa cells. The cytologic hormonal evaluation is usually based on the determination of the degree of maturation and glycogen storage of the squa-

mous cells exfoliated from the intact surface of the vaginal epithelium. Therefore, *it is most important* that the smear should be taken from a healthy vaginal mucosa and not from the cervix with an ectropion or an ulcerated area of the vagina. This is a sensitive qualitative method rather than a quantitative one.

Only the well-preserved squamous epithelial cells, shed singly or in loose clusters, should be evaluated. The columnar and the inflammatory cells, the epithelial cells that have exfoliated in sheets, and the histiocytes should not be counted. The endocervical epithelium, other than demonstrating an increase in mucus secretion toward the middle of the cycle, shows few menstrual changes.

To minimize the action of many nonhormonal factors that can affect this determination, the evaluation of a series of daily vaginal smears is preferred to the examination of one single specimen. The information obtained from one smear is often irrelevant.

CYTOLOGIC INDICES

The cytologic assessment of the hormonal condition of a patient is given by different types of indices. All are based on recognition and exact typing of the epithelial cells exfoliated from the surface of the stratified squamous vaginal mucosa. If a smear is too inflammatory and a repeat specimen is not possible after the proper treatment, the exfoliated urethral cells in the first portion of a voided urine specimen can be examined for this evaluation. The terminal part of the urethral mucosa is as sensitive to hormonal changes as is the vaginal mucosa. This method can also be used in the cytologic hormonal assessment of the male.

As summarized and illustrated in Figure 28, the most common indices used at the present time are as follows:

Maturation Index (Fig. 29)

In determining the maturation index (MI), the percentages of the basal, intermediate, and superficial cells are presented as a three-part ratio with the basal cells stated first, the intermediate cells second, and the superficial cells third; for example, MI = 80/20/0 indicates that 80 percent of the squamous cells counted are parabasal in type, 20 percent intermediate in type, and no superficial cells are seen. The results are read as "shift to the right," which indicates an increase of estrogenlike effect; "shift to the left" signifies an atrophic effect; and "shift to midzone" means a progesteronelike effect.

Karyopyknotic Index or Cornification Index

The percentage of squamous epithelial cells with sharp, squared-off cytoplasmic edges and with pyknotic nuclei is given in relation to all the other mature squamous epithelial cells that possess vesicular nuclei, regardless of the type or staining quality of the cells. The parabasal cells are not counted.

Eosinophilic Index

The number of mature squamous cells with pink eosinophilic cytoplasm, regardless of their nuclear appearance, is compared with the number of

55

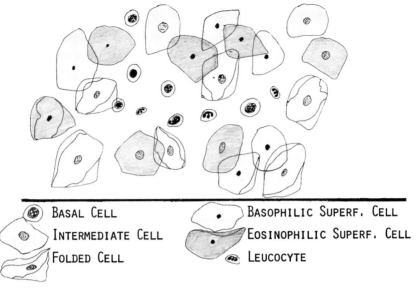

BASAL CELL   BASOPHILIC SUPERF. CELL

INTERMEDIATE CELL   EOSINOPHILIC SUPERF. CELL

FOLDED CELL   LEUCOCYTE

Figure 28
The most common indices:

Number of epithelial cells	25	Folded cell index	8/12
Karyopyknotic index	9/11	Crowded cell index	6/14
Eosinophilic index	7/13	Superficial cell index	9/16
Maturation index	5/11/9	Maturation value $0 + 5.5 + 9 = 14.5$	

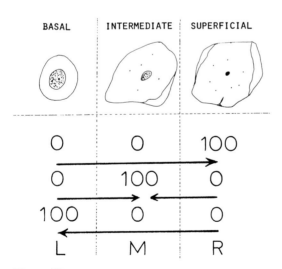

Figure 29
The principle of determining the maturation index for cervical vaginal smear.
L = shift to left, M = shift to middle, and R = shift to right.

other mature squamous cells with blue (basophilic) cytoplasm, and the result is given in the form of a ratio. The parabasal cells should be omitted from the count.

This eosinophilia should be a true one and not the result of a degenerative artifact, as may occur when a smear is air dried before its fixation. It is wise to check the staining of the cytoplasm of the neutrophils in the smear. If their cytoplasm is pink-orange instead of the normal pale blue-green, the eosinophilia of all the cells is probably artificial and no count should be made.

Folded Cell Index

The number of mature squamous cells (superficial or intermediate in type) with folded cytoplasmic rims is compared with the number of mature cells with a flat, nonfolded cytoplasm, regardless of the type of nucleus and the cytoplasmic staining reaction.

Superficial Cell Index

This index resembles the karyopyknotic index, except that all intact squamous cells with any type of nuclei are counted. Only the superficial cells, regardless of the staining quality of their cytoplasm, are evaluated and given in relation to all the other squamous cells present in the smear.

Crowded Cell Index

The number of mature squamous cells that are clustered together in groups of five or six cells is compared to the number of similar cells found singly or in groups of fewer than three or four.

Maturation Value

This method is especially useful for providing hormonal evaluation data to a computer. Each parabasal cell is counted as 0, each intermediate cell is counted as 0.5, and each superficial cell as 1. The addition of all the values given to the first 100 epithelial cells is recorded.

PITFALLS IN CYTOLOGIC HORMONAL EVALUATION

Inflammation

Genital inflammation may produce a slight hypermaturation of the vaginal mucosa. This effect is in direct proportion to the intensity of the inflammation. An artificial shift of the MI to the right may be present when the inflammation is of long duration. Furthermore a large number of inflammatory cells can mask the epithelial cells or produce severe cellular degenerative changes.

In the presence of inflammation, no hormonal evaluation should be attempted. A repeat should be recommended following antiinflammatory treatment.

Erosion

When the vaginal mucosa is eroded, the basal cells, normally located at the bottom of the stratified squamous epithelium, are then at the surface, and their exfoliation will create the impression that there is false cytologic atrophy of the mucosa. The shape of these parabasal cells (repair cells) varies more than the shape of atrophic parabasal cells (see p. 113).

Faulty Fixation and Staining Defects

This pitfall occurs when the smear is too thick or when an improper staining technique has been used. The result can be artificial eosinophilia and distortion of the epithelial cells. Therefore, no attempt should be made to evaluate these distorted cells.

Cytolysis of the Cells

When more than 50% of the cells show degenerative changes or if there is a large number of stripped nuclei, no hormonal evaluation should be attempted.

Lack of Information

The time of the collection of the smear in relation to the previous menstrual period (PMP) and the last menstrual period (LMP) should be known for the proper hormonal evaluation of a smear. Similarly, when chemotherapy and radiotherapy have been given, this should be mentioned to the cytopathologist; otherwise the hormonal evaluation cannot be relied upon.

A combination of the effect of two or three different hormones is difficult to assess.

NORMAL CYTOHORMONAL AVERAGES

Newborn (Table 3)

MI = 0/90/10 ± 10.   The increased number of intermediate cells, often containing glycogen in their cytoplasm, in the vaginal smear of the newborn is the result of the persisting effect of maternal hormones in the infant's blood. These intermediate cells are often irregular in shape and may exfoliate in sheets. A small number of superficial cells with abundant, semitransparent, and wrinkled cytoplasm may also be seen. Very little cellular debris, bacteria, lymphocytes, or mucus can be found in the first four to eight days, but slowly, during the second and third weeks, there is an increase in the number of leukocytes and microorganisms. The background of the smears progressively becomes what is termed *dirty*. After the first week, due to the progressive decrease of maternal hormones, some basal cells begin to appear, and the intermediate cells progressively become more degenerate and often are seen in lysis. The occasional anucleated squamous cells seen at this time originate from the degeneration of these intermediate cells. Fresh and old red blood cells and an excess of mucus may also be present.

Infancy (from 3 Weeks to Puberty) (Table 3)

MI = 80/20/0 ± 20.   The vaginal smears are scanty in cells and contain mainly parabasal type cells. The cytoplasm is dense and cyanophilic; the nuclei are large, with a thin, regular membrane. The chromatin is finely

Table 3.   Cytohormonal Averages

	Maturation Index	Maturation Value	Variation (±)
Newborn	0/90/10	55	10
Infancy	80/20/0	10	20
Preovulatory	0/40/60	80	10
Postovulatory	0/70/30	65	15
Menopause	0/80/20	60	20
Postmenopause	50/50/0	25	40
Estrogen therapy	0/10/90	95	10
Progesterone therapy	0/90/10	55	10
Androgen therapy	20/80/0	40	10

granular and uniformly distributed. A variable number of leukocytes is also present. These smears are very similar to the cells found in the late post-menopausal period (atrophic vaginitis) except for their smaller size. This cellular pattern persists until two to four weeks before puberty and the beginning of the menses.

Menstrual Age (Reproductive Period) (Table 3)

MI = 0/60/40 ± 20.   The cytologic changes of the vaginal smear of a woman at the different stages of her cycle are less dramatic than those found in mice or rats, where anucleated eosinophilic squamous cells are the only component cells during estrus, and basal cells and polymorphonuclear cells are the main cells found during the remainder of the cycle. In the smears of most women, superficial and intermediate cells are always present, but no basal cells should be found if the mucosa is intact. During the preovulatory time (third to fourteenth day of the cycle), the average maturation index is 0/40/60 ± 10. Toward the eighth day of the cycle, the cyanophilic intermediate cells gradually increase in size with progressive cytoplasmic eosinophilia and nuclear pyknosis. Simultaneously there is a progressive decrease in the number of leukocytes and the amount of mucus, except when secondary erosion or inflammation is present.

During ovulation and postovulatory time (fifteenth to nineteenth days), the maturation index becomes approximately 0/70/30 ± 15. The number of superficial and eosinophilic cells progressively decreases, while the number of intermediate cells increases, and the cells show a tendency to clump together. With the increased number of Döderlein's bacilli, cytolysis of the intermediate cells appears, and the background of the smear becomes dirty and granular. During the menstrual flow, the maturation index varies enormously (0/60/40 ± 20). There is a marked increase in the number of fresh and old red blood cells, cyanophilic intermediate cells, histiocytes, mucus, and leukocytes. Endometrial and stromal cells can be readily seen exfoliating singly, in clusters, or in acinic formation. To evaluate properly the physiologic variation of the maturation of the cells, a series of daily smears should be taken and placed in fixative to be stained and evaluated all at the same time in order to decrease some of the artificial factors that can influence the determination of the maturation level (staining and fixation artifacts). The report may then indicate if ovulation has occurred, as well as the level of estrogenic effect.

Menopause (Table 3)

In the early stage of the menopause, the maturation index varies greatly, averaging 0/80/20 ± 20 (shift to the middle). The exfoliated superficial and intermediate cells become progressively smaller with some decrease in their staining capacity. Döderlein's bacilli multiply and produce cytolysis of the intermediate cells and a moderate increase in the number of stripped nuclei.

In the postmenopausal period, two to six years following the cessation of menstruation, a progressive decrease of the estrogenic activity occurs with a corresponding increase in the number of parabasal cells in the smear. These parabasal cells are mainly seen in sheets in the scraped specimens. Often, parabasal cells covering the surface of the mucosa seem to be air dried, in spite of correct precautions in making the smear, due to the decrease in the normal vaginal moisture. Glycogen may be found in the cytoplasm of these cells. They are differentiated from the glycogen-containing navicular cells of pregnancy by their round shape, in contrast to the elongated, boatlike shape of the navicular cells. There is also a variable increase in the number of leukocytes and Döderlein's bacilli, which lyse the cytoplasm of some of the intermediate squamous epithelial and endocervical cells to produce a variable number of stripped nuclei.

Finally, late in the postmenopausal period, the smears of some women may become completely atrophic. The maturation index is then 100/0/0 ± 10. The size of the parabasal cells varies, but the shape remains round or oval with mild irregularities. The cyanophilic cytoplasm may show degenerative changes in the form of vacuolization. No glycogen is present. Some of the basal cells have small, dark nuclei, often pyknotic, with scanty cytoplasm stained bright orange. The background of the smear at this stage is usually dirty and granular. Hemolyzed fresh and old red blood cells may be found. When exfoliated traumatically in sheets, the basal cells can be distorted into short, spindlelike shapes; they are not to be confused with the spindle cells of an invasive carcinoma, which are longer and more keratinized.

A differential diagnosis can be made by giving the patient a small amount of estrogen. If these basal spindle cells are benign, they will disappear from the smear. If they persist, they are probably atypical.

Some of the basal cells in an atrophic smear may join to form a syncytial multinucleated cell that differs from foreign-body giant cells in variation in the size of the nuclei and the absence of phagocytized debris.

An increased estrogenic effect can be seen in some obese, hypertensive, or diabetic women even in the absence of any hormone-producing disease of the ovaries. Similarly, in about 15% of women who have passed the age of 80, an unexplained increase in the number of superficial cells in the smear can be found.

EFFECT OF EXTRINSIC HORMONES ON VAGINAL CYTOLOGY

For the study of the specific action of a hormone on the vaginal epithelium, the hormone should preferably be given to a woman with an atrophic mucosa.

Estrogen (Fig. 30A) (Table 3)

MI = 0/10/90 ± 10.   Estrogen produces a proliferation of all the layers of the squamous epithelium with an increase in the amount of intracellular glycogen due to an increase in protein and nucleic acid synthesis, which results in proliferation of the basal cells by increasing mitosis and shortening the resting period of the cells. The parabasal cells progressively enlarge to become intermediate in type. These in turn continue to enlarge and mature into eosinophilic superficial cells. The number of leukocytes in the smear decreases. Most of these changes in the vaginal mucosa are not directly related to the amount of estrogen administered. For example, 5 mg of estradiol will produce the same changes at almost the same rate as the administration of 25 mg of the same substance. Similarly, this effect is independent of the method of administration. There is no way to determine from looking at a smear the amount of estrogen given. Although some of the changes occur very early, the maximum effect is not reached until seven to ten days. A high estrogenic effect in a vaginal smear is recognized by a clean background and increased eosinophilic and karyopyknotic indices.

Progesterone (Fig. 30B) (Table 3)

MI = 0/90/10 ± 10.   Progesterone produces a proliferation of the intermediate layer of the squamous epithelium and the predominance of clusters of intermediate cells in the smear. The amount of intracellular glycogen in these cells increases. The number of leukocytes moderately decreases and the number of Döderlein's bacilli in the background, if present, increases. This effect is not directly related to the amount of progesterone given. The result and the rate of these changes are similar if the patient has received 30 mg or 100 mg of the hormone. The maximum effect of progesterone is reached in an average of four to twelve days after administration. When progesterone and estrogen are combined, the rate of exfoliation of intermediate and superficial cells increases. These cells also exhibit a tendency to fold on themselves and clump together. Progesterone given to a patient with an atrophic vaginal epithelium resulting from bilateral oophorectomy will have little effect.

Androgenlike Hormones (Table 3)

MI = 20/80/0 ± 10.   In case of masculinizing tumors (arrhenoblastoma, adrenal rest tumor), the excessive androgen produced may be detected by the presence in the vaginal smear of numerous large parabasal cells with

61

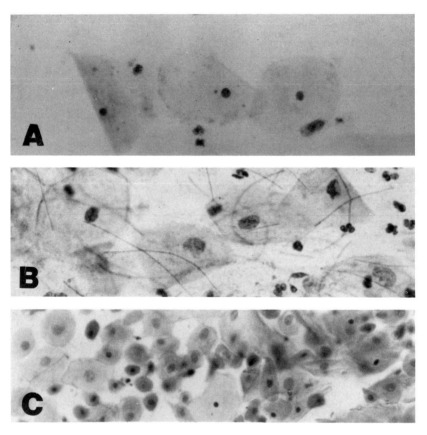

Figure 30
Vaginal smear showing (A) estrogenlike effect, (B) progesteronelike effect, and (C) atrophy.

central, almost pyknotic, nuclei and a dense, glycogen-containing cyto-plasm. If these hormones are administered to a postmenopausal patient with an atrophic mucosa, a proliferation of the basal and intermediate layers occurs, but without maturation into a superficial or keratinized layer. There is a gradual increase in the number of intermediate exfoliated squamous cells. The number of basal cells diminishes, while the number of intermedi-ate cells increases. These intermediate cells, usually lying singly, are often irregular in shape, have a highly cyanophilic cytoplasm that is rich in gly-cogen deposits, and have a round or oval, vesicular nucleus. Leukocytes and background debris decrease. On the average, six to eight days are re-quired for the maximum effect to appear following the administration of 50 to 75 mg of testosterone, for example. If an excessive amount of the andro-genic hormone is given (more than 200 mg), an estrogenic type of prolifer-ation rather than atrophy may occur with an increase in the number of superficial and intermediate cells in the smear. If androgen and estrogen are given at the same time, a shift to the middle in the maturation index generally occurs.

Adrenocortical Hormones

The administration of this type of hormone has the same action as the administration of progesterone, except that the dosage required is two to three times greater and little storage of glycogen occurs.

Gonadotropic Hormones

This type of hormone has no direct action on the proliferation of the vaginal epithelium, but it acts on the ovaries to increase the production of estrogen or progesterone or both. If the ovary is atrophic or missing, a slight proliferation of the intermediate layer can occur in the vaginal epithelium as the result of the stimulation of the adrenal cortex.

ACTH

There is disagreement concerning the effect of the administration of adrenocorticotropic hormone (ACTH) on the vaginal smear. Usually there is a mixed proliferative reaction, and the response is secondary to the presence of a functioning adrenal cortex.

Digitalis

The administration of this drug will produce a shift to the middle with proliferation of the intermediate cells in a premenopausal woman, but the same drug given for two years or longer to a postmenopausal patient or to a patient with a decreased liver function will cause a shift of the maturation index toward the right with an appreciable increase in the number of superficial cells. The molecular similarity of digitalis to estrogen-progesterone may explain their similar effect.

EFFECT OF NONHORMONAL FACTORS ON VAGINAL CYTOLOGY

Two smears showing the same indices do not necessarily indicate a similar hormonal status.

A bacterial or *Trichomonas* infection will produce an artificial eosinophilia.

Döderlein's bacillus, which affects primarily the intermediate cells, produces stripped nuclei by cytolysis, preventing the possible differentiation of the cells into the superficial type. The true estrogenic effect cannot be evaluated in such a case. For this reason, antiinflammatory treatment (sulfa or antibiotic drugs) should be given to such patients before the evaluation.

Degeneration of the epithelial cells can produce irregularity in maturation.

The presence of cervical erosion, ectopy, or a foreign body will produce an increased number of parabasal cells in the vaginal smear.

A smear taken after irradiation will show an increased number of basal cells resistant to estrogenic stimulation.

Avitaminosis (vitamins A and C) can cause a shift to the right or to the middle by the production of a generalized hyperkeratosis of the mucosa.

63

ABNORMAL CYTOHORMONAL AVERAGES

Primary Amenorrhea

This condition, in which menses have never occurred, can be the result of an ovarian or pituitary gland dysfunction or of a congenital abnormality. Cytology may help in the diagnosis of the cause by the evaluation of the maturation index, the determination of the presence of the female chromosomal mass (Barr bodies), and the detection of the absence or occurrence of ovulation.

PITUITARY DYSFUNCTION (DWARFISM, NEOPLASM) will produce infantilism as a result of hypogonadotropism and adrenal hormone insufficiency. The vaginal mucosa is atrophic, and the smear in such a case shows mainly small parabasal cells similar to the cells found in the smear from normal infants. $MI = 100/0/0 \pm 5$.

OVARIAN DYSFUNCTION (AGENESIS, NEOPLASM) will produce a partial vaginal mucosal atrophy with a maturation index shift to the middle, $MI = 20/80/0 \pm 20$. This is a result of the persisting normal activity and hormonal production of the adrenal cortex, which produces a proliferation of the intermediate layer in the squamous vaginal epithelium.

VIRILIZING ADRENAL HYPERPLASIA can be the cause of a primary amenorrhea. The maturation index will show androgenic changes with a preponderance of intermediate cells.

TESTICULAR FEMINIZATION SYNDROME demonstrates an increase in the number of superficial cells in the vaginal smear but without any cyclic variation, $MI = 0/20/80 \pm 15$. In these cases it is critically important to collect daily smears instead of only one specimen.

CONGENITAL ABNORMALITIES of the uterus (atresia or absence of the vagina or uterus) are such that the vaginal mucosa, when present, has a normal cyclic ovulatory change and the maturation index is within the normal limits. The female chromosomal mass (Barr body) is normal.

THYROID DYSFUNCTION (HYPOTHYROIDISM, CRETINISM) may produce vaginal atrophy, $MI = 80/20/0 \pm 20$. The sex chromosome Barr body is normal.

CHROMOSOMAL ABNORMALITIES are found, for example, in Turner's syndrome (testicular feminization), in which usually no ovulation occurs. An abnormal sex chromosome pattern and a consistently increased estrogen-like effect (shift to the right in the MI) occur.

Secondary Amenorrhea

This condition exists when the menses cease after variable lengths of normal cyclic menstruation. The reasons for secondary amenorrhea are complex and there is no typical cytohormonal pattern. The need for a series of daily smears over a period of time is important in evaluating the presence or absence of ovulation. Depending on the cause of the amenorrhea, the vaginal mucosa will vary from highly proliferative to atrophic. One should always consider the possibility that there is a physiologic reason for the amenorrhea (pregnancy, lactation, menopause, hysterectomy, irradiation, or ovariectomy) before investigating a pathologic cause.

UTERINE, ADNEXAL, OR SYSTEMIC DISEASES may be of inflammatory, diabetic, anemic, or psychosomatic origin. The vaginal smears are generally not affected, and the maturation index will show virtually any pattern within the normal limits.

ENDOCRINE LESIONS may cause a secondary amenorrhea that is often reflected by abnormal changes in the maturation of the vaginal mucosa.

The Chiari-Frommel Syndrome. This is a disorder of the hypothalamus that triggers the production of excess lactotropic hormones (amenorrhea, galactorrhea, and obesity). The vaginal smears show persistent atrophy, $MI = 100/0/0 \pm 5$.

Hypopituitarism. This condition is often the result of a postpartum pituitary necrosis (Simmonds' disease). The vaginal smear progressively becomes atrophic with a corresponding increase of the small single parabasal cells, $MI = 80/20/0 \pm 18$.

Acromegaly (eosinophilic adenoma of the anterior pituitary gland). The maturation index shows a marked shift to the middle with a preponderance of enlarged intermediate cells, $MI = 0/100/0 \pm 10$.

Feminizing Ovarian Tumors (luteal cyst, granulosa cell tumor, and thecoma). These lesions produce an excessive amount of estrogen, which results in a preponderance of superficial cells in the vaginal smear with an increase of eosinophilia and karyopyknosis, $MI = 0/0/100 \pm 5$. No cyclic changes can be seen in the repeated smears.

Virilizing Ovarian Tumors (arrhenoblastoma and ovarian hilar cell tumor). These lesions produce an excess of androgenlike hormone. The smear will show a preponderance of intermediate cells, $MI = 10/80/0 \pm 10$. No cyclic variation is detectable. Less than 5% of superficial cells can be seen. The removal of the tumor usually produces an abrupt and rapid (five days) return to normal of the hormonal cellular pattern.

Stein-Leventhal Syndrome. There is no characteristic pattern of the maturation index. Androgenlike effects or a combination of estrogenlike and androgenlike effects may be present, $MI = 0/60/40 \pm 30$. No cyclic fluctuation can be observed.

Castration. For a period of a few days after the ablation of the ovary, there is an increase of the basal cells, $MI = 90/10/0 \pm 10$. Then, progressively, there is an increase of intermediate cells, and the smear begins to show an androgenlike effect with a preponderance of single intermediate cells with cyanophilic, often degenerate, irregularly shaped, pale cytoplasm. An increased amount of glycogen is often present. These changes result from increased compensatory hormonal secretion by the adrenocortical glands. If these adrenal glands do not react, the smear remains atrophic. No cyclic changes are present.

Menorrhagia, Metrorrhagia, and Menometrorrhagia

The determination of the cause of excessive bleeding during menses (menorrhagia) or between the menses (metrorrhagia) or a combination of both (menometrorrhagia) in a woman can often be helped by the determination of the vaginal epithelial maturation. In this instance it is very important to evaluate several vaginal smears instead of only one. A persisting hypermaturation of the vaginal mucosa (MI shifted to the right) will indicate an excess of estrogenlike effect, which may cause an endometrial hyperplasia and vaginal bleeding. The increase in estrogen can be the result of:

1. Functioning ovarian tumor. Thecoma and granulosa cell tumors;
2. Nonfunctioning neoplasm. An unexplained pseudoestrogenic effect is sometimes seen in an endometrial, ovarian, or breast carcinoma;
3. Excessive administration of estrogenlike compounds;
4. Endometrial polyp and hyperplasia;
5. Certain systemic diseases.

Cytology of Pregnancy

During pregnancy the cyclic changes cease, and there is a cellular proliferation of the basal and intermediate layers that causes the thickening of the vaginal squamous epithelium. This is partially the result of the secretion of chorionic gonadotropin by the placenta and of progesterone and estrogen by the ovaries. The moderate estrogenlike effect seen early in pregnancy is gradually replaced by a progesteronelike effect. The pregnancy changes in a vaginal smear can be seen early (seven to eight weeks) and persist until 5 to 10 days before delivery. Although exfoliative cytology is sometimes used as a pregnancy test, it should not be. Other, more sensitive methods exist.

GENERAL CHANGES

In the first three months the karyopyknotic index progressively decreases to a level between 20% and 10% and remains constant until before delivery with few exceptions, although 8% of the patients with eventual normal term delivery of a healthy baby may not show detectable pregnancy changes in their smear and will have a continuously high maturation index. The maternal maturation index is not influenced by the sex of the fetus.

The background of the smear is usually partially covered with cellular debris and granular deposits that result from the increased serum protein exudate, cellular lysis, and endocervical hypersecretion.

The exfoliated intermediate cells have a tendency to clump and cluster together.

The incidence and the number of inflammatory cells, Döderlein's bacillus, *Leptothrix,* and *Trichomonas* increase.

Variable amounts of small histiocytes can normally be found in the early and late stages of pregnancy. When found in very large numbers and in the presence of foreign-body giant cells, the possibility of a threatened abortion should be considered.

The presence of increased fresh and old red blood cells does not necessarily indicate trouble because most of the time they are the result of the scraping of the fragile and congested cervical mucosa.

An increased number of stripped nuclei indicates lysis of the intermediate cells by Döderlein's bacilli, even in the absence of inflammation.

PARABASAL CELLS AND PREGNANCY CHANGES (FIG. 31A)

A true cellular hypertrophy occurs in 40% of the cells. The size of the parabasal cells can increase twofold or threefold without being atypical. The consistency and density of the cytoplasm and nuclei remain the same, and they are not diluted. This differentiates true hypertrophy from the degenerative edematous hypertrophy of the cell. The latter is rarely seen in a normal pregnancy smear, except when inflammation and erosion are also present.

The nuclei of the normal parabasal cell may increase in size, and multi-nucleation often occurs.

The chromatin increases in granularity and often clumps regularly and becomes prominent. This clumping should be differentiated from the malignant type, which is more irregular and has pointed projections.

The nucleoli remain central and spherical but also become more prominent. Their number can vary normally from 1 to 4. The cytoplasm is slightly more basophilic and more abundant.

It is sometimes difficult to differentiate the cellular changes of pregnancy from the cells of a squamous basal cell hyperplasia.

INTERMEDIATE CELLS AND PREGNANCY CHANGES—
NAVICULAR CELLS (FIGS. 31B, 35B)

Pregnancy causes an accumulation of glycogen in the cytoplasm of parabasal and intermediate squamous cells, giving them an elongated, boatlike shape. The cells are called *navicular cells,* but the presence of these cells in a vaginal smear is not always diagnostic of current pregnancy. They are also found in menopause, during a moderate estrogenic deficiency, or in an increase of progesterone or of androgenic hormones. In the last cases they have a tendency to be round and dispersed instead of being elongated, boat-shaped, and clustered, as in pregnancy.

The cells stain unevenly, pale blue to green toward the middle and more deeply colored toward the periphery.

The cytoplasm is delicate and contains large amounts of glycogen deposits in its endoplasm (yellow-to-brown granules). There is a moderate increase in the nuclei-cytoplasm ratio.

The margin of the cytoplasm has a tendency to curl, giving the impression of a thick, heavy border.

The nuclei are often eccentric, round, oval or elongated in shape with a moderate increase in their size. Generally vesicular in type, the nuclei may on occasion appear hyperchromatic or pyknotic.

Navicular cells might not be created by the administration of any single hormone, but may require the presence or absence of several different hormones that occur physiologically during pregnancy.

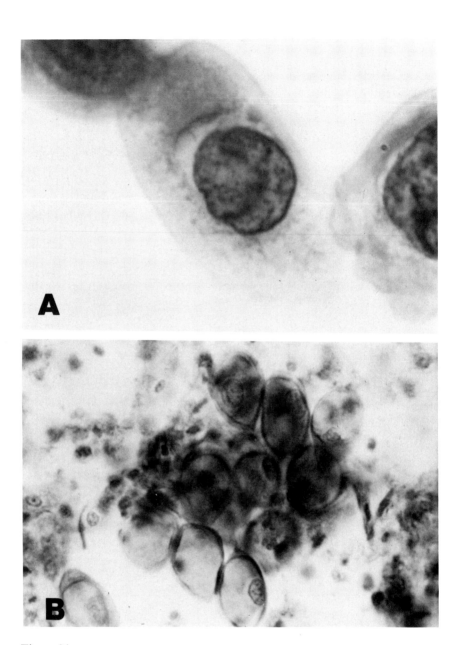

Figure 31
Pregnancy changes in normal (A) basal cells (× 1300) and (B) intermediate cells. (× 450)

TROPHOBLASTIC CELLS (FIG. 32)

These cells are rarely found during normal pregnancies. They are more common in the first and third trimesters. Associated with other changes, their presence in the first trimester of a pregnancy may indicate a threatened abortion in about 6% of the cases. When seen in the late stage of the pregnancy, they probably originate from a partial premature separation of the placenta.

They can exfoliate as mononucleated cells (easily confused with reactive basal cells) or multinucleated giant cells (syncytial trophoblastic cells).

They vary in size and shape from oval to polygonal, and from 20 to 100 μ.

The cytoplasm is abundant, thin or thick, basophilic (60%) or acidophilic (40%).

The cytoplasm can contain occasional vacuoles with leukocytes.

The cytoplasmic borders are sharp, smooth, distinct, and heavy. These are important criteria that differentiate them from endocervical and other multinucleated giant cells.

The cellular staining is dense and fairly even, differentiating them from foreign-body giant cells.

The number of overlapping rather than molding nuclei varies from 1 to 20.

The shape of the nuclei is often irregular and varied. In 90% of the cases, they have a well-delineated, heavy border.

The increased amount of chromatin is irregularly clumped (60%) but evenly distributed throughout the nucleus.

The nucleoli are absent in 70% of the cells and prominent in about 30%.

In postpartum smears these cells indicate the possibility of retained products of conception if they are present four weeks after delivery. They can be confused with anaplastic malignant cells, except for the regularity of the chromatin, their smooth nuclear membrane, and the abundance of their dense cytoplasm.

POSTPARTUM SMEAR (FIG. 33A)

Soon after delivery, in a majority of women the rate of exfoliation of the epithelial cells increases. Most of these cells are round or oval with small pointed projections and look like a transitional stage between the navicular and normal parabasal cells. It is thought that the explanation of the presence of these cells is trauma during delivery. There is also an increase of polymorphonuclear leukocytes and histiocytes. After the fifth day, if the patient is not lactating, a progressive increase of eosinophilia occurs. The cytoplasm of the intermediate cells becomes cyanophilic, the amount of glycogen decreases, and vacuolization appears. The nuclei become denser and more eccentric, and the cells begin to resemble the normal superficial cells. About 10 to 25 days following delivery, the smears return to their usual appearance in nonpregnancy, provided the patient is not lactating.

LACTATION SMEAR (FIG. 33B)

In about 85% of the cases, the mucosa of the vaginal cervix remains atrophic. An increased number of parabasal cells, which have round or oval

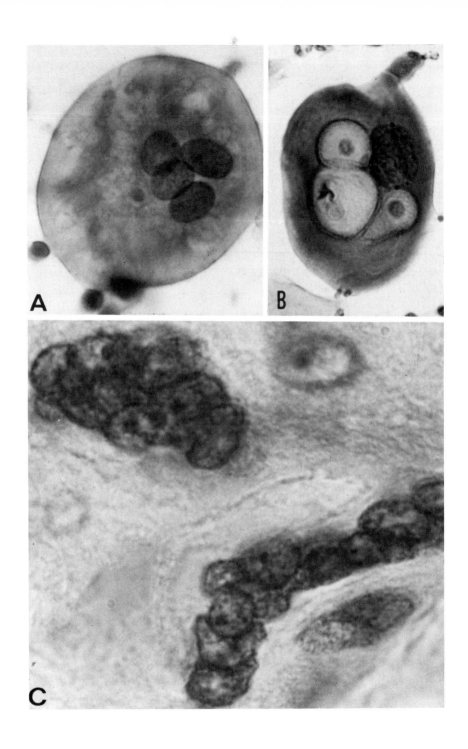

dense nuclei, glycogen-rich cytoplasm, and a thickened cellular outline, remain in the smear as long as the cyclic ovarian function is not resumed. A progressive disappearance of the intermediate and navicular cells occurs due to degeneration.

FOLATE DEFICIENCY (FIG. 34)
Folic acid deficiencies (anemias) are found mainly in postmenopausal women, in young, pregnant patients, and in preeclampsia. They have been blamed for abnormal placental location (abruptio placentae). Folate deficiency can be suspected in a pregnant patient when binucleation and multinucleation, nuclear molding, cellular hypertrophy, and discrete cytoplasmic vacuolization are observed in the squamous and endocervical cells of her vaginal smear. These changes, which are similar to those of early irradiation effects, should not be mistaken for mild squamous atypia (metaplasia and squamous dysplasia).

THREATENED ABORTION
If the fetus dies in utero, a marked shift to the left of the maternal vaginal maturation index (preponderance of basal cells) occurs in about 90% of the cases. These changes last only 4 to 10 days and are succeeded by an increased number of eosinophilic superficial cells. In the absence of pregnancy changes in a smear, a marked shift to the left should always make one consider the possibility of a threatened abortion or an intrauterine fetal death, and the patient should be followed very closely. The background of the smear generally will show increased amounts of fresh and old red blood cells, mucus, and leukocytes. No Döderlein's bacilli are seen. When anucleated, yellowish, pale, and poorly stained squamous cells are found in the smear of a woman during late pregnancy, this may indicate the rupture of the fetal membrane. Cytology seems to be more reliable for indicating fetal loss in early pregnancy than it is for the complications of late pregnancy (preeclamptic toxemia).

CYTOLOGY AND PROGNOSIS OF THE
TIME OF DELIVERY OR ABORTION (FIG. 35)
The hormonal changes that precede delivery by 7 to 10 days have been used to predict correctly, in 75% of cases, the time of birth. These changes

---

Figure 32
(A) Trophoblastic multinucleated cell seen in the vaginal smear of a woman with a retained placenta. Note the abundant vacuolated cytoplasm with its extra sharp borders and the multiple oval nuclei overlapping each other. (× 750) (B) Degenerate trophoblastic mononucleated cell seen in the vaginal smear of an aborting woman. Note the thick, vacuolated cytoplasm and the irregularities of the nuclear membrane and the chromatin, which may be confused with atypia. (× 750) (C) Trophoblastic cells seen in the histologic section of the placenta from the patient whose smear appears in A. Note the similar abundant cytoplasm and multiple overlapping nuclei with irregular nuclear membrane and chromatin. (× 750)

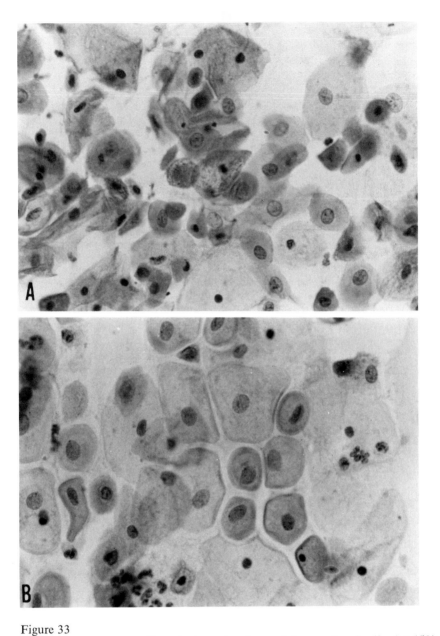

Figure 33
(A) Postpartum smear. Note the irregular shape of the parabasal cells. (× 450)
(B) Lactation smear. Note the round-to-oval shape of the parabasal cells. (× 450)

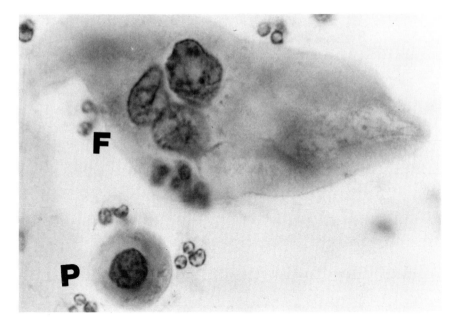

Figure 34
Parabasal cell (F) showing folic acid deficiency. Compare its size to the non-affected parabasal cells (P). (× 450)

have also been used, with disputed results, to predict the results of labor-induction attempts in an over-term pregnancy. The smears taken before delivery can be divided into the following categories:

1. Smear of a woman who is not at term (Fig. 35A). An increase in the clumping and cluster formation of the folded intermediate cells is observed. There is an increase in the number of navicular cells. There is a decrease in the eosinophilic index (less than 1%). There is a decrease in the pyknotic index (less than 10%). In 60 to 80% of the cases, a smear that presents these not-at-term characteristics will indicate that the patient, except for an accident, will not deliver within two weeks and also will not respond well to labor induction if she is over term.
2. Smear of a woman who is at term (Fig. 35B). There is a decrease in the number and size of cellular clusters. There is a decrease in the number of navicular cells (fewer than 45%). There is an increase in the eosinophilic index (above 10%). There is an increase in the pyknotic index (about 20%).

In about 80% of pregnant patients, the presence of such a smear indicates that delivery can be expected within the next two weeks.

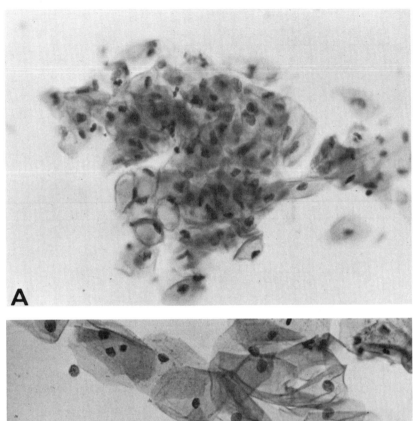

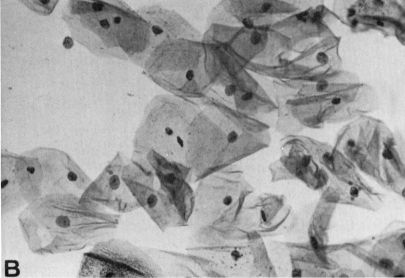

Figure 35
(A) Type of smear that indicates that term has not approached. Note the cellular clumping and the numerous navicular cells. ($\times$ 450) (B) Type of smear that indicates that term is approaching. The patient gave birth spontaneously 2 days later. Note the absence of cellular clusters and navicular cells and the increase of the number of pyknotic cells. ($\times$ 450)

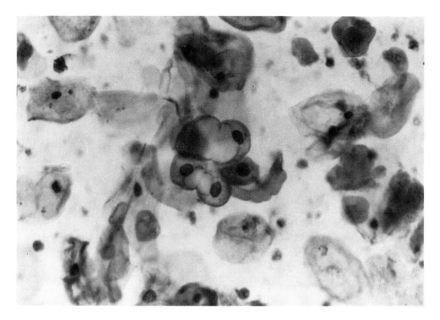

Figure 36
Aspirate amniotic fluid at week 30 of pregnancy. (× 450)

Amniotic Fluid (Fig. 36)

The amniotic fluid may be aspirated for a variety of purposes, including the determination of fetal maturation, the determination of fetal sex, fetal chromosome analysis, and the induction of labor.

There are several cytologic aspects that may be observed:

1. Before 16 weeks few epithelial cells are seen;
2. Anucleated cells and superficial squamous cells, either pink or blue, from various fetal surfaces (skin and oral mucosa) are related to the estriol concentration rather than to fetal maturity;
3. Increased numbers of hypermature squamous cells do not indicate hypermaturation of the fetus;
4. Low cuboidal cells (parabasal cells) originate from the amnion before 30 weeks of gestation;
5. Intermediate squamous or transitional cells with vesicular nuclei originating from buccal, anal, and urinary mucosas of the fetus may indicate fetal maturity;
6. Tall columnar cells, derived from the amnion, may be seen toward the end of the gestation period (after 30 weeks).

TUMORS OF PREGNANCY

Hydatidiform Mole

A cystic and hydrophilic swelling of the placental chorionic villi with a moderate degree of hyperplasia and anaplasia of the chorionic epithelium can fill up the uterine cavity and prevent the development of the fetus. Grossly it resembles a mass of thin-walled translucent, cystic, grapelike vesicles. The vaginal smear in such cases shows an MI persistently shifted to the middle with changes resembling those of normal pregnancy. In addition, clusters of single or multinucleated trophoblastic cells similar to the ones seen during abortion may be present in the vaginal smear.

Choriocarcinoma

A neoplastic change of the trophoblastic elements can occur during a normal or abnormal pregnancy or can originate from teratogenous tissues. It is a malignant tumor that rapidly metastasizes, causing multiple small, bloody metastases, mainly in the lung. The vaginal smear may show persisting pregnancy changes or, more often, a strong estrogenlike effect with an increased amount of red blood and inflammatory cells. Numerous trophoblastic cells, single or in clusters, showing variable degrees of anisokaryosis can be seen. Their nuclei have irregular membranes and contain coarse, irregularly distributed chromatin. Nucleoli are multiple and irregular in size and shape. On the basis of these cells alone, it is difficult to differentiate between a choriocarcinoma and a hydatidiform mole. In both cases the cells appear highly malignant. The detection of the continuous presence in the smear of changes like those in pregnancy can be helpful in the follow-up of the patient with such a tumor after chemotherapy or curettage.

CERVICAL CANCER DETECTION AND PREGNANCY

The superficial squamous dysplastic cells and the large-cell type of squamous in situ cancer cells, having an abundant cytoplasm, may look worse in the smear of a pregnant woman. Therefore the cytologic interpretation of the degree of atypia should be more conservative. The cells originating from a poorly differentiated carcinoma or an in situ squamous carcinoma, having a scanty cytoplasm, remain unchanged. The rate of their exfoliation often increases, and their appearance is not influenced by the pregnancy. The ease of cytologic detection of cervical carcinoma during pregnancy should be about the same as in nonpregnant women. Pregnancy at any stage is not a contraindication for having a vaginal smear.

References and Supplementary Reading

Arrighi, A., and Terzano, G. Urinary cytology during pregnancy. *Acta Cytol.* 3:298–304, 1959.

Babrinos, M. L., and Estraliades, M. C. Vaginal cytology in primary amenorrhea. *Acta Cytol.* 16:376, 1972.

Baltas-Alcalais, C., and Drossos, C. Les frottis vaginaux du nouveau-né à terme et du prémature. *Arch. Anat. Pathol.* 15:91, 1967.

Bechtold, E., Prior, J. T., and Reicher, N. B. Trophoblastic cells in vaginal and endocervical smears; problems concerned with their recognition and interpretation. In *Transactions of the Fifth Annual Meeting of the Inter-Society Cytology Council,* 1957. Pp. 35–36.

Bercovici, B., Deamant, Y., and Polishuk, W. Z. A simplified evaluation of vaginal cytology in third trimester pregnancy complications. *Acta Cytol.* 17: 67, 1973.

Borrow, M. L., Farrel, W., and Oppenheim, A. Cytologic study of 2,000 cases of incomplete abortion. In *Transactions of the Sixth Annual Meeting of the Inter-Society Cytology Council*, 1958. Pp. 99–105.

Britsch, C. J., and Azar, H. A. Estrogen effect in exfoliated vaginal cells following treatment with digitalis. *Am. J. Obstet. Gynecol.* 85:989–993, 1963.

Brunori, I. L. Effect of digitalin on the vaginal epithelium of elderly women. *Riv. Ital. Ginecol.* 49:261–272, 1965.

Butler, B., and Taylor, D. S. The postnatal smear. *Acta Cytol.* 17:237, 1973.

Casadel, R., et al. A cytologic study of the amniotic fluid. *Acta Cytol.* 17:289, 1973.

Castillo, E. B. del, Argonz, J., and Galli Mainini, C. Smears from the female urethra and their relationship to smears of the urinary sediment. *J. Clin. Endocrinol. Metab.* 9:1362–1371, 1949.

Chiaffitelli, H. S. de, and Dominquez, S. Vaginal cytology in patients treated with steroid combinations. *Acta Cytol.* 14:344, 1970.

Dallian, G. F., and Nuovo, V. M. Sur 160 cas de grossesses dites prolongées. *Arch. Anat. Pathol.* 15:115, 1967.

Gladwell, P., et al. Amnioscopy of late pregnancy with fetal membrane and decidual cytology. *Acta Cytol.* 18:333, 1974.

Haam, E. von. The vaginal smear during the luteal phase of the normal menstrual cycle. *Acta Cytol.* 6:282, 1962.

Haour, P., and Cardon, J. P. Etude de l'urocytogramme dans la stérilité masculine. *Arch. Anat. Pathol.* 15:41, 1967.

Holmquist, N. D., and Danos, M. The cytology of early abortion. *Acta Cytol.* 11:262, 1967.

Horava, A., et al. The exfoliative cytologic characteristics of the endometrium in health and disease. *Clin. Obstet. Gynecol.* 4:1128–1158, 1961.

Huisjes, M. D. Cytologic features of liquor amnii. *Acta Cytol.* 12:43, 1968.

Jing, B. J., Kaufman, R. H., and Franklin, R. R. Vaginal cytology for prediction of onset of labor. *Am. J. Obstet. Gynecol.* 99:546, 1967.

Keebler, C., and Wied, G. The estrogen test, an aid in differential cytodiagnosis. *Acta Cytol.* 18:482, 1974.

Kitay, D. Z., and Wentz, W. B. Cervical cytology in folic acid deficiency of pregnancy. *Am. J. Obstet. Gynecol.* 104:931, 1969.

Leeton, J. Vaginal cytohormonal studies in late abnormal pregnancy. *Acta Cytol.* 11:410, 1967.

Lichtfus, C. Influence de test aux oestrogènes sur le frottis vaginal en fin de grossesse. *Arch. Anat. Pathol.* 15:109, 1967.

Liu, W., et al. Normal exfoliation of endometrial cells in pre-menopausal woman. *Acta Cytol.* 7:211–214, 1963.

Liu, W., et al. Cytologic changes following the use of oral contraceptives. *Obstet. Gynecol.* 30:228, 1967.

Masukawa, T. Vaginal smears in women past 40 years of age, with emphasis on their remaining hormonal activity. *Obstet. Gynecol.* 16:407–413, 1960.

Meisels, A. Le diagnostic cyto-hormonal durant la grossesse. *Laval Med.* 34:551–560, 1963.

Meisels, A. Computed cytohormonal findings in 3,307 healthy women. *Acta Cytol.* 9:328, 1965.

Meisels, A. The menopause: A cytohormonal study. *Acta Cytol.* 10:49, 1966.

Moracci, E. Valeur pratique de la numération des cellules intermédiaires du frottis vaginal comme méthode de diagnostic cytologique de l'activité lutéale. *Arch. Anat. Pathol.* 15:53, 1967.

Moraes-Ruehsen, M. D., and Masukawa, T. The irrigation smear: Use in hormonal cytology. *Acta Cytol.* 9:307, 1965.

Mussey, E., and Decker, D. G. Intraepithelial carcinoma of the cervix in association with pregnancy. *Am. J. Obstet. Gynecol.* 97:30, 1967.

Navab, A., Koss, L. G., and LaDue, J. S. Estrogen-like activity of digitalis. *J.A.M.A.* 194:30, 1965.

Nyklicek, O. The significance of parabasal cells in the vaginal smear in prolonged pregnancy. *Acta Cytol.* 7:131–132, 1963.

Nyklicek, O. Vaginal cytology and amnioscopy in prolonged pregnancies. *Acta Cytol.* 16:48, 1972.

O'Morchoe, P. J., and O'Morchoe, C. C. Method for urinary cytology in endocrine assessment. *Acta Cytol.* 11:145, 1967.

Pundel, J. P. Normal vaginal cytology during pregnancy. *Acta Cytol.* 3:211, 1959.

Pundel, J. P., and Van Meeusel, F. *Gestation Cytologie Vaginale.* Paris: Masson et Cie, 1952.

Rahman, D., and Zaman, H. The vaginal smear in pregnant women. *Acta Cytol.* 7:287, 1963.

Rakoff, A. E. Vaginal cytology of endocrinopathies. *Acta Cytol.* 5:153, 1961.

Rakoff, A. E. A summary of the writings of George N. Papanicolaou on hormonal cytology. *Acta Cytol.* 6:532, 1962.

Rezende, J., et al. Vaginal cytology study of placental insufficiency. *Acta Cytol.* 14:78, 1970.

Romberg, G. H. The diagnostic accuracy of hormonal evaluation by means of endometrial cytology. *Acta Cytol.* 2:620–622, 1958.

Rosenblatt, R., Baechler, C., and Volet, B. Le frottis vaginal et urinaire au cours du travail d'accouchement. *Arch. Anat. Pathol.* 15:168, 1967.

Rubio, C. A., Hvalec, S., and Pareja, A. The value of cytohormonal studies (karyopyknotic index) for detecting recurrence of carcinoma. *Acta Cytol.* 11:176, 1967.

Ruiz, L. M. Vaginal cytology during delivery. *Acta Cytol.* 9:337, 1965.

Rutledge, C. E., Jr., Christopherson, W. M., and Parker, J. E. Cervical dysplasia and carcinoma in pregnancy. *Obstet. Gynecol.* 19:351, 1962.

Sedlis, A. Cytology screening of prenatal patients in a municipal hospital. *Acta Cytol.* 7:224, 1963.

Sen, D., and Langley, F. Vaginal cytology as a monitor of fetal well-being in early pregnancy. *Acta Cytol.* 16:116, 1972.

Slate, J. A., and Merritt, J. W. The significance of carcinoma in the pregnant patient. In *Proceedings of the First International Congress of Exfoliative Cytology,* Vienna, 1961. P. 128.

Smith, M. R., Figge, D. C., and Bennington, J. L. The diagnosis of cervical cancer during pregnancy. *Obstet. Gynecol.* 31:193, 1968.

Sol, J. R. del, and Rohrback, C. The effects of progestogens on the atrophic epithelium. *Acta Cytol.* 6:231–234, 1962.

Sora, P. Correlative studies on vaginal, urethral, oral smears and smears from the urinary sediment during normal pregnancy. *Acta Cytol.* 3:305, 1959.

Soszka, S., and Wisniewski, L. Cytologic evaluation of fetal death and an attempt to determine the time of its occurrence. *Acta Cytol.* 11:403, 1967.

Soule, S. D. The practical value of vaginal cytology in pregnancy: I. Cytological prediction of the fate of early pregnancy. *Acta Cytol.* 8:364–367, 1964.

Spujt, H. J., et al. Exfoliative cytology during pregnancy for detection of carcinoma of the cervix. *Obstet. Gynecol.* 15:19, 1960.

Stern, E., et al. Correlation of vaginal smear patterns with urinary hormone excretion. *Acta Cytol.* 10:110, 1966.

Stoll, B. A.   Vaginal cytology as an aid to hormone therapy in postmenopausal cancer of the breast. *Cancer* 20:1807, 1967.

Stone, D. F., et al.   Estrogen-like effects in the vaginal smears of postmenopausal women. *Acta Cytol.* 11:349, 1967.

Teter, J.   The use of selected cytologic indices for evaluation of estrogenicity of synthetic compounds. *Acta Cytol.* 16:367, 1972.

Van Leeuwen, L., Jacoby, H., and Charles, D.   Exfoliative cytology of amniotic fluid. *Acta Cytol.* 9:442, 1965.

Waard, F. de, and Oettle, A. G.   A propos des facteurs exogènes et endogènes dans les états oestrogéniques postménopausiques. *Arch. Anat. Pathol.* 15:26, 1967.

Wachtel, E.   The cytology of amenorrhoea. *Acta Cytol.* 10:56, 1966.

Wachtel, E.   The prognostic significance of the karyopyknotic index after radical treatment for cancer of the female genital tract. *Acta Cytol.* 11:35, 1967.

Wied, G. L. (Moderator).   Symposium on hormonal cytology. *Acta Cytol.* 12:87, 1968.

Zidovsky, J.   The significance of parabasal ("postnatal") cells in the vaginal smear in prolonged pregnancy. *Acta Cytol.* 5:393, 1961.

Inflammatory Diseases of the Female Genital Tract

4

The two most common symptoms of genital inflammation, pruritus (itching) and leukorrhea (vaginal discharge), are the most frequent reasons for a woman seeking medical advice.

Inflammation is the change that occurs in a living tissue after an injury, providing the injury has not killed the tissue outright. According to its intensity and persistence, the inflammation may be described as acute, subacute, chronic, mild, moderate, or marked. According to the region involved, an inflammation of the female genitalia is called vulvitis, vaginitis, cervicitis, endocervicitis, endometritis, myometritis, salpingitis, oophoritis, or pelvic inflammatory disease (PID). Infection is an inflammation produced by living organisms (viruses, bacteria, or fungi).

Symptoms of an Acute Inflammation

The following classic changes are usually present in the inflammation of the genital organs.

TUMOR. The swelling is due mainly to the accumulation of fluid in the interstitial tissue (edema).

CALOR. The localized increase of temperature of the affected tissue is the result of the dilated blood vessels and the increased rate of blood circulation, especially at the beginning of the inflammation.

RUBOR. The redness of the mucosa is the result of the dilation of the capillaries and venules and also of the regional increase, by stagnation, of the amount of blood (congestion).

DOLOR. The pain and itching are the result of irritation of the terminations of the genital nerves.

DESTRUCTION OR CHANGES OF FUNCTIONS. The endocervical glandular epithelium, for example, may overproduce secretions (leukorrhea).

Etiology of Inflammation

The inflammation is the result of injurious agents, which can be:

1. Living, such as microorganisms, bacteria, fungi, viruses, or protozoa (infection);
2. Nonliving, as in traumatic (laceration), thermal (cauterization), irradiation (x-ray), electrical (burns), and chemical (caustic) inflammation.

Types of Inflammation

The type of genital inflammation often varies with the age of the patient.

NEWBORN
The vaginal mucosa is sterile at birth, but in a few hours colonies of Döderlein's bacilli begin to invade it to help in the maintenance of the physiologic acidity of the vagina. They may produce a mild infectious inflammation. The other types of inflammation are rare.

CHILDHOOD
The vaginal mucosa, being normally atrophic, offers little resistance to an inflammation. The following, in decreasing order of frequency, are the main etiologic factors:

1. Trauma, usually secondary to the introduction of a foreign body or after a sexual assault;
2. Nonspecific bacterial vaginitis;
3. Gonorrheal infection, which can be venereal or nonvenereal in transmission;
4. Pinworm infection, secondary to an anal lesion;
5. Viral infection, the most common virus being the genital herpes simplex type II;
6. Allergic systemic diseases, as in measles, scarlet fever, blood dyscrasia;
7. Congenital abnormality, for example when an opening (fistula) exists between the vagina and the rectum, permitting the fecal material to contaminate the vagina.

Trichomonas or *Candida* infections are rare.

MENSTRUAL AGE
The main causes of inflammation here are:

1. Trichomonas infection, which is the most common, the incidence varying from 15 to 85% in women, depending on their living standards and sexual promiscuity;
2. Fungal infection, the most common fungus being *Candida;*
3. Traumatic injuries (postcoital, childbirth, foreign body);

4. Nonspecific bacterial infections (streptococci, staphylococci);
5. Gonorrheal infection;
6. Viral and chlamydial infections (herpes simplex and TRIC agent; e.g., chlamydial inclusion vaginitis);
7. Allergic reactions;
8. Other venereal diseases;
9. Systemic diseases, such as diabetes or blood dyscrasia.

POSTMENOPAUSAL

The main causes of vaginal inflammation during this period are:

1. Hormonal (atrophic vaginitis);
2. Nonspecific bacterial infections;
3. Chronic irritation (leukoplakia and neurodermatitis);
4. Fungal invasion and trichomoniasis.

Vaginocervical Inflammation

CAUSES AND PREDISPOSING FACTORS

Pregnancy

The pregnancy-related congestion, edema of the cervix and vagina, and the absence of a thick, protective, keratinized cellular layer on the surface of the squamous mucosa, combined with an increase of vaginal moisture (exudative serum and increased endocervical secretion) make the lower genital tract more susceptible to a secondary infection during pregnancy.

Variable amounts of neutrophils are normally found in most vaginal smears taken during early or late pregnancy. In the postpartum period, because of the frequency of lacerations resulting from childbirth and of the atrophy of the vaginal mucosa in the lactating period, an inflammatory smear is the rule rather than the exception.

Traumatic Injury

The most common causes of acute or chronic traumatic genital irritation are sexual relations, uterine prolapse, presence of an intragenital foreign object (pessary), and vaginal irrigation by a caustic substance for hygienic or abortive purposes. A trauma, by producing an erosion or ulceration of the mucosa, decreases its resistance to infection. Large numbers of inflammatory cells, histiocytes, and occasional foreign-body giant cells are generally seen in the smears. The presence of squamous basal (repair) cells with bizarre pseudopodlike cytoplasmic irregularities confirms this diagnosis.

Vaginal pH

A decrease of the vaginal acidity, which is normally maintained by Döderlein's bacilli, increases the chances of inflammation.

Epithelial Atrophy (Senile Vaginitis)

The decrease of available estrogen thins the epithelium and flattens the rugal folds of the vagina, making it more susceptible to traumatic injuries.

83

Similarly the absence of the protective superficial cells of the vaginal mucosa in postmenopausal women invites a secondary bacterial infection. When the vaginal secretion becomes alkaline with the decrease of the intracellular glycogen and of the number of Döderlein's bacilli, the growth of pathogenic microorganisms is enhanced. Colonies of bacteria are often seen in the smears.

A large number of basal and inflammatory cells, predominantly mononuclear in type, in the smear from an atrophic vaginitis, can often mask or distort tumor cells, which may result in false-negative cytologic interpretations. Such an atrophic smear submitted for cancer detection should always be repeated during the 10 days following local estrogenic and antibiotic therapy.

Congenital or Acquired Abnormalities

Any obstacle to the normal menstrual flow increases the chance of genital infection and allows accumulation of secretion and debris in a distended endometrial cavity. The presence of this stagnant secretion invites a secondary infection. An abnormal communication between the vagina and the urinary tract or rectum also produces contamination of the vaginal secretion by pathogenic microorganisms from these organs (postirradiation fistula).

Metabolic Abnormalities

The genital mucosa of patients with various metabolic disorders, diabetes, or hypothyroidism is more susceptible to infection (fungi and bacteria) than that of normal subjects. The infection is also usually more severe.

Irradiation

Radiotherapy may produce an epithelial atrophy that is highly susceptible to secondary infection. This atrophy is not as sensitive to estrogen as the one resulting from hormonal deficiency. In addition, irradiation can produce an extensive ulceration, necrosis, and granulation of the genital mucosa with frequent secondary inflammation. The presence of large numbers of neutrophils, foreign-body giant cells, and debris can create pitfalls in the detection of recurrent carcinomas by masking the diagnostic cells.

Postcryosurgery

Cryosurgery, which involves freezing of the cervix, is one of the treatments now used for premalignant lesions. The vaginal smear taken a few hours afterward will contain a large number of leukocytes, blood, and cellular debris. Some of the superficial and intermediate squamous cells will show perinuclear halos with increased density of the cytoplasm. Degenerative nuclear changes in the form of chromatolysis and karyorrhexis may be seen one week after freezing and may last about five weeks. Single or multiple, large, cytoplasmic vacuoles, which often compress the nuclei, are present in the parabasal cells.

Vitamin Deficiency

Deficiency of vitamins A or C produces abnormalities in the keratinization of the vaginal mucosa, making it slightly more susceptible to inflammation and infection.

Neoplasia

A space-occupying neoplasm can act as a foreign body that produces a local reactive inflammation. Furthermore, the atypical epithelial keratinization of the squamous carcinoma and the frequency of ulceration are causes of inflammation. Except for cases of early in situ carcinoma, in which the background of a smear can be clean, variable amounts of inflammatory cells are almost always found in the smear of a patient with an invasive neoplasm.

CYTOLOGY OF THE INFLAMMATORY SMEAR

Large amounts of leukocytes (neutrophils) are usually present in the smears during the acute or semiacute inflammatory stage. These leukocytes are mixed with various amounts of plasmocytes and lymphocytes in chronic inflammation.

Large pools of lymphocytes may also be present in the smears in cases of chronic follicular cervicitis if there is rupture of a cervical lymphoid follicle.

Variable numbers of histiocytes and foreign-body giant cells can also be found.

Variable amounts of poorly stained, gray particles of necrotic cytoplasmic-nucleic debris and degenerate inflammatory cells can be seen. In acute inflammation, clusters of this degenerate cellular debris, because of its usual deep basophilia, can present a pitfall in the diagnosis of a malignancy.

In the background of the smear, there is generally an increased amount of granular precipitate and protein deposits (serum) stained gray to pink.

Fresh and fragmented old red blood cells mixed with fibrin are also usually present.

Because of the increase in the maturation and keratinization of the squamous mucosa, due to inflammatory irritation, there is a corresponding increase in the exfoliation of keratinized superficial cells. Similarly, when an erosion or ulceration of the mucosa is present, the number of exfoliated basal or repair cells increases, making a cytologic hormonal evaluation (maturation index) unreliable. The inflammation may also produce general eosinophilia of the cells, which renders the determination of an eosinophilic index useless. The smear should be reevaluated after antiinflammatory treatment.

Other degenerative and reactive changes of the epithelial cells during irritative inflammation (which include cellular hypertrophy, vacuolization, perinuclear halo, and nuclear pyknosis) can simulate precancerous atypia and may constitute a pitfall leading to possible false-positive interpretations.

The specific living infective agent (e.g., *Trichomonas,* bacteria, or fungi) can be readily seen in most of the smears. Special stains can be used if necessary.

Bacterial Infection (Nonspecific Vulvovaginitis)
Bacteria are unicellular plant organisms, devoid of chlorophyll and re-producing asexually (fission). Some produce no apparent injury to the host (normal flora); others are pathogenic, either as a result of direct invasion and the destruction of the cells and tissues, or as a result of the production of toxic substances. Bacterial infection is one of the most common causes of leukorrhea in children. The clinical symptoms are often similar to those of trichomoniasis. The background of the vaginal smear can show colonies of single or multiple species of organisms living in symbiosis (mixed infection).

DÖDERLEIN'S BACILLUS–*Lactobacillus* (FIGS. 39A, 40A).   The presence of numerous immobile, semitransparent, facultative anaerobic, intracellular or extracellular small rods is diagnostic. The basophilia of the epithelial cells increases. A cytolytic effect produces a number of naked or stripped degen-erate nuclei from the endocervical or intermediate squamous cells. A mod-erate number of Döderlein's bacilli can be seen in the background of most smears. They help in maintaining the acid vaginal pH by transforming glycogen into lactic acid. In a postmenopausal smear, their increased pres-ence can suggest the possibility of diabetes mellitus. Döderlein's bacilli, which are seen in about 60% of normal vaginal smears, as well as in smears from cases of carcinoma in situ and dysplasia, are found in only 10% of the smears from patients with early invasive squamous carcinoma and in less than 1% of the smears from those with advanced cancers.

Hemophilus vaginalis.   Small, rodlike, gram-negative coccobacilli are thought to be one of the most frequent causes of nonspecific vulvovaginitis. The infection is diagnosed by the presence of masses of *Hemophilus vagi-nalis* in the background of the smear that cover the surface of large, vaginal squamous epithelial cells (*clue cells*). The organism measures about 0.5 μ in diameter by 1.5 μ in length and may occasionally be very pleomorphic.

GONOCOCCUS (FIG. 37A).   Gonorrheal vaginitis, a venereal infection, is most frequently seen during childhood and young adulthood, and it causes an abundant purulent exudate. It often involves the urethra and the vulvo-vaginal glands. The vaginal smear is covered by large numbers of inflam-matory cells with a predominance of monocytes in the early stage and polymorphonuclear leukocytes in the later stages. Eosinophilic diplococci can be found in and outside the cytoplasm of the epithelial cells and the leukocytes.
    There is a need to differentiate the phagocytized cocci from the other phagocytized debris in histiocytes and the smaller intracytoplasmic inclu-sions that result from a TRIC or chlamydial infection (inclusion con-junctivitis). The latter are found mainly in epithelial cells, rather than in neutrophils, as in the case of gonorrhea. Some diplococci in mosaic arrange-

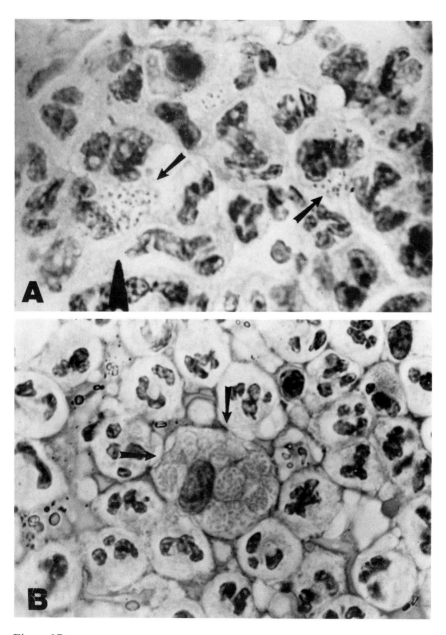

Figure 37
(A) Vaginal smear from a patient with gonorrheal vaginitis. Note the intracyto-plasmic diplococci. ($\times$ 1300) (B) Vaginal smear in a case of granuloma in-guinale. Note the numerous intracytoplasmic Donovan bodies. ($\times$ 1300)

ment may be seen attached to the surface of the cytoplasmic membranes of degenerate endocervical and squamous metaplastic cells.

Gonococci are often the cause of secondary adnexal inflammation, resulting in acute pelvic inflammation, abscess formation, or fallopian tube stricture, which can be the cause of infertility or ectopic pregnancy.

MICROCOCCI (FIG. 40B). Characteristic single cocci (staphylococci), diplococci (gonococci), or chains of cocci (streptococci) can be found inside or outside the cytoplasm of leukocytes and epithelial cells. The presence of micrococci may produce a purulent vaginal discharge, more commonly seen during pregnancy than at other times. The infected cells are often enlarged, and their cytoplasm has curled edges and contains multiple degenerative vacuoles with healthy polymorphonuclear inflammatory cells (polys). A large number of polys or large colonies of these microorganisms can hide the presence of an atypical cell, making the detection of a possible neoplasm difficult. One should not hesitate to ask for a repeat smear after adequate therapy in such a case.

GRANULOMA INGUINALE (FIG. 37B). This venereal infection is caused by a gram-negative, encapsulated coccobacillus (*Donovania granulomatis*). The scrapings of the margins of the ulcerogranulomatous lesions of the moist skin and the mucosa of the genital, rectal, urethal, and inguinal areas may be diagnostic. The bean-shaped, pink-orange organisms (Donovan bodies), measuring 1 to 2 μ, with occasional prominent polar bodies that give them the classic safety-pin appearance, are clustered in multiple cystic spaces in the cytoplasm of large (25–90 μ diameter) mononuclear macrophages. Two to 10 of these intracytoplasmic cysts (vacuoles) often mold against each other and push the nuclei toward the periphery of the cells. The smears will also contain large numbers of polymorphonuclear leukocytes, cellular debris, and protein granule deposits in the background. Such a smear should be screened carefully for cancer cells because of the reported association of granuloma inguinale with squamous cell carcinoma.

TUBERCULOSIS (FIG. 38). Tuberculous inflammation of the genital tract is difficult to diagnose by exfoliative cytology. Occasional multinucleated giant *Langhans' cells* may be found in cervical smears and other specimens, which may be recognized by the characteristic location of their nuclei toward the periphery of the cytoplasm. The multiple (more than 30) round-to-oval nuclei often overlap but do not mold. No nucleoli are seen. The abundant, pink, dense cytoplasm of the cells does not contain any vacuoles or foreign inclusions. It is most important to differentiate between these giant Langhans' cells and similar-appearing foreign-body giant cells whose nuclei are more central and whose cytoplasm may contain ingested debris and vacuoles. An acid-fast stain of an air dried smear may show the elongated and diagnostic red tuberculous organisms in the midst of blue epithelial and inflammatory cells.

Occasional *epitheloid cells* may be seen with variable, transparent, vesicular cytoplasm and elliptical nuclei that contain finely and evenly dis-

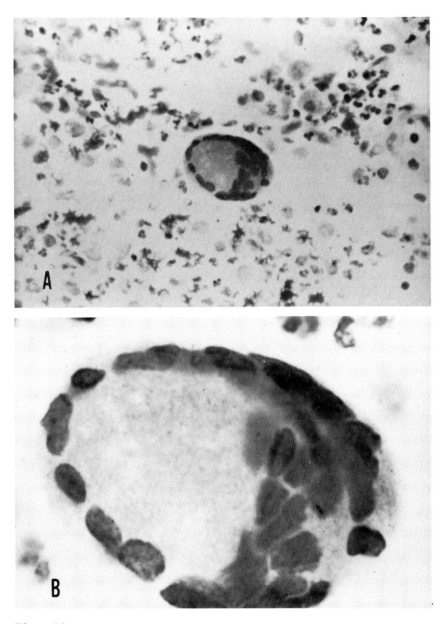

Figure 38
Tuberculous Langhans' cells in a cervical smear. Note the peripheral arrangement of the nuclei and the cytoplasm that is void of any ingested material.
(A: × 250; B: × 1300)

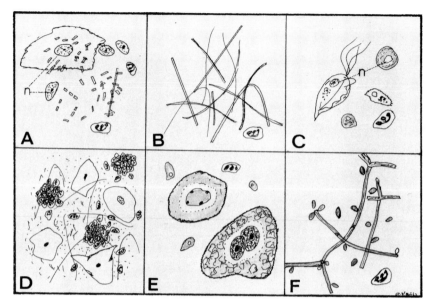

Figure 39
The cytologic findings in various genital inflammations (n = nucleus). (A) Döderlein's bacilli. (B) *Leptothrix vaginalis.* (C) *Trichomonas.* (D) *Trichomonas* infection in symbiosis with *Leptothrix.* (E) Effect of *Trichomonas* on the epithelial cells. (F) *Candida.*

tributed chromatin. They can be hard to differentiate from some histiocytes and endocervical cells.

Protozoal Inflammation

TRICHOMONAS INFECTION (FIGS. 39E, 41). In some instances patients who have no clinical complaints exhibit large numbers of trichomonads in the clean smear without any increase in inflammatory cells (infestation). In most patients, however, *Trichomonas* is the cause of leukorrhea (greenish-yellow discharge) soreness, odor, and irritation. Grossly, the vagina and the cervix are hyperemic with multiple hemorrhagic granulation (strawberry cervix). It is estimated that about 25 to 50% of the adult population is infected, the most common means of transmission being venereal. No definite association between the presence of this inflammation and cervical anaplasia has been ascertained.

The following changes provide an indication, but not a proof, of *Trichomonas* inflammation. The organism should always be identified for a definite diagnosis.

The structures, described as resembling formations of BB shot, result from an agglomeration of polymorphonuclear cells around a parasite (Fig. 41A). The presence of such multiple small, cellular clusters, which are easy to recognize even under low magnification, indicates the presence of *Trichomonas* in 85% of cases.

The cytoplasm often shows eosinophilic discoloration and irregular (worm-eaten) membranes. These cellular changes may be seen in all layers of the mucosa.

A general hypertrophy of the intermediate and basal epithelial cells with prominent perinuclear halos often occurs. These changes should not be confused with similar findings in epithelial anaplasia.

The nuclei of the parabasal cells enlarge and become either hyperchromatic or hypochromatic with smudged chromatin clustering next to a hazy nuclear membrane.

The occurrence of multinucleation in the endocervical, basal, and intermediate squamous cells increases (Fig. 41C).

A symbiosis with *Leptothrix* is common (Figs. 39D and 40C).

The presence of sheets of reactive basal cells with prominent nucleoli usually indicates mucosal erosion or ulceration.

The secondary orangeophilia of the cytoplasm of the basal, intermediate, and superficial squamous cells renders unreliable the eosinophilic index count that is used for hormonal evaluation.

An increased number of phagocytes and multinucleated foreign-body giant cells, which contain variable amounts of cellular debris, can be seen on occasion.

TRICHOMONAS ORGANISMS (FIGS. 39C, 41, 42). The size varies from 5 to 20 μ, depending on the strain, rather than on the age of the organism.

Pear-shaped, the cellular contours in the routine vaginal smear are very poorly defined.

The cytoplasm is poorly stained, very hazy, and pale gray to blue or pink.

The nucleus, which is round to oval, semitransparent, and slightly basophilic, should be definitely recognized before interpreting the small grayish blob as a trichomonad. (It is the smallest nucleus that can be found in a smear.)

The cytoplasm of *Trichomonas* can contain reddish-brown granules, which indicate a slight phagocytic power, permitting the cell to ingest debris of red blood cells as occurs in the case of amoebiasis.

No flagella can be seen in a vaginal smear because they are lost during smearing or fixation.

Trichomonads are usually found outside the epithelial cells, but on rare occasions they can be cannibalized in the cytoplasm of epithelial or phagocytic cells.

Toxoplasma gondii (FIG. 43). The infection may be congenital or acquired. The *toxoplasma* organism is an intracellular parasite, crescent to oval in shape, which measures 3 to 7 μ and stains red-orange with the Papanicolaou stain. Toxoplasma may be found in the cytoplasm of almost any epithelial or nonepithelial cell of an infested individual. In a vaginal smear, clustered dark orange-staining toxoplasma can be easily seen in the middle of a large number of inflammatory cells and cellular debris, but they are often confused with artifacts. The characteristically elongated, slightly

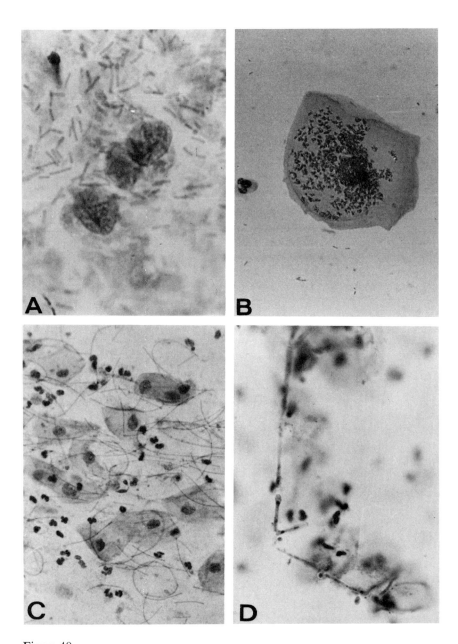

Figure 40
(A) Döderlein's bacilli and stripped nuclei. (× 1300, before 5% reduction)
(B) Micrococci covering a superficial cell (clue cell). (× 750, before 5% reduction) (C) *Leptothrix* in symbiosis with *Trichomonas*. (× 450) (D) *Candida
albicans,* showing round (spores) and long (*hyphae*) bodies. (× 450)

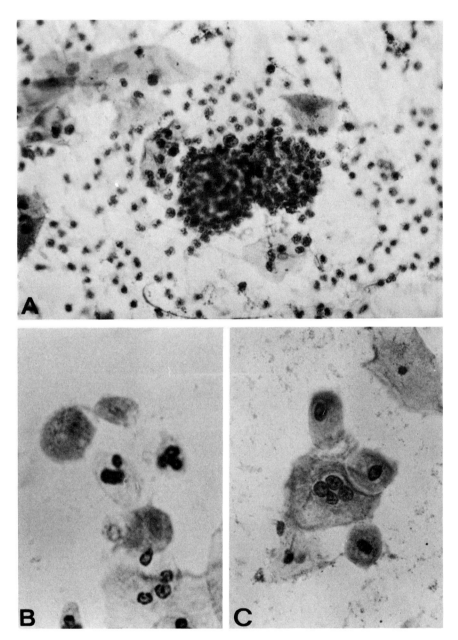

Figure 41
Trichomonas inflammation. (A) Agglomeration of polymorphonuclear cells. In 85% of the cases, they indicate the presence of *Trichomonas*. (× 450) (B) *Trichomonas vaginalis*. Note the pale, oval nuclei and the hemosiderin granules in the cytoplasm. (× 1300) (C) *Trichomonas* effect. Note the multinucleation and the perinuclear halo of the basal cells. (× 450)

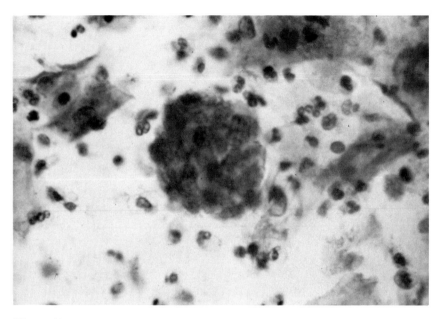

Figure 42
Agglomeration of trichomonads simulating an acinar formation. ($\times$ 450)

curved parasites that are contained in cytoplasmic vacuoles of the infected cells are easier to recognize under high magnification. The degenerative rupture of such cells and the liberation of the parasite explain their occasional presence in the background of the smear.

Enterobius vermicularis. The eggs of *Enterobius vermicularis* (pinworm) can occasionally be seen in a vaginal smear. They can be recognized, even under low magnification, as numerous oval-shaped structures measuring 25 by 60 μ. With high magnification, bright lavender- to orange-stained larvae can be seen enclosed within the refractive double-layered shell of the ova. Some of the eggs have a longitudinal indentation. Their presence should be reported.

Entamoeba histolytica. This parasite can be found in the smears of women living in areas where this disease is endemic. The cells measure 15 to 60 μ and are recognized by their round-oval shape, hazy cytoplasmic borders, and gray basophilic cytoplasm containing intact or fragmented red blood cells. The nucleus is small. The background of the smear contains basophilic granular deposits and large numbers of neutrophils.

SCHISTOSOMIASIS (FIG. 44). The infestation of the vagina and the cervix by *Schistosoma* has been reported in certain regions of Africa and other tropical areas where the disease is endemic. The diagnosis is based on the presence of blue-purple, large (100 by 60 μ) ova, which are recognized by

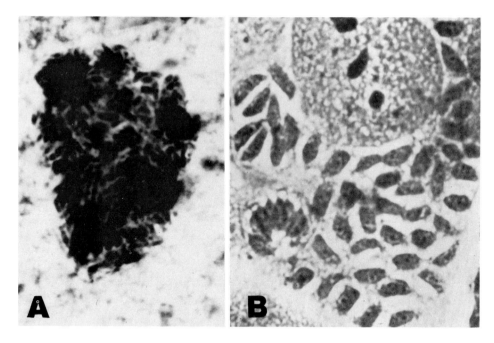

Figure 43
Toxoplasma gondii. (A) In a vaginal smear. (× 450) (B) In a tissue culture. (× 1300)

their refractive, chitinous shells and granular, segmented eosinophilic miracidia. Oval in shape, they may have a single lateral spur (*Schistosoma mansoni*) or a short tail (*Schistosoma haematobium*). Several degenerate ova and empty egg shells are often found scattered throughout the smear. Their characteristic shape and size make them hard to miss, even under low magnification.

Fungal Inflammation

Candida (FIGS. 39F, 40D). Most of the various fungi infecting the skin can also be found in vaginal smears. One of the most common, especially during pregnancy, is *Candida albicans (Monilia)*. Prolonged antibiotic treatment or diabetes favors its growth. In contrast to trichomoniasis, pruritus is the most common symptom with *Candida* infestation, while leukorrhea (white discharge) is not so frequent. The *Candida* fungus can be present in a woman without any symptoms or without an increase of inflammatory cells in the smear (infestation). It can also coexist with a *Trichomonas* infection. The cells are found in the smear as budding yeast (round bodies) or hyphae (long bodies). The diagnosis and differentiation of the hyphae from other elongated structures, such as cotton or nylon fibers and pollens, is based on the uniform staining (yellow to red), the regular

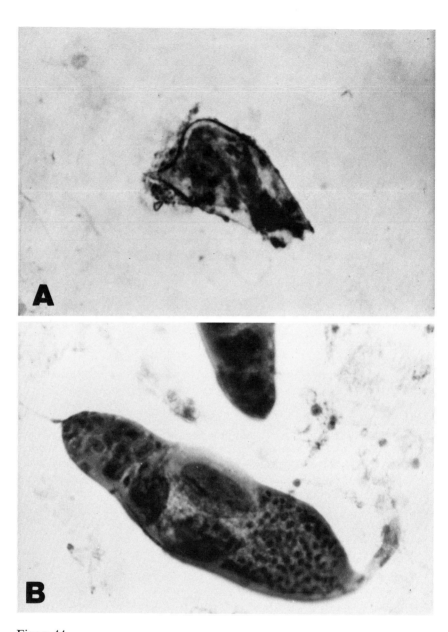

Figure 44
(A) *Schistosoma mansoni* (× 250) and (B) *Schistosoma haematobium* (× 250) in a vaginal smear. Note the lateral spur in *S. mansoni* and the short tail in *S. haematobium*.

segmentation of the mycelia, the perfect parallelism of their margins, and the presence of sharply defined pear-shaped round bodies or spores, singly or in clusters, and red to brown in color.

Leptothrix VAGINITIS (FIGS. 39B, 40C). The clinical symptoms of *Leptothrix* vaginitis have been sketchily studied. It is thought that these organisms are able to produce some clear vaginal discharge and mild pruritus, which often goes unnoticed by the patient. This filamentous fungus, similar to *Actinomyces,* is often mistaken for a bacterium. It should not be confused with *Candida,* which is larger and segmented, with Döderlein's bacillus, which is shorter and narrower, or with some species of *Corynebacterium* or *Listeria.*

Their number and size vary; they are usually elongated and very thin (hairlike structures) with rare branching.

They are poorly stained and grayish in color.

No spore is ever formed. They reproduce by segmentation.

They can be found singly or in clusters, often in symbiosis with *Trichomonas,* and are more frequently seen in the vagina of the younger patient than of the older patient.

Their incidence increases during pregnancy.

They can be the cause of some mild degenerative changes seen in the endocervical and squamous basal cells.

Viral Inflammation

Most of the viruses infecting the buccal mucosa have the capacity to infect the vaginal mucosa. Cytologic recognition of various typical viral changes in the cells of routine vaginal smears is the most readily available method of detecting genital viral infection. These viral infections may have venereal, neoplastic, or perinatal significance. The most common morphological changes of the exfoliated cells infected by the different viruses can be summarized as follows:

NONSPECIFIC STRUCTURAL CHANGES COMMON TO ALL VIRAL INFECTIONS. Different stages of viral cellular changes may be seen in any single vaginal smear because all the cells are not simultaneously infected by the virus.

The usual regular sequence of these changes is:

1. Hypertrophy of the cytoplasm or of the nucleus or both;
2. Disturbance of the texture of cytoplasmic ground substance (from its normal granular composition to a hyaline appearance);
3. Enlargement of the nucleolus followed by its distortion and disappearance by lysis;
4. Development of an amorphous or particulate, single or multiple intracytoplasmic or intranuclear inclusion body surrounded by a more or less prominent halo;
5. Late ballooning type of cytoplasmic and nuclear degenerative vacuolization.

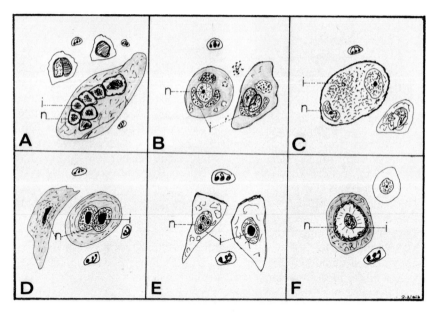

Figure 45
Cellular changes in viral genital inflammation (n = nuclei; i = inclusion). (A) Herpes simplex. Note the intranuclear inclusion and multinucleation. (B) Inclusion vaginitis. Note the cluster of intracytoplasmic inclusions. (C) Granuloma venereum. Note the cluster of intracytoplasmic inclusions. (D) Condyloma acuminatum. Note the hyperkeratotic cells and basophilic intranuclear inclusions. (E) Adenovirus. Note the intranuclear inclusions. (F) Cytomegalic inclusion infection. Note the intranuclear inclusions.

HERPES SIMPLEX (FIGS. 45A, 46A). The gross lesion can be vesicular, ulcerative (deep or shallow), or occult. It can be confined to the cervix, vagina, or vulva.

The clinical symptoms resulting from a genital herpes inflammation can be very severe with fever and enlargement of the regional lymph nodes, mainly in patients with primary infection. In other patients, the condition can be clinically undetectable. The virus is often transmitted during sexual intercourse, and the majority of the lesions ordinarily disappear without any treatment in one to two weeks in non-pregnant patients, though more slowly in pregnancy. The main danger is in the transmission of the virus to the newborn, especially when the infection occurs during the last stage of pregnancy in a mother who does not have sufficient time to produce the necessary protective antibodies. The diagnostic exfoliated cells in a vaginal smear are typical and easy to recognize.

The alterations appear in the younger cells, the parabasal cells of the squamous epithelium, or the reserve cells of the endocervical glandular epithelium.

The first visible structural change is a true hypertrophy of the cytoplasm and the nucleus.

An irregular perinuclear halo is often seen early at this stage, but this

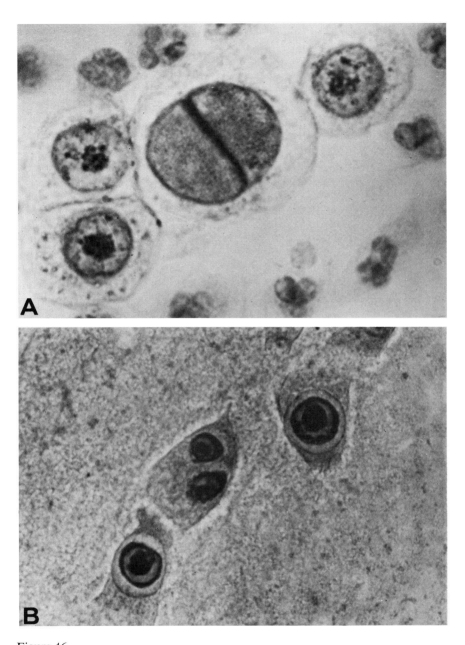

Figure 46
(A) Infected basal squamous cells in the cervical smear of a patient with a virologically proved herpetic cervicitis. Note the intranuclear granular elementary bodies of the large binucleated central cell, indicating an early infection, and the granular intranuclear inclusions of the smaller cells, indicating a late stage of infection. ($\times$ 1300) (B) A vaginal smear showing endocervical cells infected with cytomegalic inclusion virus (CMV). Note the intranuclear inclusion, the halo, and the compressed chromatin. ($\times$ 1300)

change is probably a fixation artifact and is common to most of the enlarged exfoliated epithelial cells.

From its normal granular composition, the cytoplasm becomes hyaline in appearance, denser, and more basophilic (purple in color).

Multinucleation, probably resulting from amitotic division or the fusion of several cells, is associated with the formation of the diagnostic giant cells.

The cells differ from the multinucleated foreign-body giant cells by their characteristic nuclear molding without overlapping and by variations in nuclear size and shape.

The nuclear chromatin loses its granularity, then clumps and marginates, leaving a bland, amorphous eosinophilic zone in the center. The nuclear basophilia gradually decreases except for the chromatin clumps adhering to the inner surface of the nuclear membrane.

Single, acidophilic inclusion bodies then develop in the center of the nuclei and are soon surrounded by more or less prominent halos.

Ballooning cytoplasm with multiple vacuolization or nuclear degeneration usually follows, indicating an irreversible cellular injury.

The clinical significance of genital herpesvirus infection may be summarized as follows:

1. There is a need to differentiate the viral disease from other nonviral infections, especially other venereal diseases;
2. Primary vulvar herpes causes some morbidity (fever, painful lesion, enlarged inguinal nodes);
3. In the pregnant patient, the virus may be transmitted to the newborn at delivery;
4. There is some evidence of a correlation with abortion;
5. There is strong circumstantial evidence suggesting that the virus may play a role in cervical carcinogenesis.

INCLUSION VAGINITIS (FIGS. 45B, 47). Inclusion vaginitis, caused by an organism of the *Chlamydia,* lymphogranuloma venereum group (TRIC agent) is one of the most common causes of ophthalmic infection of newborns. The genital infection produces few or no clinical symptoms, except for vaginal discharge and lesions of the mucosa ("cobblestone" type of superficial cervicitis). Often the infection is limited to the external os and transitional zone of the cervix. The smear usually shows a number of inflammatory cells with a predominance of neutrophils and occasional eosinophils. Large numbers of single hyperactive basal squamous or reserve cells with clusters of numerous perinuclear, basophilic, granular, cytoplasmic inclusion bodies are diagnostic. The cytoplasm of these cells usually has an amorphous appearance with multiple degenerative vacuoles. In cases of acute infection, after the rupture of a vacuole in the late stage of the disease, reddish-brown inclusion bodies can occasionally be seen as masses outside the infected cells in the background of the slide. Often these inclusion bodies are difficult to differentiate from colonies of microorganisms.

CONDYLOMATA ACUMINATA (FIG. 45D). This cauliflowerlike, warty growth, covered by hyperkeratotic epithelium, is generally accepted as be-

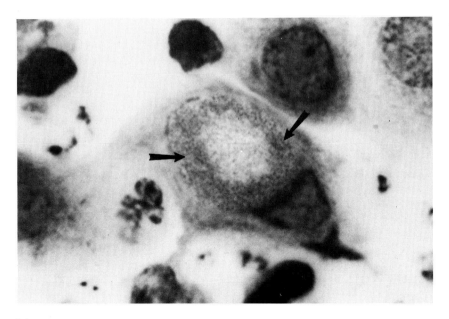

Figure 47
Inclusion vaginitis. Note the numerous (Giemsa-stained) granular inclusions (*arrows*) in the cytoplasm of the parabasal cell. (× 1300)

ing of viral origin, although the virus has not been isolated. To obtain diagnostic cells, the scraping should be energetic. The diagnostic large basal or intermediate squamous cells appear swollen and edematous with thick, pale basophilic cytoplasm having poorly defined cytoplasmic membranes.

The enlarged nuclei show little variation in shape, being round or oval with thick, regular nuclear membranes.

The nucleolus is usually prominent, round, and regular.

Occasional basophilic inclusions combine with the nucleoli, giving them the appearance of large, irregular nucleoli.

Perinuclear halos, ballooning cytoplasmic degeneration, and multinucleation with nuclear molding are other changes observed in the infected cells.

Superficial squamous atypical cells with hyperkeratosis are often present.

ADENOVIRUS (FIG. 45E). Genital infection by an adenovirus is rare and gives no clinical symptoms. The diagnosis is usually made first during the examination of a routine Papanicolaou smear. The infected cells are mainly basal endocervical in type, mononucleated, and shed singly or in clusters. Their cytoplasm is not affected and looks normal. Their nuclei are often enlarged and contain multiple small (early stage of infection) or large single (late stage of infection) basophilic amorphous inclusions. A definite periinclusion halo, which is necessary for differentiation from an abnormally prominent nucleolus, should always be present.

CYTOMEGALIC INCLUSION DISEASE (FIGS. 45F, 46B). This rare infection can be primarily localized in the genital tract, or it can be systemic. In the smear, the changes are limited to young columnar secretory endocervical cells. They are hypertrophic with scanty but thick purple-stained cytoplasm. The nucleus is enlarged with the chromatin compressed to the periphery. Single, large intranuclear granular inclusions with prominent halos are diagnostic. The detection of this disease becomes important in pregnancy because of the possible transmission of the virus to the fetus.

Inflammation of the Upper Genital Tract

ENDOMETRITIS

The diagnosis of an infection in the endometrial cavity can be made with a smear from an endocervical or endometrial aspiration or scraping. Endometritis can be the result of the invasion of the endometrial cavity by bacteria, retained gestational tissue, an endometrial polyp, or a myoma.

It is often secondary to the introduction of a foreign object in the uterine cavity, as in criminal abortion or the use of an intrauterine contraceptive device. For a short period immediately following the insertion of the device, reactive endometrial cells in the midst of increased inflammatory cells (polymorphonuclear cells) and histiocytes may appear on the smear in 10% of the cases. In the long run, the majority of the intrauterine devices (IUD's) are well tolerated; they do not affect the menstrual cycle, and their presence cannot be detected by cytology. Similarly, no relation between the IUD and endometrial or cervical cancer has been demonstrated.

The persistence of increased numbers of histiocytes in the smears of some patients with an IUD has suggested that conception may possibly be prevented by the phagocytic and enzymatic action of these macrophages upon the sperm.

Puerperal Endometritis

This type of inflammation can follow an abortion (spontaneous or induced) or a full-term delivery (normal or abnormal). It is usually the result of bacterial invasion. Its severity depends on the intensity of local reaction and the virulence of the infecting organism. In some cases, gestational products may be retained, resulting in vaginal bleeding. The aspirated endometrial smear shows a large amount of mixed inflammatory cells, an abundance of fragmented cytoplasm and nuclei, and an increased number of histiocytes and foreign-body giant cells. Some fresh and old blood and possibly trophoblastic elements can be present. These trophoblastic giant cells should be differentiated from the multinucleated foreign-body giant cells by the irregularity of their nuclei, which occasionally mold and have marked chromatin clumps and prominent nucleoli.

Atrophic Endometritis

This inflammation can occur during a severe deficiency of estrogenic hormones. The endometrial aspiration is scanty in epithelial cells, but abundant in round lymphocytes and plasma cells. Metaplastic squamous cells with semikeratinized abundant cytoplasm can be seen. The background will contain some fresh and old blood as well as a moderate amount of cellular debris and protein deposits.

Tuberculous Endometritis
This type of inflammation is difficult to diagnose by cytology. If, however, it is suspected, a special acid-fast stain of the endometrial aspiration smear should be made.

It is associated with renal tuberculosis in 10% of the cases. Streaks of clustered, small histiocytes, some with irregular elongated nuclei, may be indicative of the infection. Occasional multinucleated Langhans' giant cells can be found.

ADNEXAL INFLAMMATION AND PELVIC INFLAMMATORY DISEASE
Vaginal and endometrial aspiration smears are not as helpful in the diagnosis as is a posterior cul-de-sac needle aspiration, which will usually confirm the diagnosis by showing a large increase in inflammatory elements mixed with reactive mesothelial cells.

Cellular Degeneration and Necrosis

Decrease or cessation of the cellular functions usually results in the death of a cell. In general this process occurs gradually, except when it is produced artificially, as in the rapid protoplasmic coagulation during alcohol fixation. The amount and type of degenerative changes seen in a cell will vary with the nature and extent of the injury. Degeneration affects all cells, benign or malignant, epithelial and nonepithelial.

Rapidly growing malignant neoplasms tend to degenerate because growth outdistances blood supply, thus providing poor nourishment for the cancer cells. Degenerative changes may mask or distort some of the malignant features of the tumor cells, rendering their recognition difficult.

In benign cells degeneration can mimic atypical or malignant features, and these may cause some of the false-positive diagnoses. On the other hand, early degenerative changes may sometimes enhance the malignant characteristics of the exfoliated tumor cells (e.g., hypertrophy and hyperchromatism), thereby making their detection easier.

No diagnosis of malignancy should be based on a cell with features of advanced degeneration. Such a diagnosis should not be based on an abnormal nucleus, whatever the degree of its abnormality, if the cytoplasm is not intact. In poorly differentiated carcinoma the presence of a very scanty but intact cytoplasm may be ascertained only by high-power magnification examination (oil immersion) of the cells.

When the irreversible and persistent degenerative changes do not destroy the cell, they may be transmitted to the descendants of that cell, as is seen in the case of irradiation injury, in which the same degenerative irradiation changes in the exfoliated daughter cells may be found 20 to 30 years after the injury.

REASONS FOR DEGENERATION

Normal Aging Process

This is seen in the normal, mature, superficial squamous cell, which shows black- or brown-pigmented inclusions in its hyperkeratinized cytoplasm.

Inflammatory Conditions

The extent and the type of cellular changes depend on the nature and virulence of the inflammatory agent. Cellular hypertrophy and perinuclear halos are often present.

Metabolic Degeneration

Cloudy swelling, fatty degeneration, and hyalin degeneration are often seen in the exfoliated cells.

Trauma

This can be acidic, caustic, thermal, or mechanical in origin.

Insufficient Blood Supply

This condition produces cellular starvation, as in the case of infarction or in rapidly growing neoplasms.

CHANGES IN CELLULAR DEGENERATION (FIG. 48)

Cytoplasmic Changes

VACUOLIZATION. The vacuoles are usually multiple, vary in size and shape, and surround the nucleus. They often contain well-preserved, healthy polymorphonuclear cells. The vacuolar margins are delicate but well defined. They can, when overdistended, rupture and leave only scanty, frayed cytoplasm still attached to the bare or naked nuclei, that needs to be differentiated from the scanty but intact cytoplasm of certain malignant cells.

CLOUDY SWELLING. This condition is the result of an accumulation of excess fluid in the cytoplasm and nuclei. The size of the cells increases two to three times, and the cytoplasmic background looks granular, cloudy, and fuzzy. The density and staining intensity of the cells decreases. The cellular borders become fuzzy, distended, smooth, and poorly defined. This cloudy swelling can occur when the cells have been poorly preserved in weak fixative solution or when the patient has taken a vaginal douche before her smear. This artificial increase in cellular size should not be confused with the true cellular hypertrophy in which the density of the cell remains the same or even increases, as is seen in neoplastic cells.

Cytoplasm may rupture to release an enlarged, swollen, naked nucleus without any attached fragments of cytoplasm. This is different from the previous mode of degeneration (vacuolization), in which the size of the nucleus is not increased and fragments of cytoplasm are retained.

PERINUCLEAR HALO. This is often produced artificially during fixation. It is an indication that the cytoplasm, nucleus, or both, have enlarged and that

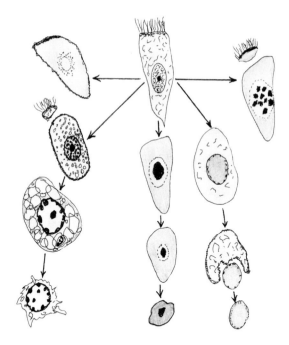

Figure 48
The most common type of degeneration of a ciliated columnar cell.

a part of the nucleus and cytoplasm has shrunk on itself during fixation and dehydration, leaving an apparent empty space between the endoplasm and nuclear membrane. The perinuclear halo is often the result of degenerative hypertrophy (cloudy swelling), but it can also be found in cellular atypia in which a true cellular hypertrophy exists. It is frequently seen in the basal and intermediate squamous cells during an inflammatory condition (*Tricho-monas* or viral infection). The halos are seldom found in small cells. They should be differentiated from the true halo that can surround a viral inclusion.

LOSS OF CELLULAR ORGANELLES.   Very early in the degenerative processes, some of the nonvital specialized organelles of the cytoplasm disappear, such as the cilia (ciliocytophoria).

IRREGULAR SHAPE.   The cytoplasm can become very irregular in shape with pseudopodlike projections. Most of the time, this irregularity indicates a cellular attempt to heal after an injury that allows the survival of the cell (repair cells after trauma and irradiation injury). These changes should not be confused with the cytoplasmic irregularities of a neoplastic cell, which often are the results of a too-rapid cellular growth in a too-narrow space. The loss of portions of the cytoplasm may also produce an apparent irregularity of cellular shape.

POLYCHROMASIA. The staining reaction of the cytoplasm of a degenerate cell can vary from one cell to the other, or even within an individual cell. Pseudoeosinophilia, for example, occurs as a result of poor fixation or inflammation. Because of this, the staining reaction of the cytoplasm alone cannot be used as a criterion for diagnosis of the nature of a cell. All the deep orange-stained cells in a smear, for example, are not necessarily atypical. It is helpful to compare the color of a cell to the color of the cytoplasm of a leukocyte, which normally is transparent, pale blue.

HYPERKERATOSIS. Some of the cells will accumulate an abundance of keratin in their cytoplasm during their slow degeneration. This accumulation of keratin causes them to appear deep orange and glassy with the Papanicolaou stain. The hyperkeratosis can be the result of natural aging or of cellular atypia, and it may even persist after the lysis of the nucleus, as seen in the hyperkeratotic cells of leukoplakia. These changes should be differentiated from the rapid, abnormal accumulation of the cytoplasmic keratin, which is seen in the neoplastic cells of a well-differentiated squamous cell carcinoma.

CYTOLYSIS. The homogenization and vacuolization of the cytoplasm during degeneration can be followed by the fragmentation and the production of stripped nuclei. This is best illustrated by the action of Döderlein's bacilli on the intermediate squamous cells in a vaginal smear. The gray fragments of cytoplasm present in the background of such a smear may be differentiated from the similar-looking *Trichomonas* by the absence of a nucleus.

MULTINUCLEATION. Occasionally, multinucleated, degenerative giant cells in syncytial formation are produced when the cytoplasm fuses, as seen in smears of atropic mucosa, degenerated endocervical cells, some viral inflammations, and after irradiation. The cytoplasmic borders are often poorly defined, and the numerous nuclei vary in shape and size, overlapping each other without molding, which differentiates them from other multinucleated cells.

Nuclear Changes (Fig. 49)

CHROMATIN CLUMPING. During the first stages of degeneration the chromatin has a tendency to clump together to form multiple spherical particles. These clumps are variable in size but regular in shape (round). They have a tendency to migrate toward the center or the periphery of the nuclei to form a cartwheel appearance. These chromatin clumps are differentiated from the malignant ones by the fuzziness of their contours and the regularity of their shape.

KARYOPYKNOSIS (FIGS. 49A, B). The nucleus may shrink up on itself, stain darker, and render the identification of the chromatin structure impossible. This shrinkage may be the result of either a slow or a rapid degenerative change. The slow process is best seen in the nucleus of benign or atypical superficial squamous cells and is the result of normal aging and maturation.

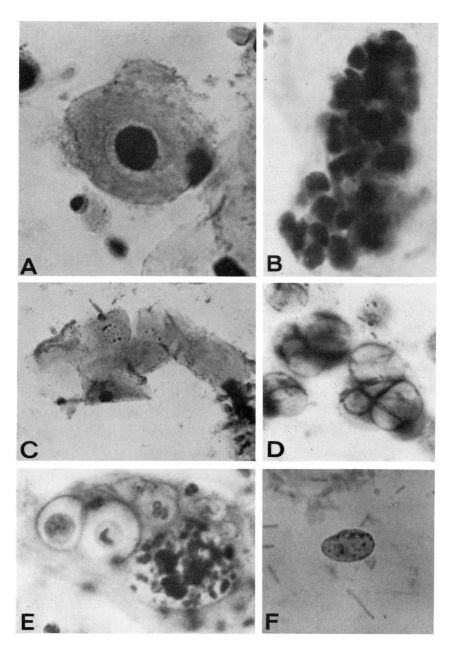

Figure 49
Cellular degenerative changes. (A) Nuclear pyknosis and perinuclear halo in a degenerate basal squamous cell. (× 1300) (B) Nuclear pyknosis and cytolysis of a cluster of reserve endocervical cells. (× 750) (C) Karyolysis of intermediate squamous cells. (× 450) (D) Karyolysis of endocervical cells. (× 1300) (E) Karyorrhexis and cytoplasmic vacuolization of a squamous cell. (× 1300) (F) Cytolysis of an endocervical cell producing a stripped nucleus. (× 1300)

The nuclear contour of the gradually occurring pyknosis is usually very sharp, in contrast to the fuzziness of the nuclear contour that results from a rapid degeneration like that seen in the pyknotic nuclei of degenerated columnar, squamous, and inflammatory cells that have been subjected to a violent irritation. This degenerative karyopyknosis should be differentiated from the irregular pyknosis of certain cancer cells that is the result of an abnormal hyperchromatism (nuclear pyknosis of invasive squamous cancer cells).

NUCLEAR EDEMA (FIG. 49). In cloudy swelling, the nucleus can become larger and its content become diluted, giving the impression of hypochromatism. In the vaginal smear this is seen as a result of poor fixation, mainly in the young squamous or columnar basal cells. Cells showing changes resulting from poor fixation should be differentiated from those with true nuclear hypertrophy, in which an increase of the chromatin occurs. A perinuclear halo is often seen.

KARYORRHEXIS (FIG. 49E). The nucleus condenses and breaks into several round or oval masses, variable in size, which often scatter in the cytoplasm after the rupture of the nuclear membrane. This change is usually the sign of a rapid degeneration like that seen in the death of the cell during radiotherapy.

KARYOLYSIS. As a result of cloudy swelling, hyperkeratosis, or other degeneration, the nuclei of some cells may disappear. A ghost—or empty contour of the nucleus—may persist. These anucleated cells, if squamous, can be differentiated from the anucleated cells of leukoplakia, amniotic fluid, and skin contaminants by their other severe degenerative changes.

NUCLEAR VACUOLIZATION. Occasional, small, round, single or multiple vacuoles can be found in the nuclei, compressing the chromatin. This condition usually indicates the rapid death of the cell from irradiation injury.

STRIPPED NUCLEI (FIG. 49F). In a cell with delicate cytoplasm even a mild injury (such as the trauma of smearing the cells on the slide) may separate the nuclei from their cytoplasm, producing variable numbers of stripped nuclei, as seen in the smears of endocervical or atrophic vaginal mucosa. These round or oval nuclei will often show few other degenerative changes. Their chromatin usually remains finely granular and stains uniformly gray-blue. Their nuclear membranes are distinct, sharp, and regular.

Cellular degeneration increases in the presence of benign but irritative, space-occupying lesions or in the case of ulcerated, infiltrating, malignant carcinoma. One can seldom find a vaginal smear in a patient with an invading carcinoma without having degenerative cellular debris in its background. Frequently, degeneration may occur also as a result of therapy, such as electrocautery, irradiation, or broad-spectrum antibiotic therapy. Cellular degeneration may become a pitfall in diagnosis, either by imitating malignant cells or by blurring the structural detail of the diagnostic cancer cells.

References and Supplementary Reading

Alousi, M. A., et al.   Microinvasive carcinoma and inflammatory lesions of the cervix uteri: Histologic and cytologic differentiation. *Acta Cytol.* 11:132, 1967.

Berry, A.   A cytopathological and histopathological study of bilharziasis of the female genital tract. *J. Pathol. Bacteriol.* 91:325, 1966.

Braya, C. A., and Teoh, T. B.   Amoebiasis of the cervix and vagina. *J. Obstet. Gynaecol. Br. Commonw.* 71:299, 1964.

Carvalho, G.   Döderlein bacilli in vaginal smears of post-menopausal women. *Acta Cytol.* 9:244, 1965.

Christian, R. T., et al.   Viral study of the female reproductive tract. *Am. J. Obstet. Gynecol.* 91:430–436, 1965.

Colmenares, R. F., and Naib, Z. M.   Significance of lymphocytic pools in the routine vaginal smear. *Obstet. Gynecol.* 26:909, 1965.

Douglas, C. P.   Lymphogranuloma venereum and granuloma inguinale of the vulva. *J. Obstet. Gynaecol. Br. Commonw.* 69:871, 1962.

Fontanes de Torres, E., and Bribiesca, L.   Cytologic detection of vaginal parasitosis. *Acta Cytol.* 17:252, 1973.

Gondos, B., et al.   Cytology changes in cervical epithelium following cryosurgery. *Acta Cytol.* 14:386, 1970.

Highman, W. J.   Cervical smears in tuberculous endometritis. *Acta Cytol.* 16:16, 1972.

Heller, C.   *Neisseria gonorrhoeae* in Papanicolaou smears. *Acta Cytol.* 18:338, 1974.

Hoffman, H., and Frank, M. E.   Microbial burden of mucosal squamous epithelial cells. *Acta Cytol.* 10:272, 1966.

Hulka, B. S., and Hulka, J. F.   Dyskaryosis in cervical cytology and its relationship to trichomoniasis therapy. *Am. J. Obstet. Gynecol.* 98:180, 1967.

Hunter, C. A., Long, K. R., and Schumacher, R. R.   A study of Döderlein's vaginal bacillus. *Ann. N.Y. Acad. Sci.* 83:217, 1959.

Ishisama, A., et al.   Cytologic studies after insertion of intrauterine contraceptive devices. *Acta Cytol.* 14:35, 1970.

Josey, W., et al.   Genital herpes simplex infection in the female patient. *Am. J. Obstet. Gynecol.* 96:493, 1966.

Kraus, G. W., and Yen, S. S. C.   Gonorrhea during pregnancy. *Obstet. Gynecol.* 31:258, 1968.

Langlinais, P. C.   *Enterobius vermicularis* in a vaginal smear. *Acta Cytol.* 13:40, 1969.

Naguib, S. M., Comstock, G. W., and Davis, H. J.   Epidemiologic study of trichomoniasis in normal women. *Obstet. Gynecol.* 27:607, 1966.

Naib, Z. M.   Exfoliative cytology of viral cervico-vaginitis. *Acta Cytol.* 10:126, 1966.

Naib, Z. M., Nahmias, A. J., and Josey, W. E.   Cytology and histopathology of cervical herpes simplex infection. *Cancer* 19:1026, 1966.

Nasiell, N.   Histiocytes and histiocytic reaction in vaginal cytology. *Cancer* 14:1223, 1961.

Rad, M.   Enterobius vermicularis in cervical smears. *Acta Cytol.* 14:466, 1970.

Ross, L.   Incidental finding of cytomegalovirus inclusions in cervical glands. *Am. J. Obstet. Gynecol.* 95:956, 1966.

Sagiroglu, N., and Sagiroglu, E.   The cytology of intrauterine contraceptive devices. *Acta Cytol.* 14:58, 1970.

Scarpelli, D. G.   The cytology of degenerative and inflammatory lesions. *Acta Cytol.* 5:206, 1961.

Stern, E., and Longo, L. D.   Identification of herpes simplex virus in a case showing cytological features of viral vaginitis. *Acta Cytol.* 7:295–299, 1963.

Trussell, R. E.   *Trichomonas vaginalis and Trichomoniasis.* Springfield, Ill.: Thomas, 1947.

Varga, A., and Browell, B.   Viral inclusion bodies in vaginal smears. *Obstet. Gynecol.* 16:441–444, 1960.

Wied, C. L.   Interpretation of inflammatory reactions in vagina, cervix and endocervix by means of cytologic smears. *Am. J. Clin. Pathol.* 28:233, 1957.

Yen, S. S., Reagan, J. W., and Rosenthal, M. S.   Herpes simplex infection in the female genital tract. *Obstet. Gynecol.* 25:479–492, 1965.

Youssef, A. F., Fayad, M. M., and Shafeek, M. A.   The diagnosis of genital bilharziasis by vaginal cytology. *Am. J. Obstet. Gynecol.* 83:710, 1962.

Benign and Precancerous Lesions of the Female Genital Tract 5

The various types of benign but atypical epithelial cells found in the genital smear can be correctly recognized most of the time as originating from a specific lesion. The cytopathologist, using slightly different criteria, can usually give a cytologic interpretation of the nature of the atypia similar to the one rendered by the histopathologist, who examines tissue sections.

A Class III report from the classification used by Dr. Papanicolaou is no longer sufficient. It provides very little help to the clinician in the management of his patient, and it may be overused as a convenient way to hide the ignorance of the screener.

Atrophy
(Figs. 50B, 51A, 52A)

This condition usually reflects a physiologic or pathologic estrogen deficiency and is not a precancerous change. The maturation index is shifted to the left.

The presence in the smear of an excess of small, round or oval, squamous parabasal cells is diagnostic. In natural exfoliation they shed singly; in traumatic exfoliation, however, they are often found in large sheets.

The cytoplasm of these cells is scanty (relatively high nuclei-cytoplasm ratio), dense, basophilic, or cyanophilic, and it contains occasional multiple small vacuoles. The cytoplasmic membrane is usually smooth, even, and regular.

The nuclei are centrally located and round or oval. The amount of chromatin is slightly increased but is usually finely granular and evenly distributed. The nuclear membrane is regular, sharp, and delicate. The nucleoli, when present, are minute and round.

Multinucleation can occur from the fusion of several cells, forming syncytial giant cells. These cells can be differentiated from the foreign-body giant cells by the variation of the size, shape, and structure of their nuclei.

The background of the smear generally shows an increased amount of degenerate cellular debris, granular protein deposits, and fresh and old blood with an increased number of inflammatory cells (neutrophils and small histiocytes).

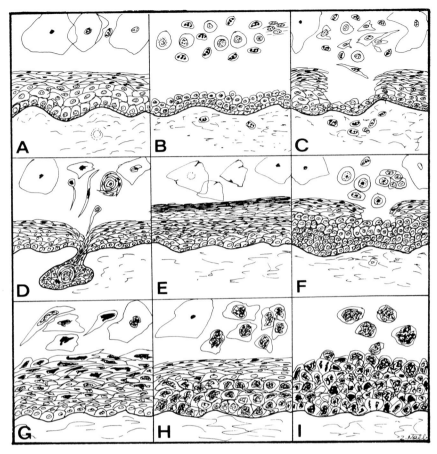

Figure 50
Histology of common benign lesions of the cervix. (A) Normal cervical epithelium. (B) Atrophy. (C) Erosion. (D) Squamous metaplasia. (E) Leukoplakia.
(F) Basal cell hyperplasia. (G) Superficial cell dysplasia. (H) Basal cell dysplasia. (I) In situ carcinoma.

On occasion, aborted keratinized basal cells (small round basal cells
with pyknotic nuclei and orange cytoplasm) can be seen.

Similarly, numerous traumatically exfoliated endocervical stripped nuclei—singly or in clusters or sheets—can be found in the smears of postmenopausal women.

The presence of large numbers of basal cells may become a handicap
in the detection of poorly differentiated, invasive squamous or in situ carcinoma. In such a case, the smear should be repeated after estrogen therapy.
A majority of the benign basal cells will disappear and will be replaced by
superficial cells, while the dysplastic or tumor cells will persist unchanged
and become easier to detect.

In some atrophic smears, large numbers of multinucleated foreign-body

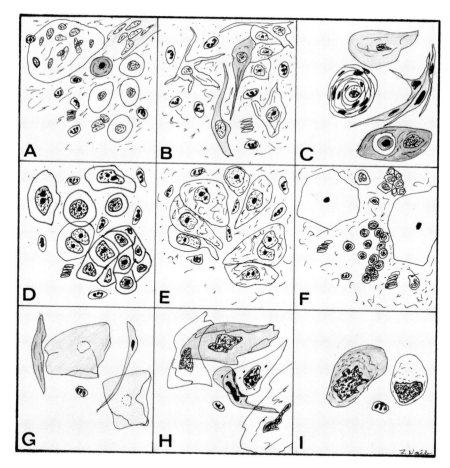

Figure 51
Cytology of benign lesions of the female genital tract. (A) Atrophy. (B) Erosion (repair cells). (C) Squamous metaplasia. (D) Basal cell hyperplasia. (E) Reserve cell hyperplasia. (F) Endometrial hyperplasia. (G) Leukoplakia. (H) Superficial cell dysplasia. (I) Basal cell dysplasia.

giant cells may be present. Keratinized vaginal superficial squamous cells may be distinctly seen phagocytized in their cytoplasm. It appears that the histiocytes of the patient mistake the normal superficial squamous cells for foreign bodies and try to neutralize them by ingestion. The significance of this abnormal autoimmune reaction has not yet been evaluated.

Erosion or Ulceration—Repair Cells
(Figs. 50C, 51B, 52B)

The regional, partial, or total loss of the vaginocervical squamous epithelium is commonly the result of genital trauma or acute inflammation (e.g.,

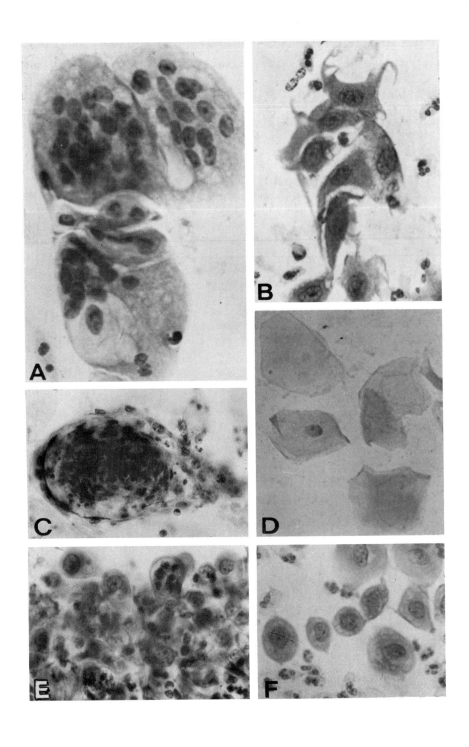

due to *Trichomonas*). The diagnostic cells are often associated with the repair or healing reaction of the tissues. They are common in smears taken from patients after hysterectomy or radiotherapy, or from those who have had recent cautery, cryosurgery, biopsy, or delivery. The epithelial erosion may coexist with a carcinoma, necessitating a more careful screening of the specimen that contains such diagnostic benign repair cells.

The smears are indicative of inflammatory conditions (large number of neutrophils are present) and have a dirty background because of the large amounts of granular serum protein deposits and cellular debris. Fresh and old blood cells are also found. Histiocytes may be present in about 20% of the cases.

Often the so-called repair or spider cells are abundant. They exfoliate singly (25%) or in clusters or sheets (75%). They appear as large, hyperactive basal cells with numerous cytoplasmic, pseudopodlike projections similar to the ones found in irradiated cells. They are seen undergoing varying degrees of differentiation. Generally adequate, the cytoplasm can be dense or semitransparent, basophilic (60%), or eosinophilic (40%). The nuclei are enlarged (mean value $30 \pm 5 \ \mu^2$), and they vary in size from one cell to the other. Usually vesicular, they are centrally located, have a heavy nuclear membrane, and show a variable degree of hyperchromatism. Their chromatin is finely granular with chromocenters in 60% of the cells and coarsely granular and irregularly distributed chromatin in the remainder. Their nucleoli are often large, spheroidal, and prominent, varying in number and size. Multinucleation (1–3) is occasionally seen, the different nuclei being uneven in size, shape, and structure, overlapping each other, rather than molding. Mitotic figures may be found in about 5% of the cases.

These repair cells are often found in large sheets and can be difficult to differentiate from the cells exfoliated from an endocervical polyp or a low-grade endocervical adenocarcinoma or sarcoma.

Ectropion

When the columnar endocervical cells extend beyond the external os and appear as a red circular zone in the portio vaginalis, the ectropion may

---

Figure 52
(A) Large syncytium of basal cells in an atrophic smear. Note the occasional prominent nucleoli and absence of ingested debris. ($\times$ 450) (B) Repair or spider basal cells seen in vaginocervical mucosa erosion or ulceration. Note the pseudopodlike projections of the cytoplasm. ($\times$ 450) (C) Mature squamous metaplasia. Central squamous formation like a pearl surrounded by flattened endocervical cells. ($\times$ 125) (D) Leukoplakia. Note the anucleated, hypermature, polygonal squamous cells. ($\times$ 350) (E) Squamous basal cell hyperplasia. Note the dense cytoplasm, large nucleus, coarse chromatin, and prominent nucleoli. ($\times$ 350) (F) Reserve cell hyperplasia. Small basal cell with vacuolated cytoplasm. ($\times$ 450)

grossly resemble an erosion or ulceration. Such lesions are more commonly seen in the postpartum period.

There is an increased number of intensely stained endocervical cells that appear singly or in sheets. They are generally well preserved, normal looking, or hyperplastic.

The abundant cytoplasm of these cells is often dense, acidophilic, or distended by large vacuoles that contain mucous secretions.

When in sheets, the cells show an intensive molding with a pronounced distortion of their shape which makes them look more atypical than they actually are.

Their nuclei may be moderately enlarged and hyperchromatic with prominent, red, occasionally multiple nucleoli. Their chromatin is coarsely granular but uniformly distributed, a criterion useful in differentiating them from an endocervical adenocarcinoma cell.

Contrary to what is seen in the case of erosion, no parabasal repair cells are present. It is also hard to differentiate these cells from the ones scraped from an ulcerated endocervical glandular polyp.

A moderate increase of leukocytes, histiocytes, fresh and old red blood cells, and cellular debris may obscure the background of the smear.

Squamous Basal Cell Hyperplasia
(Figs. 50F, 51D, 52E, 53A, C)

This is a reversible lesion occurring as a reaction to regional irritation. It has very little tendency to progress into dysplasia or in situ carcinoma. The normal equilibrium between the birth of new cells and the exfoliation of old ones is disturbed, resulting in an abnormal increase in the number of parabasal layers. These atypical basal cells may be covered with normal epithelium, which will prevent their detection by cytology. Frequently, however, the surface is eroded and the atypical basal cells desquamate readily, especially if the cervix has been energetically scraped.

Numerous hypertrophic basal cells, uniform in shape (round or oval) but variable in size (15–40 μ), may shed singly, in clusters, or in large sheets.

The cells' basophilic, dense, uniform cytoplasm is usually adequate, round or oval with irregular and poorly defined borders.

The centrally located nuclei are large and spherical with well-defined, regular, thick borders. The chromatin is darkly stained and coarsely clumped.

The nucleoli are prominent, massive, spheroidal, and usually single, but they may vary in number (2–4).

The cells can be multinucleated (2–3 nuclei), variable in their size and shape, molding, or overlapping each other.

The background of the smear shows an increased number of inflammatory leukocytes, fresh and old red blood cells, and deposits of cytoplasmic debris and protein.

The discovery of such reactive basal cells in routine vaginal smears in-

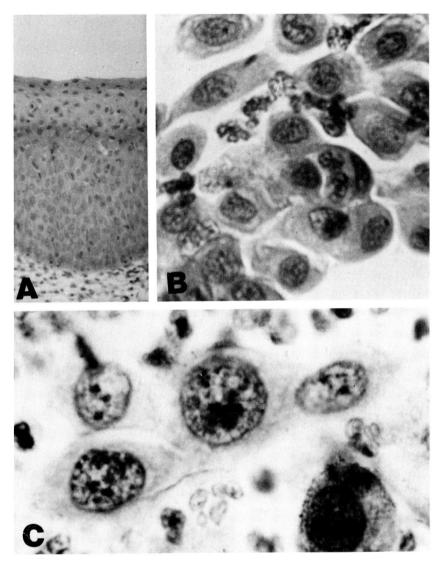

Figure 53
Basal cell hyperplasia. (A) Histologic section. (× 125) (B) Reserve cell hyperplasia. (× 450) (C) Squamous parabasal cell hyperplasia. (× 1300)

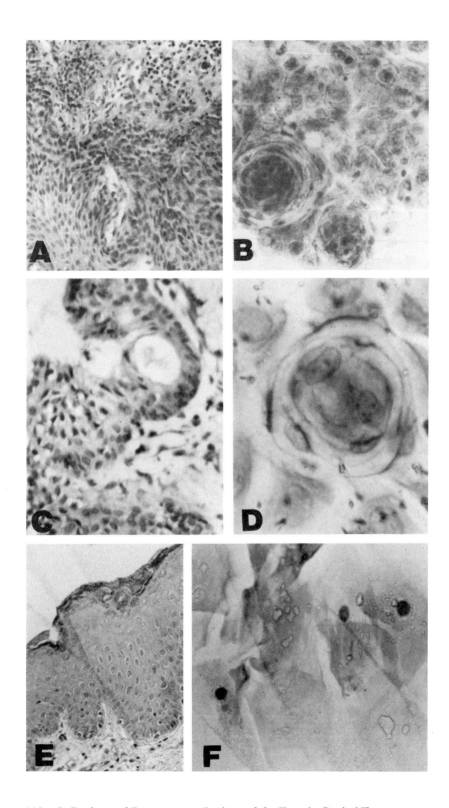

dicates the need for a repeat smear in the next two to three months after the removal of the source of irritation.

Reserve Cell Hyperplasia
(Figs. 51E, 52F, 53B)

Reserve basal cell hyperplasia is evidenced by a proliferation in the endocervical columnar mucosa of the small basal cells, which have the potentiality to develop into either squamous or glandular epithelium. The diagnostic cells are more abundant in the cervical and endocervical scraping specimens than in the vaginal pool smears.

The cells are oval or elongated in shape and exfoliate either in loose clusters or, mainly, in sheets. They resemble the cells from squamous basal cell hyperplasia, except for their cytoplasm, which is semitransparent and vacuolated instead of being dense. The cytoplasm stains unevenly, is basophilic, and often contains mucin droplets in vacuoles.

The nuclei are enlarged, centrally located, and round or oval in shape. The chromatin is homogeneous, coarsely granular, and regularly distributed. The nucleoli are prominent, round, and usually single.

Some of the cells show degenerative swelling and hypertrophy, which result from cellular edema, with a dilution of their structures.

The significance of these lesions is similar to that of squamous basal cell hyperplasia, and the process is usually reversible. The absence of dysplastic cells and the regularity of the smooth nuclear membrane of these cells differentiate them from the cells of poorly differentiated or in situ carcinoma.

They are differentiated from reactive histiocytes by their round, central nuclei and the occasional molding of the cells against one another.

Mature Squamous Metaplasia, Epidermization, and Squamous Prosoplasia
(Figs. 50D, 51C, 52C, 54A, B)

Under this classification we consider those areas of glandular epithelium (mainly of the endocervix but also of the endometrium) that are transformed and replaced by mature stratified squamous epithelium. This can be the result of chronic inflammation, hormonal changes, repeated pregnancies, or congenital abnormalities.

The malignant potentiality of these cells is very low, being only slightly higher than that of the normal epithelium. The presence of metaplasia increases the length of the squamous-columnar junction, which is considered a weak zone, one from which carcinoma preferentially arises.

As a result of scraping, numerous mature intermediate or superficial

Figure 54
Squamous metaplasia. Mature: (A) Histology. (B) Cytology. Immature: (C) Histology. (D) Cytology. (E) Leukoplakia histology. (F) Cytology. Note the anucleated keratinized cells. ($\times$ 450)

squamous cells may be shed singly, but they are more commonly found in sheets and often in pearl formation.

The cytoplasms are abundant, stain deep orange, and have keratohyalin granules, occasionally in concentric rings. Their shape is irregular, varying from one cell to the other; they often show pseudopodlike projections, indicating that these cells have multiplied and grown in a narrow space (i.e., the endocervical gland lumen). The cytoplasmic borders are well defined and sharp.

The nuclei are often irregular in shape, single or multiple (2–3) with moderate variation in their size. In the vesicular nuclei, the chromatin pattern is coarsely granular and uniformly or irregularly distributed, depending on the severity of the lesion. Hyperchromatism is common with occasional irregular nuclear pyknosis. The nucleoli are minute and most often cannot be seen.

These cells must be distinguished from those of well-differentiated, low-grade, squamous invasive carcinoma, in which the cytoplasmic keratosis and nuclear abnormalities are more extreme.

Immature Squamous Metaplasia
(Figs. 54C, D)

In immature squamous metaplasia, areas of the endocervical and endometrial glandular epithelium are replaced by young, squamous, reactive parabasal cells. They can exfoliate singly, but mainly they exfoliate in tight sheets or groups (85%).

Their shape is irregular, angular, round, oval, or spinous, and they mold tightly against one another.

The cytoplasm is scanty to moderate in amount, irregular in shape, and has indistinct borders. It can stain deep blue, pink, or orange. Occasionally, the cytoplasm contains multiple small vacuoles in which mucus or glycogen granules, staining yellow to brown, can be demonstrated.

The enlarged nuclei are centrally located, vesicular, and often irregular in shape as a result of the pressure of the neighboring cells, and they have heavy borders. The chromatin is coarsely granular and often irregularly distributed. The nucleoli, when present, are prominent, as seen in basal cell hyperplasia (20%).

The presence, in a young woman, of sheets of deeply stained parabasal cells, irregularly shaped and strongly molded against each other, but without a gross cervical ulceration, suggests the presence of an immature squamous metaplasia. Since these lesions can be found associated with an in situ carcinoma, the smear with such cells should be more carefully screened. They are more commonly seen in the smears of pregnant women. A close follow-up of the patient is indicated until the lesion disappears.

Leukoplakia—Hyperkeratosis, White Patches
(Figs. 50E, 51G, 52D, 54)

This hyperkeratinization of the superficial layer is often the result of chronic irritation or hormonal disturbance and is most commonly seen in cases of uterine prolapse. Clinically, the whitish plaques may become grossly more apparent just before menstruation.

The smear will contain an increased number of hypermature, polygonal, often anucleated, cells shed singly (80%), in tight sheets, or in pearl formation. The presence of these hyperkeratinized cells does not necessarily indicate an increase of estrogenic effect, which is a pitfall in hormonal evaluation by cytology.

Their cytoplasm is abundant, angular, thin, stained deep orange or yellow with a ground-glass appearance, and contains keratinized granules but no glycogen.

Their nuclei are completely missing in a majority of the cells (85%). Occasionally, an empty space in the cytoplasm indicates the location of the lysed nucleus. In a few cells, small retained pyknotic nuclei may still be detected.

Because of their staining characteristics, these cells can easily be differentiated from the anucleated semitransparent cells of fetal origin found after the rupture of the membrane or from anucleated cells present as contaminants from the perineum or skin.

The leukoplakia, particularly if situated toward the squamocolumnar junction, is associated, in about 10% of the cases, with a cervical dysplasia or in situ carcinoma. The smears with such anucleated cells should be particularly carefully screened.

The cells of leukoplakia are differentiated from the degenerate squamous cells with lysed nuclei by their hyperkeratinization, sharpness of the cytoplasmic border, and absence of other degenerative changes.

Endocervical and Endometrial Polyps

Polyps are more common in the postmenopausal woman and can be glandular or fibrous in type. They seldom become malignant. As a result of acute or chronic traumatic irritation, the surface often shows areas of squamous metaplasia or ulceration. Specific cytologic diagnosis is difficult, especially if the polyp is fibrous in type.

The smear is usually inflammatory with an increase in the number of neutrophils, lymphocytes, plasma cells, histiocytes, and occasional foreign-body giant cells.

The maturation index is generally shifted to the right (40%), but an atrophic smear can occasionally be seen (10%).

Fresh and old red blood cells, some ingested by phagocytes, are commonly found. The background of the smear appears dirty owing to granular serum protein precipitates.

There is an increased number of benign ciliated and nonciliated hypersecretory endocervical and endometrial cells, mixed with metaplastic eosinophilic squamous cells. They shed singly or in large sheets, forming glandular or polypoid acini. Their cytoplasm is usually abundant, and their round-to-oval nuclei seem benign. Their nuclei may show extreme variations in size and shape with coarse chromatin and prominent nucleoli. Their nuclear membrane, however, remains smooth, and their chromatin is uniformly distributed.

Endometrial Hyperplasia
(Fig. 51F)

This condition results from an overgrowth of the glandular and stromal elements and is most common in the adolescent and premenopausal woman. It often indicates the persistence of a follicle or a decrease in the amount of secreted progesterone. The bleeding may be due to a fall of the estrogen level and necrotic changes in the endometrial mucosa. The condition must be suspected when benign endometrial cells are found 10 days after the end of the last menstrual period.

Granulosa cell tumor or theca cell tumor of the ovary should be suspected when hyperplasia is diagnosed in a premenopausal woman. The smears should be repeated daily for one to two weeks for hormonal evaluation. The persistence of a vaginal hyperestrogenlike effect should always be investigated for its cause.

The smears are usually bloody but without cellular debris or protein deposits, except when cervical ulceration is also present. Döderlein's bacilli are rare.

An abundance of single, superficial, squamous cells are the main squamous epithelial components. Clusters of endometrial cells may be present with benign, slightly hyperchromatic nuclei. Isolated endometrial glandular and stromal cells may also be seen mixed with variable amounts of histiocytes. These hyperplastic endometrial cells may show variable degrees of nuclear and cytoplasmic abnormalities. Nuclei are often enlarged, but the increased chromatin is usually regular. The cytoplasm is finely or coarsely vacuolated and contains occasional polymorphonuclear cells. In some atypia, the cells are difficult to differentiate from low-grade, well-differentiated endometrial adenocarcinoma cells. They can be considered to be in situ, or the precursors of endometrial adenocarcinoma. Their presence in a postmenopausal smear warrants a dilation and curettage. The cytology of the various endometrial hyperplasias is summarized in Table 4.

Squamous Superficial Cell Dysplasia
(Figs. 50G, 51H, 55, 56)

This lesion is characterized by atypical changes that occur in the mature superficial layer of the stratified squamous epithelium rather than in the basal layer, which remains normal. This lesion comprises 25% of all the cervical dysplasias and is more commonly found in women in the second or third decade of life. Such lesions are the result of partially impaired differentiation and maturation of the squamous cells with a thickening of the epithelium.

The smear shows an increased number of atypical superficial or intermediate type squamous cells in various degrees of differentiation. The abnormality of their features may be mild, moderate, or extreme.

They shed singly (35%), in small sheets or clusters (60%), or in pearl formation (5%). The shape of the cells is round or oval (40%), distorted, polygonal, or square (60%).

Table 4. Cytology of Endometrial Hyperplasia

Type	Cystic	Adenomatous	Atypical
Number of cells	+	++	+++
Size	$90 \pm 20 \; \mu^2$	$100 \pm 20 \; \mu^2$	$120 \pm \mu^2$
Hyperchromatism	++	++	++
Irregularity of chromatin	+	++	+++
Prominent nucleoli	2%	5%	15%
Nuclear molding	+	++	+++
Nuclear pyknosis	0.5%	0.5%	2%

Their cytoplasm, usually abundant, stains deep orange and contains keratohyalin granules, which are often disposed in perinuclear concentric rings. The cytoplasmic borders are sharp and well defined.

The single or multiple (2–6) nuclei are central, enlarged, round or oval in shape, and vary in size from cell to cell (10–20 μ). The nuclear membrane, usually well defined, is smooth (60%) or irregular and wrinkled (40%). The chromatin is increased and coarsely granular with occasional irregular clumping or pyknosis (25%). The distribution of the chromatin is uniform to irregular, depending on the severity of the lesion. Nucleoli, if present, are not prominent.

The background of the smear is not remarkable, but when ulceration is present, it becomes dirty with protein and deposits of cellular debris.

If the irregularity is moderate to extreme, a cervical biopsy examination is indicated. In a woman with a cytologic diagnosis of mild superficial cell dysplasia, a close follow-up by repeated smears will be sufficient.

It has been shown that superficial cell dysplasia coexists with an in situ carcinoma in more than 25% of the cases. Furthermore, 5 to 25% of the patients with cervical superficial dysplasia eventually develop a squamous carcinoma. All these patients, if not biopsied, should be carefully followed with repeated smears until the atypical cells eventually disappear from the smear.

Squamous Basal Cell Dysplasia
(Figs. 50H, 51I, 57B, C, 58B)

With squamous basal cell dysplasia, the histologic examination of the cervical epithelium shows increased mitotic figures and abnormalities in the shapes of the cells in the lower and middle third of the stratified squamous epithelium. The lesion differs from an in situ carcinoma only by virtue of the lesser and variable degree of maturation seen in the intermediate and superficial layers. The dysplasia is the result of partially impaired differentiation of the basal squamous cells.

The dysplasias may show variable degrees of irregularity of their features and, according to their severity, they are separated into categories of

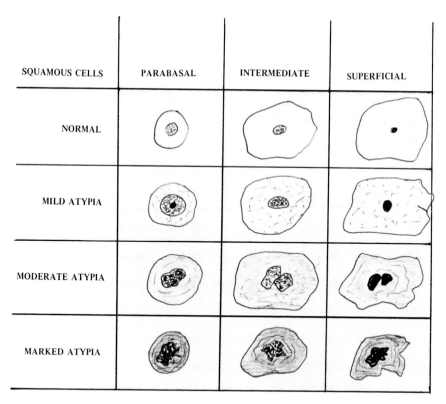

SQUAMOUS CELLS	PARABASAL	INTERMEDIATE	SUPERFICIAL
NORMAL			
MILD ATYPIA			
MODERATE ATYPIA			
MARKED ATYPIA			

Figure 55
Various cervical squamous atypia.

mild, moderate, and extreme dysplasia. Their capacity for becoming malignant increases with the degree of the irregularity of the nucleus and cytoplasm.

The diagnostic cells may shed singly (60%) or in clusters or sheets (40%). Their size varies moderately (30 ± 10 μ). Their shape is irregularly polygonal (60%), oval (20%), or round (20%).

Their cytoplasm is usually adequate (99%), thick, basophilic (50%), or eosinophilic (50%). The cytoplasm contains no glycogen or vacuoles unless degenerate. It has sharp, distinct borders.

The nuclei are central, usually single (80%) or multiple (2–3 nuclei) (20%), and often enlarged (20 ± 10 μ). They are irregular in shape (40%), oval (30%), or round (30%). Usually hyperchromatic, the chromatin is uniformly coarsely granular (90%), clumped (8%), or pyknotic with no discernible pattern (2%).

The nuclear membrane is often irregular in shape and thickness with wrinkling in 10 to 80% of the cells.

In fewer than 5% of the extremely atypical cells, there may be seen prominent, usually single, nucleoli, which are irregular in shape and size.

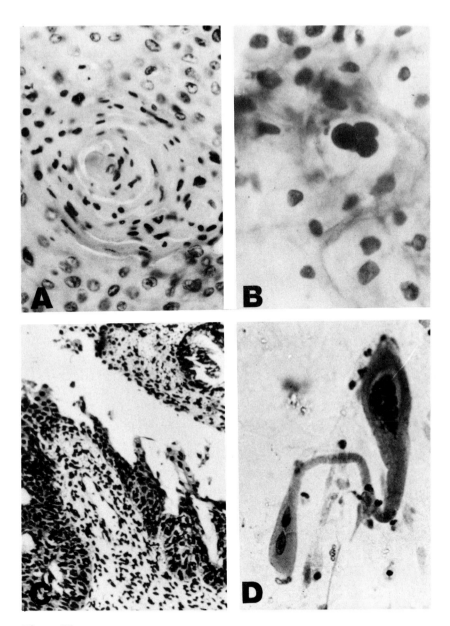

Figure 56
Superficial cell atypia. (A, B) Mild. (C, D) Moderate.

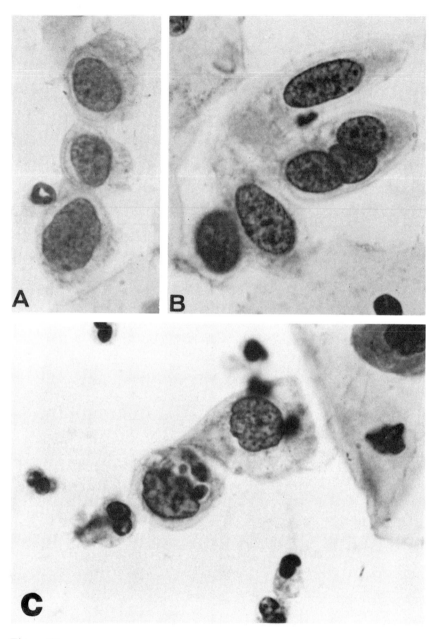

Figure 57
Squamous and reserve basal cell dyskaryosis diagnostic of basal cell dysplasia.
(A) Mild. (B) Moderate. (C) Marked. Note the progressive increase in irregularity of the chromatin and nuclear membrane and the decrease in nuclear size and amount of cytoplasm. ($\times$ 1300)

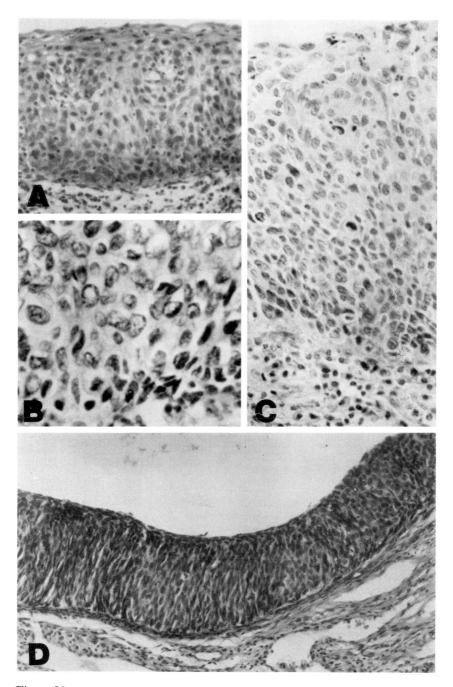

Figure 58
Histologic differences among mild basal cell atypia (A: × 150; B: × 900), (C) moderate atypia (× 450), and (D) in situ carcinoma. (× 150)

The background of the smear may be obscured by cellular debris and degenerate stripped, benign, or atypical nuclei. Döderlein's bacilli can be seen in the background of the smears in 35% of the cases.

It is frequently difficult to distinguish by cytologic or even histologic study between some cases of precancerous squamous basal cell extreme dysplasia and early malignant in situ carcinoma. The main cytologic differentiation is based on the abundance of the cytoplasm; their nuclei may look the same. This is the worst atypia of the squamous epithelium that one can encounter, and its detection by cytologic study warrants a cervical biopsy, followed by a cone, as in the case of a cervical in situ carcinoma.

References and Supplementary Reading

Anderson, W. A. D., and Gunn, S. A. Premalignant and malignant conditions of the cervix uteri. *Cancer* 20:1587, 1967.

Beutler, H. K., Dockerty, M. D., and Randall, L. M. Precancerous lesions of the endometrium. *Am. J. Obstet. Gynecol.* 86:433, 1963.

Buck, C. E., and Henderson, P. H., Jr. The histopathology and cytopathology of cervical leukoplakia. *Obstet. Gynecol.* 21:279, 1963.

Epstein, N. A. The significance of cellular atypia in the diagnosis of malignancy in ulcers of the female genital tract. *Acta Cytol.* 16:483, 1972.

Fluhmann, C. F. The histogenesis of acquired erosions of the cervix uteri. *Am. J. Obstet. Gynecol.* 82:970, 1961.

Fornari, M. Cellular changes in the glandular epithelium of patients using IUCD—A source of cytologic error. *Acta Cytol.* 18:341, 1974.

Fox, C. H. Biological behavior of dysplasia and carcinoma in situ. *Am. J. Obstet. Gynecol.* 99:960, 1967.

Fox, C. H. Time necessary for conversion of normal to dysplastic cervical epithelium. *Obstet. Gynecol.* 31:749, 1968.

Frost, J. K. Cytology of benign conditions. *Clin. Obstet. Gynecol.* 4:1075–1096, 1961.

Girolami, E., de. Perinuclear halo versus koilocytotic atypia. *Obstet. Gynecol.* 29:479, 1967.

Gouzalez-Merlo, Z., et al. Regeneration of the ectocervical epithelium after its destruction by electrocauterization. *Acta Cytol.* 17:366, 1973.

Hall, J. E., and Rosen, I. H. Significance of the Class III cervical smear. *Am. J. Obstet. Gynecol.* 79:709, 1960.

Henderson, P. H., Jr., and Buck, C. E. Cervical leukoplakia. *Am. J. Obstet. Gynecol.* 82:887, 1961.

Johnson, L. D. The pathogenesis of atypical lesions of the cervix in women of the childbearing age. *Acad. Med. New Jersey Bull.* 8:233, 1962.

Jordan, M. J., Bader, G. M., and Day, E. Carcinoma in situ of the cervix and related lesions. *Am. J. Obstet. Gynecol.* 89:160, 1964.

Kaminetsky, H. A., and Swerdlow, M. Atypical epithelial hyperplasia of the uterine cervix. *Am. J. Obstet. Gynecol.* 82:160, 1964.

Klavins, J. V. Intra-epithelial carcinoma with differentiated surface cells and dysplasia. Definition and separation of these lesions. *Acta Cytol.* 7:351, 1963.

Lambert, B., and Woodruff, J. D. Spinal cell atypia of the cervix: A clinico-pathological study. *Cancer* 16:1141, 1963.

Lerch, V., et al. Cytologic findings in progression of anaplasia (dysplasia) to carcinoma in situ: A progress report. *Acta Cytol.* 7:183, 1963.

McKay, D. G., et al. Clinical and pathologic significance of anaplasia (atypical hyperplasia) of cervix uteri. *Obstet. Gynecol.* 13:2, 1959.

Ng, A., Reagan, J., and Cechuer, L. The precursors of endometrial cancer: A study of their cellular manifestations. *Acta Cytol.* 17:439, 1973.

Nieburgs, H. E. The significance of tissue cell changes preceding uterine cervix carcinoma. *Cancer* 16:141, 1963.

Nieburgs, H. E. Tissue and cell pathology of uterine cervix dysplasia and carcinoma in situ. *Acta Cytol.* 15:513, 1971.

Peckham, B., and Greene, R. R. Follow-up on cervical epithelial abnormalities. *Am. J. Obstet. Gynecol.* 74:804, 1957.

Piver, M. S., Whitely, J. P., and Bolognese, R. J. Effect of an intrauterine contraceptive device upon cervical and endometrial exfoliative cytology. *Obstet. Gynecol.* 28:528, 1966.

Reagan, J. W., et al. Dysplasia in the uterine cervix during pregnancy: An analytical study in the cells. *Acta Cytol.* 5:17, 1961.

Reagan, J. W., and Patten, S. F., Jr. Dysplasia; basic reaction to injury in uterine cervix. *Ann. N.Y. Acad. Sci.* 92:662, 1962.

Richart, R. Colpomicroscopic studies of the distribution of dysplasia and carcinoma in situ on the exposed portion of the human uterine cervix. *Cancer* 18:950, 1965.

Sagiroglu, N. Progression and regression studies of precancer (anaplastic or dysplastic) cells, and the halo test. *Am. J. Obstet. Gynecol.* 85:454, 1963.

Saphir, O., Leventhal, M. L., and Kline, T. S. Podophyllin-induced dysplasia of the cervix uteri. *Am. J. Clin. Pathol.* 32:446, 1959.

Schmidt, A. L., and Christiaans, A. P. L. The vaginal smear pattern in cases of endometriosis. *Acta Cytol.* 9:247, 1965.

Stern, E., and Neely, P. M. Carcinoma and cervical dysplasia. *Acta Cytol.* 7:357, 1963.

Ta-wang, Y. Precancer cell change and early cancer observed in cervical smear. *Shanghai Sci. Tech. Pub.* 1:17, 1962.

Traut, H. F., and Papanicolaou, G. N. Vaginal smear changes in endometrial hyperplasia and in cervical keratosis. *Anat. Rec.* 82:478, 1942.

Varga, A. The relationship of cervical dysplasia to in situ and invasive carcinoma of the cervix. *Am. J. Obstet. Gynecol.* 95:759, 1966.

Ward, H. N., Konikov, N., and Reinhard, E. H. Cytologic dysplasia occurring after busulfan (Myleran) therapy. *Ann. Intern. Med.* 63:654, 1965.

Malignant Cells

6

The number of tumor cells present in a smear is independent of the size or stage of the neoplasm. The number of exfoliated cells varies with the type and location of the lesion and the sampling technique. For example, an occult, squamous in situ carcinoma of the cervix will often shed more recognizable diagnostic cells than an advanced, fungating cervical squamous cell carcinoma, in which large amounts of inflammation and cellular degeneration may be present. In cytology, it is easier to miss a large, ulcerated, invasive cervical cancer than a small, invisible in situ cancer.

The final cytologic interpretation should be based on more than one cell. If only one diagnostic cell is seen in the smear, the possibility of this cell's being a contaminant from another specimen should always be considered.

As often as possible, the cytologic diagnosis should be based on all the criteria found in an intact, well-preserved cell. Cellular degeneration with frequent partial cytoplasmic loss is one of the most common cytologic diagnostic pitfalls.

The structure of a suspected atypical cell should always be compared to the structure of the other similar but benign cells present in the smears.

No single morphologic criterion alone is pathognomonic for malignancy. A combination of criteria is necessary.

One should not add the malignant features of one cell to the features found in a neighboring cell, even if both cells are found side by side in the smear. Irregularity of the chromatin of one cell, for example, cannot be added to the prominence of the nucleolus of another cell to arrive at the diagnosis of adenocarcinoma. Both atypical changes should be located in the same cell. The changes found in different cells can be added only when these cells are part of a definite sheet or acinus.

It is frequently impossible to pinpoint the site of a malignant lesion by cytology alone. Malignant squamous cells in a sputum, for example, can originate from bronchial, tracheal, or buccal mucosa, or even esophageal carcinoma.

In the presence of insufficient cytologic evidence, one should not be influenced by the clinical history of the patient or by pressure exerted by the clinicians in interpreting a smear as positive or negative for cancer. To say, on occasion, that we cannot interpret the cells and to recommend a repeat smear indicates a good judgment rather than a poor one. It is permissible for about 2% of the cytologic reports to be inconclusive. An increase of this percentage (above 4%) may indicate that this inconclusive diagnosis is used by the pathologist as a wastebasket for difficult cases. Errors in the fixation and staining technique may also have a definite bearing on having a larger percentage of this type of report.

If the Papanicolaou classification is used, one should be sure that the clinician understands the meaning of these classes. One should always try to add a comment indicating what type of follow-up is recommended. The specific nature of the lesion, based on the cellular morphologic changes in the smear, should be given whenever possible.

Changes in the Nucleus

The nuclear changes are the most important criteria used for the cytologic diagnosis of a carcinoma. No single structural change is diagnostic by itself. A combination of several abnormalities is always necessary.

NUCLEAR HYPERTROPHY (FIG. 59)
To appreciate the degree of hypertrophy, the abnormal nucleus should always be compared to the size of several other nuclei of similar but benign cells. A nuclear diameter greater than 10 μ (about the size of neutrophil) should attract the attention of the screener.

The size increase of a cell in interphase should be real (i.e., an increase in the DNA content) and not the result of edematous swelling, which may occur in a poorly fixed air-dried smear or when the cells have been in contact with a hypotonic solution (cloudy swelling). One should beware of the degenerated, enlarged but pale, almost empty nuclei.

One should remember that a flattened nucleus, as illustrated in Figure 59, will seem larger than it is when viewed on a vertical axis. The need for continuous up and down focusing of the microscope cannot be emphasized enough.

The nuclei of a poorly differentiated or embryonal carcinoma, although malignant, are relatively small, as, for example, in oat-cell carcinoma or squamous in situ carcinoma.

All the nuclear enlargement is not necessarily due to malignancy. Irradiation, regeneration, chemotherapy, administration of alkalinizing agents, cautery, viral infections, and *Trichomonas* inflammation may all increase nuclear size.

An empty perinuclear halo resulting from the shrinkage of an enlarged nucleus in the dehydrating fixative is prominently seen in cases of benign nuclear hypertrophy (see Fig. 61). A slight physiologic increase of the nuclear volume can be seen during cellular mitosis just after telophase.

NUCLEAR SIZE VARIATION
This variation may be observed in the cells of most malignant neoplasms. The exceptions are few, such as low-grade endocervical adenocarcinoma or pulmonary alveolar cell carcinoma. The importance of this criterion varies with the type of suspected neoplasm. For example, it is more important in

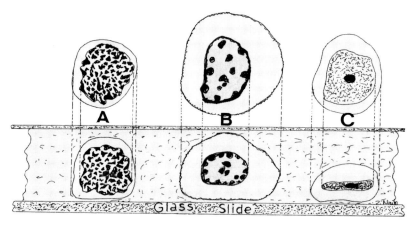

Figure 59
Differences between cancerous and noncancerous nuclear hypertrophy. (A)
Malignant hypertrophy. (B) Degenerative edematous hypertrophy. (C) Pseudohypertrophy (compressed nucleus).

the diagnosis of a well-differentiated cervical squamous cell carcinoma than in an adenocarcinoma.

This criterion is also more important when the cells are part of a sheet or acinus, rather than appearing singly, since the different single cells found in a smear often originate from different areas or tissues.

One should be wary of the pseudovariation in size resulting from the position of the cell on the slide in relation to the axis of observation.

The nuclei of some benign cells (e.g., histiocytes and endocervical cells) vary physiologically in size according to their function.

NUCLEAR SHAPE VARIATION
Variation in the nuclear shape is often the result of the rapid rate of growth of the neoplastic cells, which become crowded against each other and have to occupy a continuously narrowing space. This explains the molding of one nucleus against the other (pseudocannibalism; see Figure 65), which has led to the use of the term *birdseye* cells.

Some of the irregularities in shape are the result of abnormal mitoses producing an irregularity in the number and shape of the chromosomes.

Multinucleation may be simulated by a single, convoluted nucleus.

Whenever the diagnosis of nuclear malignancy cannot be based on the chromatin pattern—for example, because of a pyknotic nucleus—the irregularity of shape becomes the major criterion for diagnosis.

An artificial variation of the shape of the nucleus can occur in benign cells as a distortion resulting from poor cellular fixation, trauma, degeneration, cautery, or hormonal changes, but in such cases of distortion, the cells are then often surrounded by a fairly prominent halo.

By itself, irregularity in nuclear shape, if not extreme, is not sufficient for diagnosis.

HYPERCHROMATISM

By itself, this is a fairly poor criterion of malignancy. Some benign cells, such as a lymphocyte or a reactive squamous basal cell, are normally hyperchromatic. The staining intensity can also be altered by an increase in the staining time or the strength of the hematoxylin solution (artificial hyperchromatism).

Atypical hyperchromatism may be differentiated from hyperchromatism resulting from nuclear degeneration by the sharpness of the nuclear membrane and the sharpness of the chromatin clump borders. These are fuzzy during degeneration.

Hyperchromatism should not be confused with pyknosis; chromatin details are seen in the former and absent in the latter.

A mild hyperchromatism (increased quantities of DNA) is seen better with basic dyes (methylene blue) or with acridine-orange fluorescence than with the routine Papanicolaou stain.

IRREGULARITY OF THE CHROMATIN

One of the important criteria for malignancy is irregularity of the chromatin. The chromatin pattern in a normal cell in interphase, except for the sex chromosome masses, is usually very finely granular and almost invisible. In some atypia and in malignancies, the chromatin often forms irregular clumps, variable in size and shape, with multiple sharp-pointed projections.

One should differentiate irregularly distributed, malignant, abnormal clumping from the degenerative type in which the chromatin clumpings are spherical and fuzzy with a tendency to migrate toward the periphery. Furthermore, the areas between the chromatin clumping are often clean and transparent in atypical cells, whereas there is a pink granular deposit in degeneration.

In the cells of some well-differentiated carcinomas with vesicular nuclei (adenocarcinoma), the sex chromosome mass, Barr body, when present is more prominent, owing probably to the nondisjunction of the sex chromosomes during an abnormal mitosis. It is two to three times larger than the normal chromosomal mass as a result of an increase in the size or number of the abnormal sex chromosomes.

A moderate increase of regular chromatin clumping can be found in the nuclei of irradiated, regenerating, or premalignant cells. Because of this, it is sometimes difficult to differentiate these cells from malignant cells. The presence of pointed, irregular projections in malignancy, which often can be seen only with oil immersion magnification, is most important in such cases.

MULTINUCLEATION

By itself, this is a poor criterion for malignancy since it frequently occurs in certain normal cells (e.g., mesothelial, transitional, or endocervical cells). It can be useful in identifying the nature of the malignant neoplasm— for example, in recognizing the cells exfoliating from a giant cell carcinoma of the lung or a mixed mesodermal tumor of the uterus. If abnormal, the different nuclei in a cell will vary excessively in size, shape, and chromatin

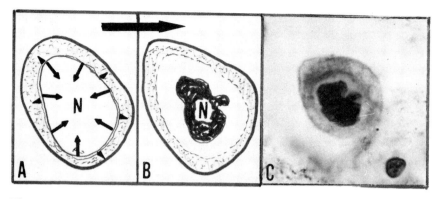

Figure 60
Explanation of degenerative nuclear membrane wrinkling and perinuclear halo formation. (A) Cell before fixation. (B, C) Retraction of the nucleus toward the center of the cell and retraction of the cytoplasm toward the periphery of the cell.

structures. This variation differentiates these cells from the multinucleated foreign-body giant cell, in which all the nuclei are of the same size and texture (sister image).

CHANGES IN THE NUCLEAR MEMBRANE
The irregularity of the shape of the nuclear membrane (indentation or extensive wrinkling) is important in the diagnosis of malignancy. Wrinkling of the nuclear membrane of any normal cell can be produced by placing it in a hypertonic solution. The nucleus gives the impression of having shrunk upon itself, and a perinuclear halo is almost always present (Fig. 60). In the presence of such perinuclear halos, the wrinkling of the nuclear membrane is less significant. In atypia, this wrinkling of the nuclear membrane is the result of the rapid and abnormal growth of the cells. The irregularity in the thickness of this membrane is often the result of the adhesion of some of the abnormal chromatic clumps to its inner surface.

The contours of the nuclear membrane in malignant cells are sharp and well defined, which differentiates them from the irregularity resulting from cellular degeneration. These atypical irregularities may be totally absent in certain malignant cells that have been exfoliated, for example, from a low-grade endocervical adenocarcinoma.

CHANGES IN THE NUCLEOLI
The variation in the size, shape, number, and staining quality of the nucleoli is often used not only in the diagnosis of malignancy but also in typing the neoplasm. The nucleoli are present and prominent in well-differentiated squamous cell carcinoma. A cell with a nucleolus larger than 5 μ should be suspected of being atypical.

To be diagnostic of a malignancy, the nucleoli should present a pronounced irregularity in size and shape, with multiple, sharp-pointed projections.

Variation in the number of the nucleoli becomes important when it is extreme and is observed in a sheet of identical cells or in acini rather than in scattered or clustered single cells. In certain types of nonmalignant cells the number of nucleoli may vary physiologically from 1 to 5.

The nucleoli can be prominent in certain normal cells, such as in endocervical cells, or during the regeneration of any cells (repair cells). In such cases, they are generally round and regular in shape.

The nucleoli must be differentiated from intranuclear viral inclusions, which usually possess a distinct periinclusion halo.

Changes in the Cytoplasm

SCANTINESS OF THE CYTOPLASM

The amount of the cytoplasm is important not only in the diagnosis of malignancy, but also in typing the nature of the neoplasm. For example, an oat-cell carcinoma of the lung is recognized mainly by its very scanty, almost nonexistent, cytoplasm. In an atypical cell, the amount of cytoplasm present should be related to the nuclear size. An increase in the nuclei-cytoplasm ratio is a good criterion of malignancy in the diagnosis of many tumors. By itself, an abnormal, naked nucleus cannot be diagnostic, no matter how irregular its shape or size. An intact cell must be found for a definite diagnosis of malignancy.

Some cancer cells, however, may have an abundant amount of cytoplasm (certain adenocarcinomas or well-differentiated invasive squamous cell carcinomas), while some benign cells have very little cytoplasm (lymphocytes).

One should be aware of an apparent cytoplasmic scantiness of exfoliated cells resulting from a cytoplasmic torsion, traumatic partial loss (frayed edges), or the position of the cells in relation to the viewing axis (Fig. 61).

CYTOPLASMIC BOUNDARIES

The cytoplasmic boundary is sharp, distinct, and regular in some malignant cells (flattened, infiltrating carcinoma squamous cell) or indistinct and heavy, as in the case of a thick, spherical, malignant cell with scanty cytoplasm (undifferentiated carcinoma). The irregularity of the cytoplasmic borders (if thin and frayed) can indicate that a portion of the cytoplasm has been lost traumatically (Fig. 62), and the apparent cytoplasmic scantiness of the cells alone is not a dependable criterion of malignancy.

CYTOPLASMIC VARIATION IN SIZE

To be a valid criterion, the cytoplasmic variation in the size of atypical cells should always be compared to the physiologic variation in size of similar but normal cell populations in the smear. The variation in cellular size is most important when seen in the component cells of a sheet or acinus. One should be aware of the pseudovariation in size that can be the result of the position of the cell in relation to the viewing axis (Fig. 63). The difference in cellular size can also be the result of other factors, such as irradiation, pregnancy, anemia, viral inflammation, cellular degeneration and regeneration.

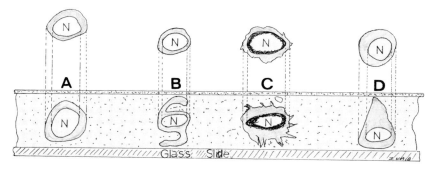

Figure 61
Real and various types of apparently scanty cytoplasm. (A) Real scantiness.
(B) Folded but abundant cytoplasm. (C) Partially lost cytoplasm. (D) Abundant cytoplasm but positioned above the nucleus.

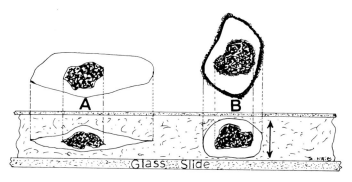

Figure 62
Relation between the sharpness of the cytoplasmic and nuclear boundaries and the thickness of the edges. (A) Sharp. (B) Fuzzy.

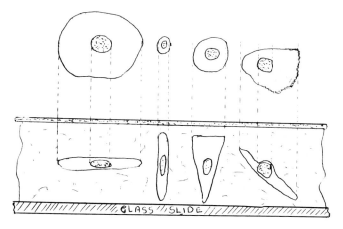

Figure 63
The apparent variation of the amount of cytoplasm according to the position of the cell in relation to the axis of observation.

137

CYTOPLASMIC VARIATION OF SHAPE

The variation of shape of the cytoplasm is diagnostic only when it is extreme. It often results from the crowding of the cells in a narrow, nonexpandable structure, as in the case of squamous metaplastic cells that are developing in the lumen of an endocervical gland. This shape variation can also be the result of mechanical distortion or of cellular regeneration (repair cells).

CYTOPLASMIC STAINING

The intensity of the staining and the color of a cell can be helpful in the diagnosis of the nature of a neoplasm. If properly done, staining becomes important. For example, an abnormal amount of cytoplasmic keratinization produces a glassy, deep orange stain, as is seen in the hyperkeratinized cells exfoliated from a well-differentiated, infiltrating, squamous cell carcinoma. Generally, the cytoplasm is basophilic in the cells of immature or poorly differentiated carcinoma.

CYTOPLASMIC INCLUSIONS

Most of the cytoplasmic inclusions are rarely diagnostic of malignancy, except in certain neoplasms, such as a melanoma (where melanin pigment is in the cytoplasm) or a mucinous adenocarcinoma (where mucin that is positive for periodic acid–Schiff stain is contained in a large secretory vacuole distending the cytoplasm). The various significant types of inclusions and vacuolization must be differentiated from the common ingested debris and degenerative vacuoles.

CYTOPLASMIC AND NUCLEAR
MEMBRANE RELATIONSHIP (FIG. 64)

In benign cells, it is very rare for the nuclear and cytoplasmic membranes to meet at more than one point (as seen in a histiocyte). When the cytoplasmic border is found tightly molded against the nucleus and touching it in more than one place, the meeting of membranes at several points becomes a very important criterion of malignancy, as is the case in the diagnosis of cervical in situ carcinoma.

Changes in the Cells as a Group

CANNIBALISM–CELLULAR PHAGOCYTOSIS (FIG. 65)

In very rare instances, some malignant cells can phagocytize another cell, usually a leukocyte. Most of the time, the cannibalism between two atypical epithelial cells (one cell appearing to be contained in a vacuole of the cytoplasm of another epithelial cell) is more apparent than real. The smaller cell is usually lying in a depression of the cytoplasm of the larger one. This indicates only that two cells have grown in a very narrow space. This pseudocannibalism can be seen in benign lesions (such as squamous metaplasia or basal cell hyperplasia) as well as in malignant ones (such as in situ or invasive squamous cell carcinoma). Pseudocannibalism indicates only the rapid growth of the cell in a confined area.

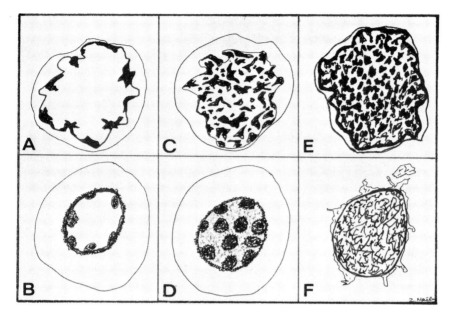

Figure 64
Similarities and differences between degenerative and malignant changes in cells. Irregularity of nuclear membrane (A) in malignancy, (B) in degeneration. Irregularity of chromatin clumping (C) in malignancy, (D) in degeneration. Irregularity of the scantiness of the cytoplasm (E) in malignancy, (F) in degeneration.

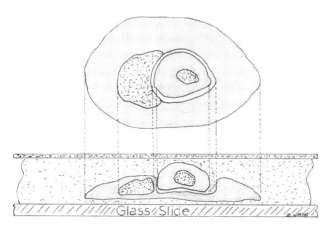

Figure 65
Explanation of the apparent cannibalism of tumor cells.

VARIATION OF SIZE AND SHAPE

The size and shape of a cell usually depend on the size and amount of its cytoplasm. Other factors are overcrowding, rapid multiplication, and the site of the cellular growth. By itself, variation in size and shape is a poor criterion of malignancy, especially if found in single adjoining cells that could originate from different areas of the mucosa. This variation becomes more important as a criterion when it is found in a sheet of cells or in acinar formation.

LACK OF CELLULAR ADHESION

Lack of mutual cellular adhesion between atypical epithelial cells is an important criterion of malignancy. It is thought to be the result of abnormalities in the desmosomes. The tumor cells tend to shed singly and in abundance. This explains why a large number of diagnostic malignant cells is often present in the smear even when the lesion is very small. There is little relation between the number of cells found in a smear and the actual size of the lesion. In certain neoplasms, the cells have a tendency to shed in tight clusters, as in endometrial adenocarcinoma. Sheets of tumor cells are usually obtained by the direct surface scraping of a neoplasm or after therapy (e.g., irradiation or cryosurgery).

MITOSIS

The number of mitoses has very little significance in exfoliative cytopathology (as opposed to histopathology). Most of the cells, even if malignant, have, by the time they exfoliate, completed their mitotic division. A cell in mitosis found in a routine smear is probably a histiocyte. An increased number of abnormal mitoses, especially if atypical, in cells found in sheets or acinar formation, is a good criterion of malignancy.

BLOOD

The presence of fresh red blood cells is meaningless in a scraping or in a specimen obtained by any other traumatic method of collecting cells.

Malignancy should be suspected when the blood is old, partially ingested by histiocytes, or present in a specimen obtained without trauma.

FOREIGN CELLULAR STRUCTURES

The presence of certain cellular structures foreign to the site investigated can be considered a very good criterion of malignancy—for example, the presence of an acinar formation in a pleural effusion or a psammoma body in a routine vaginal smear. One must be sure that the presence of these foreign structures is not due to contamination of the smear by a floating cell from another, previously or simultaneously stained slide.

DEGENERATION AND INFLAMMATION

Degeneration and inflammation are usually seen in the smears from patients with large malignant lesions. The inadequate blood supply and frictional or mechanical trauma of such a tumor mass often produces an erosion, ulceration, or necrosis. A large number of inflammatory cells (leukocytes, lymphocytes, phagocytes) and red blood cells in the presence of malignant cells usually indicates an *invasive carcinoma rather than an in situ one.* In the case of an unexplained, nonspecific inflammation in the smear of a postmenopausal woman, a neoplasm should be suspected and the smear screened more carefully or repeated after antiinflammatory treatment. If necrosis exists, lipid material may be found phagocytized by histiocytes or free in the debris of the background. A Sudan stain of such a smear may be helpful.

References and Supplementary Reading

Atkin, M. B.   Variant nuclear types in gynecologic tumors: Observations on squashes and smears. *Acta Cytol.* 13:569, 1969.

Bernard, W., and Granboulau, N.   The fine structure of the cancer cell nucleus. *Exp. Cell Res.* [Suppl.] 9:19, 1963.

Caspersson, O.   Quantitative cytochemical studies on normal, malignant, premalignant and atypical cell populations from the human uterine cervix. *Acta Cytol.* 8:45, 1964.

Coman, D. R.   Decreased mutual adhesiveness, a property of cells from squamous cell carcinomas. *Cancer Res.* 4:625–629, 1944.

Cowdry, E. V.   *Cancer Cells.* Philadelphia: Saunders, 1955.

Dellepiane, C.   Etude du noyau dans le diagnostic de la cellule cancéreuse. *Rev. Fr. Gynecol. Obstet.* 54:591, 1959.

Frost, J. K.   Gynecologic and Obstetric Cytopathology. In Novak, E. R., and Woodruff, J. D. (Eds.), *Novak's Gynecologic and Obstetric Pathology* (6th ed.). Philadelphia: Saunders, 1967.

Hinglais-Guillaud, N., Moricard, R., and Bernhard, W.   Ultrastructure des cancers pavimenteux invasifs du col utérin chez la femme. *Bull. Assoc. Fr. Cancer* 48:283, 1961.

Horvath, W. J., Tolles, W. E., and Bostrom, R. C.   Quantitative measurement of cell properties on Papanicolaou smears as criteria for screening. In *Transactions of First International Cancer Cytology Congress,* Chicago, 1956. P. 371.

Koller, P. C.   The nucleus of the cancer cell. *Exp. Cell Res.* [Suppl.] 9:3, 1963.

McMaster, C. W.   A measurement of the pattern of normal and malignant cervical cells. *Acta Cytol.* 12:9, 1968.

Patten, S. F.   *Diagnostic Cytology of the Uterine Cervix.* Baltimore: Williams & Wilkins, 1969.

Pfitzer, P., and Pape, H. D.   Characterization of tumor cell populations by DNA measurements. *Acta Cytol.* 17:19, 1973.

Reagan, J. W., Hamonic, M. J., and Wentz, W. B.   An analytical study of cells in cervical squamous cell cancer. *Lab. Invest.* 6:241, 1957.

Singleton, H., et al.   Human cervical intraepithelial neoplasia: Fine structure of dysplasia and carcinoma in situ. *Cancer Res.* 28:695, 1968.

Sprenger, E., Moore, G., and Naujoks, H.   DNA content and chromatin pattern analysis on cervical carcinoma in situ. *Acta Cytol.* 17:27, 1973.

Tolles, W. E., Horvath, W. J., and Bostrom, R. C.   A study of the quantitative characteristics of exfoliated cells from the female genital tract. *Cancer* 14:437, 1961.

Genital Squamous Cell Carcinoma 7

Uterocervical squamous cell carcinoma (Fig. 66) is still one of the most frequent cancers. Of every 100 women, 2 will eventually develop this neoplasm, and 1 may still die from it. When localized to the epithelium (in situ carcinoma), the cure rate approaches 100%, as compared to only 45% survival rate when the basement membrane and surrounding tissues and organs have been penetrated (invasive carcinoma).

The in situ cancer is also called preinvasive, noninvasive, incipient, surface, intraepithelial, intramucosal, and stage 0 squamous cell carcinoma and Bowen's disease of the cervix.

As illustrated in Figure 67, a squamous carcinoma of the cervix in the in situ stage can grow in only three directions, one side being obstructed by the intact basement membrane. The lateral expansion of the neoplasm is difficult, owing to the resistance of the normal stratified squamous epithelium, which often is found compressed in the histologic sections. This resistance reduces the space that the tumor can occupy and explains in part the premature exfoliation of the surface cells before their maturation. The in situ carcinoma can expand easily only toward the surface of the squamous epithelium or toward the lumen of the gland, where the cells are subjected to exfoliating frictional trauma. If the basement membrane has been ruptured and penetrated, as is the case in invasive carcinoma, the cells have more space in which to grow, which gives them time to mature. The exfoliated cells from the surface of an invasive carcinoma usually show these maturation changes, which help in cytologic diagnosis and typing. In short, one of the main differences, cytologically, between an in situ carcinoma and an invasive squamous carcinoma is that, in the latter, the cells have had time and space to mature. An exception is the poorly differentiated, infiltrating squamous cell carcinoma, in which the excessive increase of cellular division prevents the maturation of the cells, even in the presence of adequate growth space.

The following are good *histologic* criteria of malignancy, but they have little significance in *cytology:*

1. Loss of polarity and stratification;
2. Frequent mitoses;
3. Stroma and blood vessel invasion;
4. Involvement of the entire thickness of the mucosa.

143

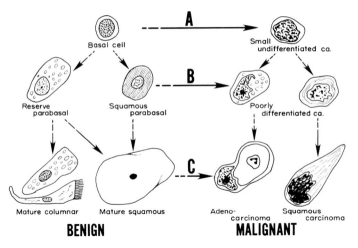

Figure 66
Maturation and differentiation of normal and malignant epithelial cells. (A) Embryonal cells. (B) Immature cells. (C) Mature cells.

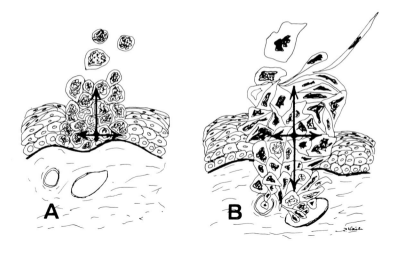

Figure 67
Direction of growth of (A) an in situ carcinoma and (B) an invasive carcinoma.

Clinical Staging of Cervical Carcinoma

Stage 0 = intraepithelial, or in situ carcinoma
Stage 1 = early stromal invasion, with the cancer confined to the cervix
Stage 2 = involving the cervix and the upper third of the vagina
Stage 3 = involving the cervix, the entire vagina, and the pelvic wall
Stage 4 = involving surrounding organs or with distant metastasis

SMALL-CELL OR POORLY DIFFERENTIATED CARCINOMA IN SITU (FIGS. 68A, 68B, 69A)

A squamous cell in situ carcinoma is an early, irreversible, malignant lesion that has not penetrated the basement membrane underlying the squamous epithelium. It does not produce metastases. The neoplasm may be multi-focal and usually arises from areas where dysplasia, anaplasia, or squamous metaplasia preexisted. It is found commonly in the portio vaginalis and the endocervical canal, mainly at the squamocolumnar junction. Grossly, the cervix appears normal in 60% of the patients and eroded in 40%. There is good evidence that at least half the patients with a detected but untreated cervical in situ carcinoma will develop an invasive cancer within four years.

TISSUE DIAGNOSIS (FIG. 68B)

The examination of a tissue biopsy specimen is needed to confirm the cyto-logic diagnosis. Since early carcinoma is a microscopic disease, visual in-spection of a normal-appearing cervix is not sufficient to rule out the neoplasm. Colposcopy may be helpful, but it is limited to visible, external os lesions. To be diagnostic, the cervical epithelium should show:

1. Loss of uniform cellular polarity and stratification;
2. Nuclear hyperchromatism with irregularity in its chromatin clumping;
3. Cytoplasmic depletion of glycogen;
4. Involvement of the entire thickness of the mucosa;
5. Frequent mitosis, with some abnormalities;
6. Epithelial localization over an intact basement membrane with no in-vasion of the subadjacent connective tissue (stroma).

GENERAL CYTOLOGIC MALIGNANT CELLULAR CRITERIA (TABLE 5)

A large number (more than 500) of diagnostic, abnormal cells of the basal cell type can usually be found in the smears of endocervical aspirations and cervical scrapings of patients with an in situ cervical carcinoma. The abundance of the exfoliated cells is related to the remarkable lack of adhe-sion between the neoplastic cells. More abnormal cells are often present in the smears of in situ cancer than in those of basal-cell dysplasia. Their number in a smear has no relation to the size of the lesion but depends on the type and location of the tumor and method used to obtain the specimen. (Direct scraping will produce more cells.) At least 10 abnormal cells should be seen before a positive cytologic diagnosis is rendered.

The tumor cells exfoliate singly in 90% of the cases. When found in small sheets or syncytial formation, they are slightly more difficult to interpret.

The neoplastic cells coexist and intermix with all types of atypical and dysplastic squamous cells in more than 90% of the smears.

The diagnostic cells show little variation in size, the average being $25 \pm 5 \mu$. They appear slightly larger during pregnancy ($30 \pm 7 \mu$).

The cell shape varies little, being oval in 80%, round in 15%, and irreg-ular in 5% of the specimens. The irregularity in shape is often an indica-

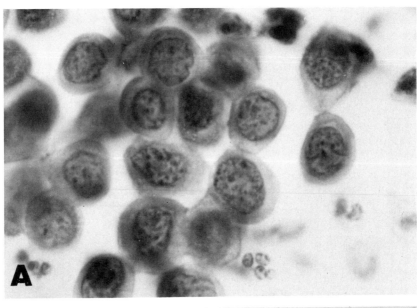

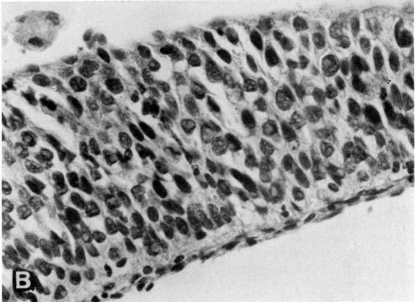

Figure 68
(A) Small cell type of in situ squamous cell carcinoma. Note the irregularity of the chromatin and of the nuclear membrane. ($\times$ 1300) (B) Strip of epithelium from cervical in situ carcinoma in the biopsy specimen of the same case. ($\times$ 450)

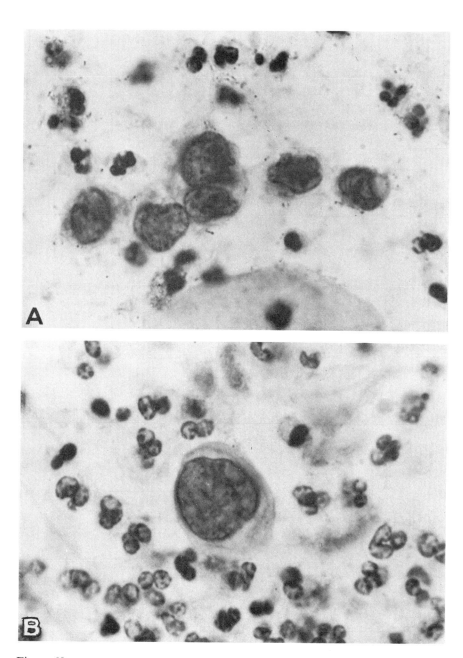

Figure 69
(A) In situ carcinoma, small cell type. Note the scantiness of the cytoplasm and the irregularity of the nuclear membrane and chromatin. (× 1300) (B) In situ carcinoma, large cell type. Note the size of the cell, the greater amount of cytoplasm, and the irregularity of the nuclear membrane and chromatin. (× 130)

147

Table 5. Differential Characteristics of In Situ Carcinoma, Basal-cell Dysplasia, and Invasive Squamous Cell Carcinoma

Criteria	Basal-cell Dysplasia		In Situ Carcinoma		Microinvasive Squamous Cell Carcinoma		Invasive Keratinizing Squamous Cell Carcinoma	
Size	$30 \pm 10\ \mu$		$25 \pm 5\ \mu$		$40 \pm 54\ \mu$		$110 \pm 80\ \mu$	
Shape	Polygonal	60%	Oval	50%	Oval	70%	Polygonal	70%
	Oval	20%	Polygonal	36%	Polyhedral	15%	Irregular	20%
	Round	20%	Round	14%	Irregular	5%	Spindle	10%
Occurrence	Single	80%	Single	90%	Single	30%	Single	55%
	Sheet	20%	Sheet	10%	Sheet	70%	Sheet	45%
Amount of cytoplasm	Adequate	99%	Adequate	10%	Adequate	50%	Adequate	90%
			Scanty	90%	Scanty	50%	Scanty	10%
Cytoplasmic stain	Basophilic	50%	Basophilic	70%	Basophilic	40%	Basophilic	30%
	Eosinophilic	50%	Eosinophilic	30%	Eosinophilic	60%	Eosinophilic	70%
Cytoplasmic vacuolization	25%		25%		15%		10%	
Nuclei-cytoplasm ratio	4:10		8:10		4:10		2:10	
Nuclear size	$20 \pm 10\ \mu$		$13 \pm 3\ \mu$		$15 \pm 5\ \mu$		$30 \pm 20\ \mu$	
Nuclear shape	Oval	80%	Oval	40%	Oval	30%	Oval	30%
	Round	10%	Round	40%	Round	30%	Round	30%
	Irregular	2%	Irregular	20%	Irregular	40%	Irregular	40%
Chromatin pattern	Granular	90%	Granular	80%	Granular	40%	Granular	25%
	Clumped	8%	Clumped	20%	Clumped	40%	Clumped	30%
	Pyknotic	2%			Pyknotic	20%	Pyknotic	45%
Multinucleation	10%		2%		15%		20%	

Nucleoli	15%	0.1%	15%	10%
Number of diagnostic cells	Less abundant	Abundant	Abundant	Abundant
Nuclear indentation (wrinkling)	40%	85%	90%	90%
Degenerate stripped nuclei	10%	5%	5%	20%
Döderlein's bacillus	35%	20%	15%	5%
Tumor diathesis	1%	5%	60%	95%

tion of extension of the neoplasm to the endocervical canal or of a possible early microinvasion.

The background of the smear is usually clean with very few leukocytes, inflammation, or red blood cells, unless an erosion exists. The few leukocytes, when present, are usually very well preserved.

CYTOPLASMIC MALIGNANT CHANGES

The cytoplasm is very scanty, and the cytoplasmic membrane often molds and fuses regionally with the nuclear membrane. The presence of this type of cytoplasm is sometimes difficult to ascertain, except when the cells are examined under oil immersion magnification. This criterion is important in differentiating malignant cells from degenerative, stripped nuclei, which are not diagnostic, no matter how atypical they appear.

The shape of the cytoplasm varies very little. Round or oval in 80%, it can be angular, polygonal, deformed, or frayed in 20% of the cells with slightly more abundant cytoplasm.

The cytoplasm stains basophilic in 70% of the cases, acidophilic or eosinophilic in 20%, and intermediate in 10%. Some of the staining reaction depends partially on the localization of the main lesion. An increase of basophilia is present when the lesion is situated high in the endocervical canal and is not exposed to air.

The cytoplasm may be uniformly dense or may have multiple, small vacuoles. This vacuolization may be in the form of a degenerative perinuclear halo, may appear as diffuse, ill-defined, lacy, multiple vacuoles, or may be present as one or two large secretory vacuoles distorting slightly the cellular shape.

When dense, the cytoplasm will occasionally contain a perinuclear, ringlike, linear condensation concentric to the nuclear membrane.

The cytoplasmic membrane is usually well preserved, smooth, and delicate.

No inclusions or glycogen deposits should be seen in the cytoplasm.

NUCLEAR CRITERIA FOR MALIGNANCY

The nucleus is enlarged and varies only moderately from one cell to the other ($13 \pm 3 \mu$). Its size is often smaller than the dysplastic nucleus.

There is little variation in the shape of the nucleus, which is oval in 40%, round in another 40%, and irregular in 20% of the cells.

The nuclear membrane shows marked irregularity in thickness and shape with indentation and wrinkling in 85% of the cells.

Multiple deep indentations on one side of the nucleus (tuliplike shape) provide an excellent nuclear criterion of severe dysplasia or carcinoma (Fig. 70).

The chromatin in about 80% of the cells is increased and is coarsely granular, resembling the chromatin seen in the prophase stage of a mitotic division. Moderately larger and irregularly distributed clumping of the chromatin can be seen in the other 20% of the cells.

The nucleolus is generally absent. The presence of prominent nucleoli in a majority of the atypical, suspicious squamous cells is evidence against the diagnosis of a small-cell squamous in situ carcinoma.

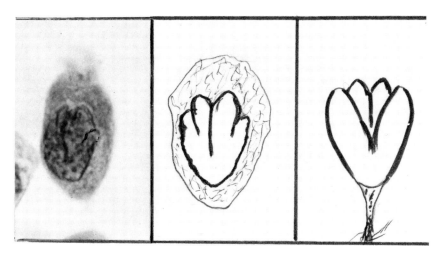

Figure 70
Deep indentations on one side of the nucleus of an in situ cancer cell make the nucleus resemble a tulip. Tuliplike formation is a good criterion of severe dysplasia or in situ carcinoma.

Multinucleation (2–3 nuclei) can be seen in less than 5% of the cells and may indicate a glandular extension of the neoplasm or coexisting dysplasia. The nuclei of these cells are uneven in size and structure.

Large-Cell or Well-Differentiated In Situ Carcinoma (Figs. 69B, 71)

This type of neoplasm is often located higher in the endocervical canal, and the best cellular sample is obtained by a direct endocervical scraping.

Numerous malignant cells, medium to large (30–80 μ) exfoliate in sheets or clusters (40%) or singly (60%), often forming stringlike alignments in the smeared mucus.

The size and shape variations are greater than those found in the small in situ carcinoma. Often these cells are triangular or have pseudopodlike elongations or tails.

The dense (80%) or delicately vacuolated (20%) cytoplasm is more abundant than in the small-cell carcinoma, but is still relatively scanty and occasionally shows a moderate degree of keratinization and eosinophilia.

The cellular borders are distinct, well defined, and regular.

The enlarged nuclei (15–40 μ) can be multiple (2–4) in 30% of the exfoliated cells.

The chromatin, usually increased, is coarsely granular (30%) or forms large, variable, irregular clumps with pointed projections (70%).

There is an increase in the number of dyskaryotic, superficial, and intermediate squamous cells.

The background of the smear is usually clean with very little inflammation or degenerative debris.

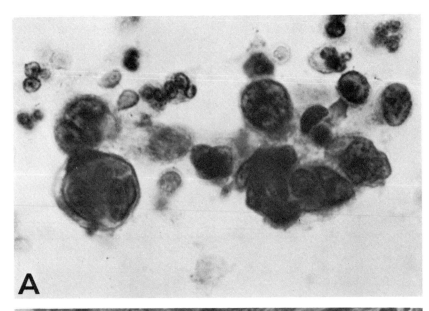

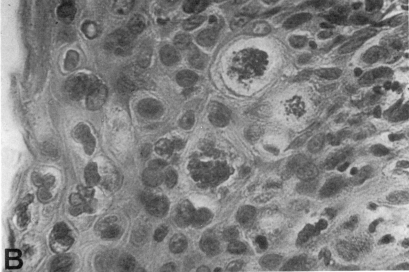

Figure 71
(A) Large-cell type of in situ squamous, cervical carcinoma cells in a smear of a 25-year-old woman. Note the size of the cells, their clustering, and irregularities. ($\times$ 1300) (B) Cervical biopsy specimen from the same patient. Note the abnormal mitosis and size of the cells. ($\times$ 750)

The exfoliated malignant cells are often underevaluated and mistaken for dysplastic basal and superficial cells. The true nature of the lesion is often shown only by the cone biopsy. Occasionally, malignant cells have pyknotic nuclei and dyskaryotic cytoplasm, which is probably due to surface differentiation and maturation of the in situ carcinoma.

IN SITU CARCINOMA WITH ENDOCERVICAL GLANDULAR EXTENSION (FIG. 72)

The possibility that there may be a glandular extension of an in situ carcinoma can be suspected by cytology in the presence of malignant in situ cells with the following changes:

1. About 50 to 90% of the malignant cells have vacuolated cytoplasm (multiple small, delicate vacuoles);
2. Large numbers of single cells or sheets of benign endocervical cells are present in the smear, showing mild-to-moderate reactive nuclear atypia;
3. The polymorphism and the variability in size of the malignant cells are increased;
4. There is an increase of cytoplasmic basophilia and a decrease of eosinophilia;
5. The nuclei are larger (15–20 μ) and show larger chromatin clumping. The incidence of prominent nucleoli is increased in up to 25% of the neoplastic cells;
6. Numerous multinucleated malignant cells (2–4 nuclei) are often present;
7. The cells tend to cluster together and mold, rather than scatter singly in the smear;
8. The background of the smears is usually dirty with an increased number of degenerate inflammatory cells, cellular debris, and protein deposits.

IN SITU CARCINOMA AND PREGNANCY

Cervical in situ carcinoma is detected by cytology with the same frequency in pregnant and nonpregnant women. There is no evidence that an in situ carcinoma diagnosed during pregnancy will have a greater tendency to regress after parturition than will an in situ carcinoma found in a nonpregnant woman.

It was once thought that the cytologic diagnosis of an in situ cervical carcinoma was more difficult during pregnancy because of the frequent association of certain epithelial pseudoatypia with the pregnancy (squamous and glandular hyperplasia and metaplasia). We are now able to recognize the true nature of these changes, and the diagnosis of an in situ carcinoma during pregnancy, based on the recognition of poorly differentiated, hyperchromatic, malignant cells with scanty cytoplasm, is as definite and easy to make as in a nonpregnant woman.

The morphologic criteria used are almost the same except that during pregnancy the nuclear size and the hyperchromatism increase slightly, the wrinkling of the nuclear membrane decreases, and there is a higher per-

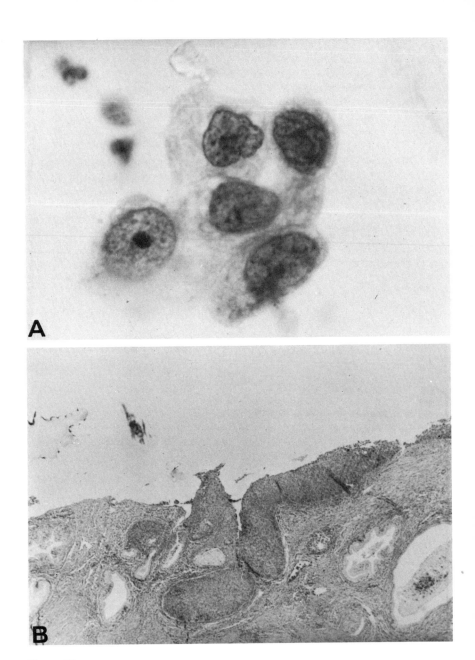

Figure 72
(A) In situ carcinoma cells indicating a probable endocervical glandular extension. Note the vacuolated cytoplasm and prominent nucleoli. ($\times$ 1300) (B)
Tissue section from the same patient showing the glandular extension of the in
situ carcinoma. ($\times$ 150)

centage of malignant cells with prominent nucleoli (15%). During pregnancy, the diagnostic malignant cells may be more numerous in the smears than at any other time. Dysplastic cells in the early stage of pregnancy may appear more abnormal because of their premature exfoliation, and the smear could be overcalled, i.e., the pathology could be overestimated.

The background of the smears frequently contains more leukocytes, cellular debris, and granular protein deposits.

IN SITU CARCINOMA WITH MICROINVASION (FIG. 73)
This term should be restricted to a cervical carcinoma that infiltrates the cervical stroma no more than 5 mm.

In spite of the fact that the basement membrane is not seen in a smear, characteristic morphological changes in the individual exfoliated malignant cells allow the cytologic diagnosis of possible infiltration.

Large numbers of malignant cells are found in sheets or syncytia about 75% of the time.

A majority of the abnormal cells are round or oval with frequent single, unilateral, short, tail-like projections of their occasionally keratinized cytoplasm.

The nuclei-cytoplasm ratio is about double that of normal parabasal squamous cells, slightly less than the noninfiltrating in situ cancer cell.

The abnormal chromatin granules are more extreme than in noninvasive in situ cancer cells.

Abnormal but small nucleoli are seen in about 20% of the malignant cells.

The background in 70% of the smears is cluttered by various amounts of cellular debris, granular protein deposits, and altered erythrocytes (tumor diathesis).

These cellular changes require a more comprehensive histopathologic study of the cervix, such as a cone biopsy examination. This is especially important during pregnancy.

IN SITU CARCINOMA IN POSTMENOPAUSAL WOMEN
Cervical in situ squamous carcinoma cells in postmenopausal smears having atrophic cellular patterns are more difficult to detect than those in smears from younger women. The diagnostic cells, often seen in syncytial groups, are difficult to distinguish from sheets of atrophic, benign, parabasal cells.

The cells are also harder to differentiate from squamous invasive carcinoma cells. The background of the smears is equally dirty, being cluttered with precipitates and debris that are not subjected to the periodic menstrual cleansing. Furthermore, a greater number of the in situ cancer cells show some degree of keratinization of their more abundant cytoplasm.

The nuclei of the in situ cancer cells are also more irregular than in the cells of younger women and are more pyknotic or hyperchromatic. Therefore, the cytologic diagnosis of an in situ cancer rendered on an atrophic smear should be very conservative. Estrogen therapy should often be suggested and the smear repeated.

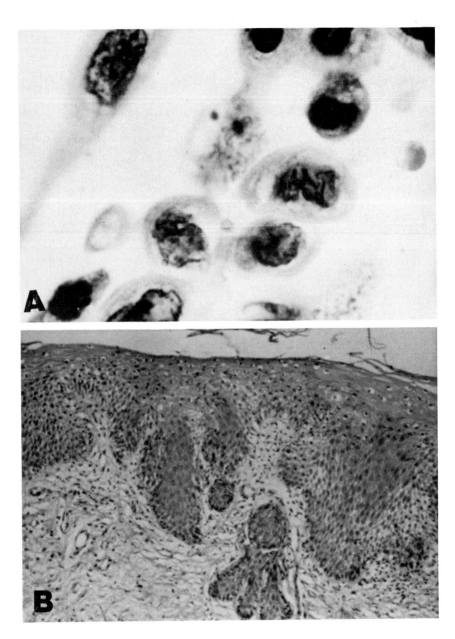

Figure 73
(A) In situ carcinoma cells indicating a probable early invasion. Note the short cytoplasmic tail of the cell in the upper left corner. ($\times$ 1300) (B) Tissue section from the same patient showing the microinvasion. ($\times$ 450)

Both histiocytes and the diagnostic in situ cancer cells usually shed singly and may have scanty cytoplasm and large nuclei. The vacuolated cytoplasm of the histiocyte may contain phagocytized debris, and often the nucleus is eccentric and touches the cytoplasmic membrane in only one place instead of several. The nuclear membrane is smooth and regular instead of wrinkled, and the chromatin clumping, although variable, is less atypical than that found in the in situ carcinoma cells.

Invasive Squamous Carcinoma

Cervical invasive squamous carcinoma may be defined as a malignant neoplasm that invades the underlying stroma of the cervical epithelium by infiltration, or destruction, or both. Squamous cell carcinoma constitutes 85% of all cervical cancers, adenocarcinoma accounts for 5%, and the remaining 10% consists of mixed adenosquamous carcinomas and other cancers and sarcomas.

Squamous cell carcinomas are graded in relation to their differentiation, which is indicated mainly by the amount of keratin present in the predominating tumor cells. A well-differentiated squamous carcinoma (grade I) will exfoliate mainly keratinized, malignant cells that stain deep orange with Papanicolaou stain, whereas a poorly differentiated squamous carcinoma (grade III) will exfoliate nonkeratinized cells that stain blue and resemble parabasal cells.

Well-Differentiated, Infiltrating Squamous Cell Carcinoma
(Figs. 74, 75, 76)

GENERAL CELLULAR MORPHOLOGY
Depending on the size, the location of the neoplasm, and the technique used, an abundant but variable number of diagnostic cancer cells may be present in the smear.

The cancer cells shed singly (50%), in clusters or sheets (30%), as a part of a structure (malignant pearl, 2%), and in syncytial formation (18%).

The size of the cells varies ($220 \pm 80 \ \mu^2$). Often, the malignant cells are about half the size of the corresponding normal superficial or intermediate squamous cells.

The shapes of the cells vary, being flat, round, polygonal (70%), tadpole (10%), spindle or pearl formation (10%), or irregular (10%), depending on the maturation of the tumor and the technique used in making the smear.

A large number of nonspecific inflammatory cells is frequently present in the smear. The absence of inflammation is an indication against the presence of an invasive squamous cell carcinoma. In most cases, in the back-

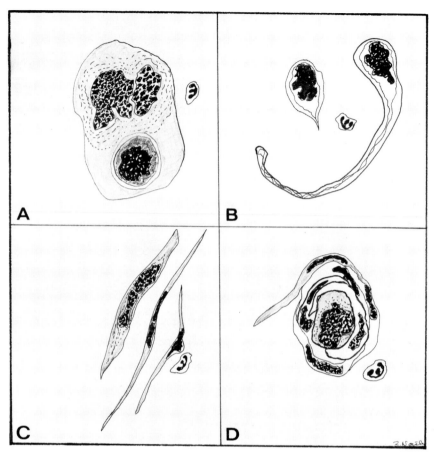

Figure 74
(A) Polygonal, well-differentiated, squamous carcinoma cell with pseudo-cannibalism of a smaller, round cancer cell. (B) Squamous tadpole cancer cells. (C) Squamous spindle cancer cells. (D) Malignant pearl.

ground of the smear there are no normal vaginal flora or Döderlein's bacilli.

Fresh and old red blood cells, fibrin, cytoplasmic debris, and granular protein deposits are almost always present in the smears of patients with fungating, ulcerated, large tumors, whereas the background remains remarkably clean in smaller lesions.

In 85% of the cases, other atypical cells, diagnostic of an in situ or squamous dysplasia, coexist with the invasive malignant cancer cells.

CYTOPLASMIC CHANGES
The amount of cytoplasm varies with the differentiation of the squamous cancer cell. In the mature carcinoma, the cytoplasm is adequate to abundant, forming approximately 75% of the total surface of the cells.

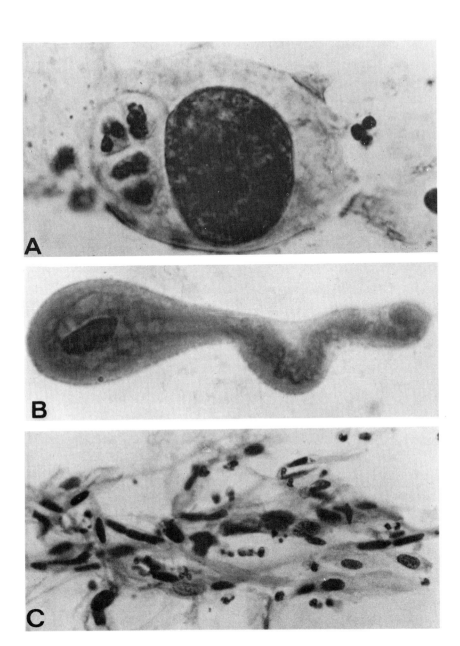

Figure 75

Well-differentiated, cervical, invasive squamous carcinoma cells. (A) Squamous, oval tumor cell with keratinized, abundant cytoplasm and large, central, hyperchromatic nucleus. Note the diagnostic irregularity of the chromatin clumps. (× 1300) (B) Squamous, tadpole tumor cell with an abundant, elongated, keratinized cytoplasm and a prominent Herxheimer's spiral. Note the eccentric, irregularly hyperchromatic nucleus. (× 750) (C) Squamous, spindle tumor cells with their keratinized cytoplasm and diagnostic, elongated, semi-pyknotic, irregular nuclei. (×450)

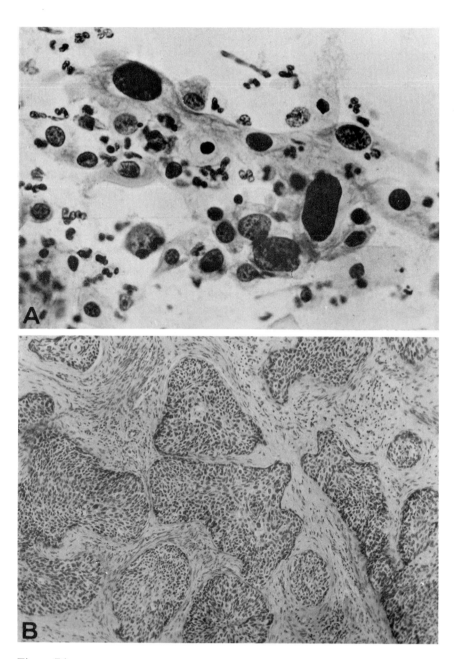

Figure 76
(A) Invasive squamous carcinoma cells (× 750) compared to the corresponding tissue section (B) of Grade IB epidermoid carcinoma of the uterine cervix. (× 230) Note the occasional prominent nucleoli in the vesicular nuclei.

The amount of intracytoplasmic keratin depends, in direct proportion, on the maturation of the neoplasm. The cells are stained deep orange–red or yellow in 40% of the cases, deep blue in another 40%, and indeterminate in color in 20%.

In about 40% of the malignant cells, the keratin is deposited in concentric, perinuclear, ringlike formations. The endoplasm is often thick and granular and can be clearly seen, in contrast to the exoplasm, which in about 70% of the hypermature malignant cells is thin and semitransparent.

The cellular borders are sharp and well limited, without the curling tendency seen in the corresponding benign cells.

NUCLEAR CHANGES

The size of the nuclei varies (mean nuclear surface 75 ± 40 μ^2). It generally ranges from 2 to 10 times the size of the normal squamous cell.

The nuclei-cytoplasm ratio is increased with the nuclei occupying about one-third of the volume of the cells, as compared to only one-twentieth in the benign squamous cells.

The nuclei show irregularity in their shape, being round (30%), oval (30%), or irregular (40%). The nuclear shape varies with the shape of the cells, being elongated when in a spindle cell and round when in a polygonal cell.

The irregularly distributed chromatin is coarsely clumped with pointed projections (50%), and is pyknotic or has severe hyperchromatism in another 50%.

An occasional nucleolus can be seen in 10% of the vesicular nuclei in which the chromatin is not obscured. This nucleolus is often enlarged, acidophilic, and pleomorphic with pointed projections.

The malignant nuclei are multiple (2–3) in approximately 15% of the exfoliated cancer cells. These nuclei are irregular in shape with occasional molding.

TADPOLE CELLS (FIGS. 74B, 75B)

The tadpole cell, an elongated cell with a rounded "head" containing its nucleus, can be benign or malignant, depending on the appearance of the nucleus.

The elongated portion of the cytoplasmic "tail" may vary in length and may be adequate to abundant. Usually heavily keratinized, it stains bright orange with Papanicolaou stain (90%).

In the tadpole cell, the endoplasm is often stretched to the tip of the tail, and its rim gives the appearance of a thin, spiderlike filament that is often twisted on itself (Herxheimer's spiral).

The end of the cytoplasmic tail is rounded and bulbous in 80% of the cases and pointed in the remaining 20%.

The nucleus is eccentric, being situated in the terminal, rounded end of the cytoplasm. Irregular and angular in shape (80%), the nuclei vary in size and have pyknotic, or coarse, unevenly clumped chromatin with occasional prominent nucleoli. The other previously described malignant criteria are usually present.

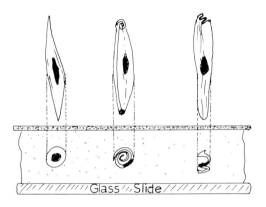

Figure 77
Difference between true squamous spindle cells and rolled-up or folded superficial squamous cells.

Tadpole cells should be differentiated from the benign metaplastic elongated columnar cells and basal repair cells, whose nuclei are often central
or toward the cellular tail and have little keratin. Their nucleoli are prominent, and their fine chromatin is regularly distributed.

SPINDLE CELLS (FIGS. 74C, 75C)
Spindle cells shed singly (50%) or as part of a sheet or a pearl.
   The cytoplasm is orangeophilic (80%), abundant, and elongated, with
a well-defined, sharp, and smooth border.
   The cytoplasmic membrane often merges with the nuclear membrane on
both sides of the cell.
   The separation between the ectoplasm and the endoplasm is usually in-
distinct, as opposed to that of the tadpole cell.
   The nuclei are enlarged and elongated, and often have pyknotic (85%)
or dark, coarsely clumped chromatin (15%).
   The nuclear membrane is irregular with indentations.
   The nuclear ends vary in shape. One end can be seen as pointed, while
the other is square, for example. This is an important criterion in differentiating them from benign spindle cells, such as smooth muscle cells and
collagen or elastic fibers.
   Spindle cells should be differentiated also from the elongated, rolled-up
or folded, thin-walled, benign superficial or intermediate squamous cells that
are occasionally seen (Fig. 77). The differentiation is made by focusing
the microscopic image up and down and studying the different cellular
levels.
   In the scraping of benign, atrophic, cervical cells, sheets of elongated
spindlelike cells can be seen. They are often orangeophilic because they are
usually partially air dried. Their differentiation from sheets of malignant
spindle cells is based on the density of the cytoplasm and the abnormality
of the nuclei of the tumor cells.

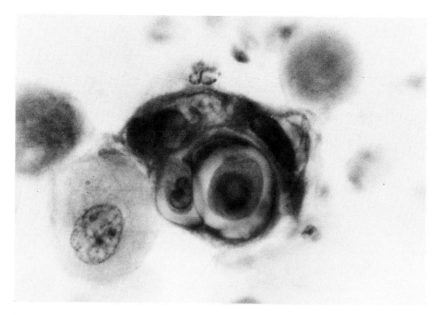

Figure 78
Malignant pearl formation in an invasive, squamous cell carcinoma. Note the nuclear abnormalities. (× 750)

PEARL FORMATION (FIG. 78)

These pearls (cluster of cells in a whorl formation) are interpreted as benign or malignant according to the aspect of the individual component cells, especially the appearance of the nuclei.

They are rarely seen, appearing in less than 1% of the smears from squamous cell carcinoma, and their presence indicates that the tumor is well differentiated.

The criteria for malignancy should be present in a majority of the component cells, not only in the central ones. This central cell, even in benign pearls, will show some cytoplasmic and nuclear abnormalities as a result of degeneration and compression.

Mitosis, when present, is helpful in the diagnosis of malignancy.

When seen with adenocarcinoma cells, pearl formations may indicate the presence of an adenoacanthoma or a mixed carcinoma. They should not be confused with psammoma bodies, which are made up of noncellular concentric rings.

Poorly Differentiated, Infiltrating, Small Squamous Cell Carcinoma
(Figs. 79, 80)

An abundance of diagnostic cells is usually found in the smear from a poorly differentiated, infiltrating squamous cell carcinoma (more than 1000). Usually single, they may also be found in loose syncytia.

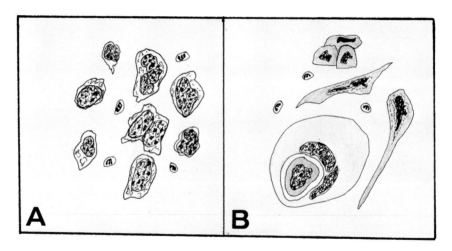

Figure 79
(A) Poorly differentiated, cervical squamous cell carcinoma. (B) Well-differentiated, cervical squamous cell carcinoma.

The cells vary moderately in size ($100 \pm 25\ \mu^2$) and shape, usually being either round or oval. Fewer than 0.5% are elongated or caudate.

The delicately vacuolated cytoplasm of these cells is basophilic (80%), scanty, and often seen as a very narrow rim. It may also be dense with concentric rings. No cytoplasmic inclusions are present.

The cytoplasmic membrane is poorly defined but distinct, differentiating these cells from those with stripped, degenerate, endocervical and parabasal nuclei.

The angular or oval-shaped nuclei are large and hyperchromatic with irregular chromatin clumps that have pointed projections. Multinucleation is common.

The abnormal nucleoli, when present (30%), are irregular and vary in size, shape, and number.

In a cluster, the cells may overlap or mold loosely against one another, with intercellular, slitlike spaces. Pseudocannibalism (birdseye appearance) is common.

Large amounts of inflammatory cells, cellular debris, and fresh and old red blood cells are almost always present (tumor diathesis). These tumor cells can be differentiated from normal endometrial stromal cells by their greater degree of anaplasia and more extreme irregularity of the nuclei. The tumor diathesis is absent in the benign condition.

Nonkeratinizing, Infiltrating Squamous Cell Carcinoma

This cancer, which probably originates from a large-cell type in situ carcinoma, is sometimes called anaplastic or large-cell, poorly differentiated carcinoma. The diagnostic cells are numerous (more than 600 cells), large (mean cell area of $200 \pm 80\ \mu$), and either single (70%) or in syncytial

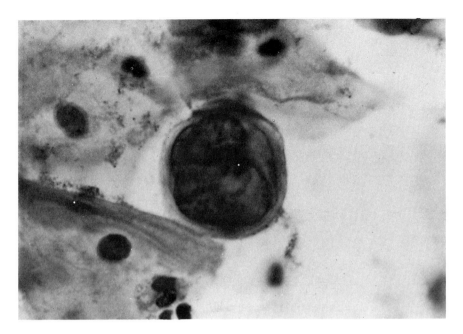

Figure 80
Poorly differentiated squamous cell carcinoma. Note the scantiness of the cytoplasm, multilobular nucleus, and the irregular chromatin clumping. ($\times$ 1300)

formation (30%). The cytoplasm. which varies in amount, is usually
cyanophilic and contains no keratin. It is fragile and often lysed, which results in the presence of numerous naked abnormal nuclei in the smears.

The nuclei are round or irregular, and large ($80 \pm 10 \mu$) with increased, coarsely granular chromatin.

The background of the smear is almost always cluttered with large
amounts of degenerate cellular debris, red blood cells, and protein deposits
(severe tumor diathesis).

These cells may be difficult to differentiate from reactive (hyperplastic)
endocervical parabasal cells, as well as from repair basal squamous cells.
All have enlarged nuclei with coarsely granular chromatin, smooth nuclear
membranes, prominent nucleoli, and variable amounts of cyanophilic cytoplasm. The benign cells are usually found in sheets, rather than singly, their
chromatin is slightly less irregular, and the background of the smear is relatively clean.

Squamous Cell Carcinoma of the Vulva

The cells exfoliated from a vulvar squamous carcinoma can contaminate a
vaginal smear, and it is very difficult to differentiate these cells cytologically
from the ones exfoliated from a cervical squamous carcinoma. They usually
show better differentiation and have more cytoplasm than the cells from a

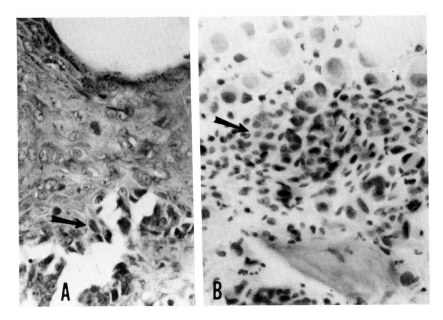

Figure 81
Histology (A) and cytology (B) of Paget's disease of the vulva.

primary cervical squamous carcinoma. Anucleation, anaplasia, and marked degeneration with air-drying artifacts are commonly found. Most of the previously described malignant features are present.

Paget's Disease of the Vulva
(Fig. 81)

This malignant dermatosis involves mainly the vulva but can be found extending into the vagina and cervix of elderly women. It produces vulvar pruritus, irritation, and discomfort.

The cytologic and histologic characteristics of this lesion are diagnosed by the presence of Paget cells, which, in histologic sections, are usually found in clusters in the basal layers of the skin. Energetic scraping of the lesion will produce an abundance of these Paget cells. They are large, parabasal cells with moderate amounts of finely granular, amphophilic cytoplasm. Melanin pigment may be seen. Their central nuclei are usually single, large, and vesicular, with moderately coarse chromatin clumping. Occasional, prominent, red, round nucleoli are present. Moderate, superficial, squamous dysplasia is frequently seen coexisting with this disorder.

References and Supplementary Reading

Alousi, M. A., et al.   Microinvasive carcinoma and inflammatory lesions of the cervix uteri, histologic and cytologic differentiation. *Acta Cytol.* 11:132, 1967.

Baugle, R., Berger, M., and Levin, M.   Variations in the morphogenesis of squamous carcinoma of the cervix. *Cancer* 16:1151, 1963.

Bredahl, E., Koch, F., and Stakemann, G.   Cancer detection by cervical scrapings, vaginal pool smears and irrigation smears, a comparative study. *Acta Cytol.* 9:189–193, 1965.

Brux, J. de, and Dupre-Froment, J.   Junctional areas in squamous cell carcinoma of the cervix. *Am. J. Obstet. Gynecol.* 93:181, 1965.

Christopherson, W. M., and Parket, J. E.   Microinvasive carcinoma of the uterine cervix. *Cancer* 17:1123, 1964.

Darby, R. E. W., and Williams, S. E.   The cytological diagnosis of carcinoma of the cervix. *N.Z. Med. J.* 64:98, 1965.

Diddle, A. W., and Williamson, P. J.   Cervical carcinoma. *Am. J. Obstet. Gynecol.* 89:975, 1964.

Dougherty, C. M.   Relationship of normal cells to tissue space in the uterine cervix. *Am. J. Obstet. Gynecol.* 84:648, 1962.

Fidler, H. K., and Boyes, D. A.   Patterns of early invasion from intra-epithelial carcinoma of the cervix. *Cancer* 12:673–680, 1959.

Hadju, S. I., and Adelman, H. C.   Anatomic distribution and grading of carcinoma in situ of the cervix. *Cancer* 19:1466, 1966.

Handy, V. H., and Wieben, E.   Detection of cancer of the cervix: A public health approach. *Obstet. Gynecol.* 25:348, 1965.

Koss, L. G., Melamed, M. R., and Daniel, W. W.   In situ epidermoid carcinoma of the cervix and vagina following radiotherapy for cervical carcinoma. *Cancer* 14:353, 1961.

Lerch, V., et al.   Cytologic findings in progression of anaplasia (dysplasia) to carcinoma in situ: A progress report. *Acta Cytol.* 7:183, 1963.

MacGregor, J. E., Fraser, M. E., and Mann, E. M. F.   The cytopipette in the diagnosis of early cervical carcinoma. *Lancet* 1:252–256, 1966.

Margulis, R. R., Ely, C. W., Jr., and Ladd, J. E.   Diagnosis and management of stage 1A (microinvasive) carcinoma of the cervix. *Obstet. Gynecol.* 29:529, 1967.

Marshall, C. E.   Effects of cytologic screening on the incidence of invasive carcinoma of the cervix in a semi-closed community. *Cancer* 18:153, 1965.

Messelt, O. T., and Hoeg, K.   Mass screening for cancer and precancerous conditions of cervix uteri: A preliminary report. *Acta Cytol.* 11:39, 1967.

Nieburgs, H. E.   Tissue and cell pathology of uterine cervix dysplasia and carcinoma in situ. *Acta Cytol.* 15:513, 1971.

Ng, A., Reagan, J., and Luider, E.   The cellular manifestation of microinvasive squamous cell carcinoma of the uterine cervix. *Acta Cytol.* 16:5, 1972.

Old, J. W., Wielenga, G., and Haam, E. von   Squamous carcinoma in situ of the uterine cervix. I-Classification and histogenesis. *Cancer* 18:1598, 1965.

Osbaud, R., and Jones, W. N.   Carcinoma in situ in pregnancy. *Am. J. Obstet. Gynecol.* 83:599, 1962.

Reagan, J. W., and Hamonic, M. J.   The cellular pathology in carcinoma in situ. A cytohistopathological correlation. *Cancer* 9:385–402, 1956.

Reagan, J. W., Seidemann, I. L., and Patten, S. F., Jr.   Developmental stages of in situ carcinoma in uterine cervix: An analytical study of the cells. *Acta Cytol.* 6:538, 1962.

Roberts, T. W., and Linkins, T.   Differentiation between intramucosal and invasive carcinoma of the cervix by cytologic smear. Is it reliable? *Acta Cytol.* 8:280, 1964.

Sedlis, A.   Cytology screening of prenatal patients in a municipal hospital. *Acta Cytol.* 7:224, 1963.

167

Tweeddale, D. N.   Cytopathology of cervical squamous carcinoma in situ in postmenopausal women. *Acta Cytol.* 14:363, 1970.

Wied, G., et al.   Cytology of invasive cervical carcinoma and carcinoma in situ. *Ann. N.Y. Acad. Sci.* 97:759–766, 1962.

Genital Adenocarcinoma and Sarcoma

8

The incidence of uterine adenocarcinoma is about one-third that of invasive cervical squamous cell carcinoma. Uterine adenocarcinoma occurs mainly in patients in their fifth or sixth decade. The histologic diagnosis is based on the distortion, stratification, and variation of shape and size of the glandular epithelium, combined with a loss of cellular polarity. When the adenocarcinoma is of a well-differentiated type, often the glandular pattern is unchanged, and the individual component cells resemble the normal columnar or cuboidal ones, except for the abnormalities of their nuclear structure. In the poorly differentiated adenocarcinoma, the glandular structure is not so apparent, and the size and shape of the individual cells vary extensively with very little resemblance to the original benign columnar cells.

Endometrial Adenocarcinoma

DETECTION RATE

According to the type of smear examined, the average detection rates are:

1. Cervical scraping only: 15%;
2. Vaginal pool smear and cervical scraping: 40%;
3. Endocervical aspiration and vaginal pool smears: 65%;
4. Intrauterine and endocervical aspiration and lavage: 90%.

The exfoliated cancer cells can remain in the uterine cavity from one day to two weeks before they appear in the vaginal pool. This explains their usual marked degeneration, which makes them more difficult to recognize and indicates the need for a direct endometrial sampling method in clinically suspect patients.

HELPFUL HINTS FOR DIAGNOSIS

The possibility of an endometrial lesion is suggested by the presence of an unexplained shift to the right of the maturation index (hyperestrogenlike effect) in the vaginal smear of a postmenopausal woman. Any other space-occupying mass in the ovary, endometrium, or fallopian tubes may also occasionally produce a similar maturation index abnormality. This mass can be a benign or a malignant tumor, an intrauterine device, or even a nonfunctioning neoplasm of the ovary.

The presence of degenerate or well-preserved benign endometrial cells in the smear of a postmenopausal woman warrants an endometrial washing or a dilation and curettage to rule out the possibility of an early or advanced but nonexfoliating uterine adenocarcinoma.

Also, the possibility of the existence of an endometrial adenocarcinoma is indicated by the presence of an increased number of small or large histiocytes, often with phagocytized necrotic debris, in the nonatrophic smear of postmenopausal women.

An increased number of foreign-body giant cells in such smears should also bring to mind the possibility of a neoplasm. These foreign-body giant cells should be differentiated from the syncytial epithelial giant cells commonly seen in the atrophic smear and not related to any endometrial lesion.

The presence of unexplained fresh and old blood or an increase in necrotic debris or fibrin in the smear from a postmenopausal woman should also be investigated.

A history of diabetes, obesity, hypertension, late menopause, or postmenopausal bleeding also increases the chances of having an endometrial adenocarcinoma.

GENERAL DIAGNOSTIC CELLULAR CRITERIA (FIG. 82)
The number of exfoliated diagnostic cells in a smear varies according to the site, type, and size of the tumor and the technique used. They can be very scanty (only a few cells) or abundant (more than 4000 cells in one smear). They are abundant when the neoplasm is necrotic, poorly differentiated, or advanced.

Although generally larger than normal, the tumor cells, with few exceptions, imitate the appearance of the benign columnar cells from which they originate.

The cells can shed in tight clumps (30%), in clusters (30%), or singly (40%).

The majority of the endometrial tumors are poorly differentiated and have scanty cytoplasm (60%).

The well-differentiated carcinomatous cells often contain enlarged secretory vacuoles distending their cytoplasm; and they are being cannibalized by leukocytes.

The background of the smear is usually dirty, containing large numbers of fresh and old red blood cells, cellular debris, fibrin, and granular protein deposits (tumor diathesis).

The diagnostic malignant cells also often show advanced degenerative changes and distortion because of the delay in their passage through the endocervix.

CYTOPLASMIC CRITERIA
The amount of cytoplasm varies from one cell to another. It is relatively scanty in relation to the size of the nucleus in 80% of the cells, especially in poorly differentiated carcinoma.

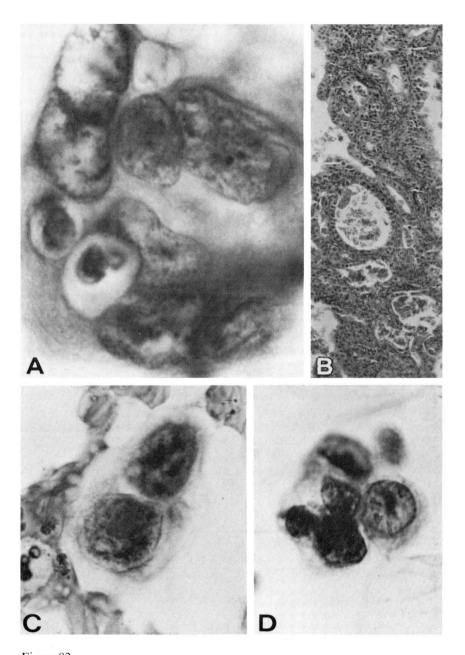

Figure 82
(A) Cluster of endometrial adenocarcinomatous cells. Note the irregularity of shape and size of the nuclei and the overlapping neutrophils. The nucleoli in these cells are not prominent. ($\times$ 1300) (B) Tissue section of the curettage specimen. ($\times$ 150) (C) Endometrial adenocarcinomatous cells with large, prominent, irregular nucleoli. ($\times$ 1300) (D) Cluster of malignant cells in the routine vaginal smear of a 60-year-old woman with an asymptomatic early endometrial carcinoma. ($\times$ 1300)

171

Although usually round, the shape of the cytoplasm varies, especially when the cells originate from a well-differentiated neoplasm.

The cytoplasmic border in 90% of the cells appears indistinct as a result of their thickness or degeneration.

The cytoplasm is transparent, vesicular with tiny vacuoles (60%), and homogeneous (30%). It may contain distorting large vacuoles, as seen in the signet ring cells (10%), in which the vacuoles compress the nuclei into a crescent shape and distend the cytoplasmic borders.

The staining of the cytoplasm is often uneven, with basophilia (60%), acidophilia (20%) or with an indeterminate stain (20%). No keratin is present.

Large numbers of leukocytes infiltrating the vacuoles can almost hide the cells. These leukocytes are well-preserved and healthy in appearance, which differentiates them from the degenerate ones found ingested in the phagocytic vacuoles.

NUCLEAR CRITERIA

The nuclei are enlarged, 2 to 6 times the size of the normal endometrial cell nucleus. They cover about 40–60% of the cellular area, with a mean area of 60 μ^2. On occasion, particularly in poorly differentiated adenocarcinoma, monstrous nuclei (about 100 μ in diameter) may appear. They indicate anaplasia and have a poor prognosis. They need to be differentiated from the giant nuclei found in certain sarcomas.

A majority of the nuclei are oval (60%) or round (10%) with a mild degree of hyperchromatism. Another 30% of the cells may show a marked variation of their shape as a result of the rapid growth in a confined space. Molding and pseudocannibalism are common. The nuclei are often eccentric, especially when pushed to one side by the presence of large, overdistended vacuoles.

The nuclear membrane is thick, distinct in 97% of the cells and thin and indistinct in the rest. The nuclear membrane is regular in outline, or it can be indented by the pressure of vacuoles, which occurs in about 12% of the cells.

The nuclei are usually vesicular with marked irregularity of the chromatin clumping in 65% of the cells and fine granularity in 35%, in direct proportion to the activity of the cells.

The chromatin clumping often is not as prominent as that seen in squamous cell carcinoma and is more uniformly scattered throughout the nucleus.

Although the nucleus is usually single, it may be multiple (2–3) in 5% of the malignant cells.

The nucleoli present in 30% of the tumor cells are moderately enlarged and vary in size, shape, and number. They appear more distinct in malignant cells that are physiologically exfoliated to the vagina (vaginal pool) than in cells obtained by aspiration or direct scraping.

PITFALLS

MENSES.   The presence of large numbers of normal stromal and glandular endometrial cells, as seen in the smears during the menstrual flow, can prevent the detection of the atypical ones. If an endometrial lesion is clinically suspected, cytology should be repeated at midcycle.

BLEEDING.   The epithelial cells may be diluted with excessive blood, exudate, or transudate, which renders the smears very scanty in diagnostic cells. In the presence of a large number of red blood cells, as seen in the endometrial lavage specimen, Carnoy's fixative may be used to lyse the red blood cells before processing the specimen. Any bloody smear from a postmenopausal woman should be carefully screened and a repeat almost always recommended.

ENDOMETRIAL HYPERPLASIA.   The presence of reactive hyperplastic endometrial cells and an increased maturation index may make the cytologic detection of an endometrial adenocarcinoma more difficult. In a postmenopausal woman this cytologic picture indicates, for the patient, a dilation and curettage or endometrial washing, even in the absence of any diagnostic cells.

ENDOMETRIAL POLYP.   Blood, histiocytes, increased pseudoestrogenlike effect and reactive endometrial cells may make the differential diagnosis between glandular polyp and low-grade endometrial adenocarcinoma almost impossible by cytology alone. The presence of such cells in postmenopausal smears indicates the need for histology (dilation and curettage) for the final diagnosis.

POSTCURETTAGE SMEAR.   The regenerating endometrial glandular and stromal cells after a curettage may vary in size and shape, have prominent nucleoli, and be mixed with reactive histiocytes. They can be the cause of a false-positive cytologic interpretation if the cytopathologist is not apprised of a recent operation.

REACTIVE ENDOCERVICAL PARABASAL CELLS.   Because of their resemblance to the cells exfoliating from a well-differentiated endometrial carcinoma (large nuclei with prominent nucleoli), the presence of numerous reactive endocervical cells in the smear can be an extreme handicap in the detection of the neoplasm.

REACTIVE SQUAMOUS PARABASAL CELLS.   Cytoplasmic keratin rings parallel to the nucleus in some of the cells aid in their recognition as squamous cells in spite of their prominent nucleoli and their enlarged hyperchromatic nuclei.

ENDOMETRITIS.   The presence of large numbers of degenerate endometrial and inflammatory cells in the smear renders recognition of the tumor cells difficult. In cases of advanced endometrial tuberculosis, the danger of mistaking the isolated reactive epithelioid cells becomes even more acute.

HORMONE ADMINISTRATION.   Administration of hormones to the patient can produce an abnormal maturation index and hyperplasia of the endometrium that can be mistaken, by the cytologist who is unaware of the hormone therapy, for the changes resulting from endometrial lesions.

REASONS FOR FALSE-POSITIVE CYTOLOGIC DIAGNOSES, AS SUMMARIZED FROM THE LITERATURE. Endometrial hyperplasia: 30%; endometrial polyp: 30%; proliferative endometrium: 18%; secretory endometrium: 12%; atrophic endometrium: 7%; endometritis: 3%.

IN SITU ENDOMETRIAL ADENOCARCINOMA
The exfoliated cells from an endometrial in situ adenocarcinoma have few distinctive morphological features and are not easily recognized as such. They cannot be differentiated from the cells exfoliating from an invasive adenocarcinoma. A majority of the cells are seen singly. Their cytoplasm is abundant and eosinophilic with no secretion, and the single nucleus is hyperchromatic with an enlarged prominent nucleolus. In a postmenopausal woman, the frequent concomitant endometrial hyperplasia and unexplained maturation index shift to the right can be helpful in detecting this lesion.

WELL-DIFFERENTIATED ENDOMETRIAL ADENOCARCINOMA (FIG. 83C)
The exfoliated cells vary in their arrangement. They can be isolated or in small, tight clusters with overlapping nuclei. The cells are often scanty in the smears.

The cytoplasm is usually abundant and cyanophilic, showing marked variation of size and shape, with distinct cellular borders.

Large secretory vacuoles, single or multiple, overdistend their cytoplasm and often contain numerous healthy neutrophils that can obscure the nuclei. *In a postmenopausal woman, every cluster of polymorphonuclear leukocytes should be examined under high magnification.* The eccentric nuclei vary in size and shape and cover usually more than 50% of the cell surface ($130 \pm 80 \ \mu^2$).

Mitotic figures are more numerous, and the nucleoli are generally more prominent than those found in cells exfoliating from the poorly differentiated adenocarcinoma.

PAPILLARY ENDOMETRIAL ADENOCARCINOMA
The majority of the previously mentioned criteria of malignancy apply in the cytologic diagnosis of a papillary adenocarcinoma. The main differences are as follows:

1. There is an increase in the number of exfoliated tumor cells;
2. A larger portion of the cells is grouped into papillary formations with an external smooth common border (50% of the cells);
3. The cytoplasm of the individual cells is scantier and better preserved than in cells from a nonpapillary adenocarcinoma;

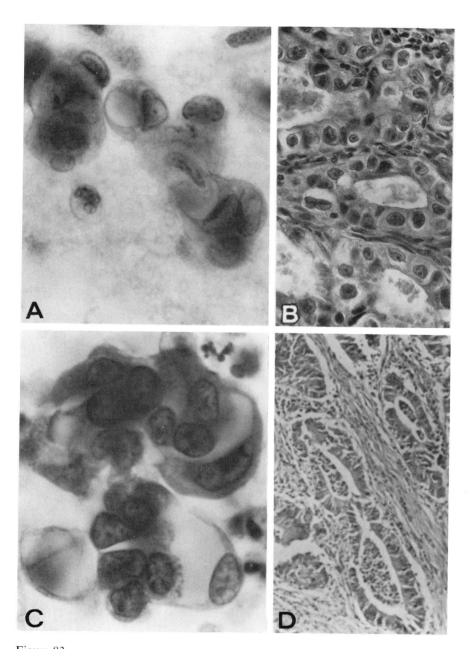

Figure 83

(A) Endometrial aspirate. Cluster of malignant cells with marked irregularity of their nuclei. (× 750) (B) Grade II endometrial adenocarcinoma in the curettage specimen. (× 450) (C) Well-differentiated endometrial adenocarcinoma. Note the large secretory vacuoles distending the cytoplasm. (× 750) (D) Grade III endometrial adenocarcinoma. (× 225)

4. The molded nuclei are more irregular in their shape with moderate hyperchromatism;
5. The nucleoli are less prominent.

POORLY DIFFERENTIATED ENDOMETRIAL ADENOCARCINOMA

The cells are crowded together in an irregular cluster. The cytoplasm is scanty, cyanophilic, often hazy with little vacuolization. The mean cell surface area is about $200 \pm 50\ \mu^2$. The round or irregular hyperchromatic nucleus is almost always single. The nucleoli are often prominent and irregular in size, shape, and number.

ENDOMETRIAL ADENOACANTHOMA

This is a rare tumor (1 per 15,000 patients) and is most commonly seen in patients between the ages of 50 and 70 years.

The diagnosis is based on a mixture of exfoliated cells that are typical of adenocarcinoma and squamous metaplasia or dysplasia, more than of frank invasive carcinoma. For a definite cytologic diagnosis, the two types of atypical cells need to be found in continuity in a cluster.

The size of the malignant endometrial cells varies ($80–120\ \mu^2$). The cells are usually found in tight clusters.

The cytoplasm is abundant (80% of the cells), acidophilic (46%), basophilic (45%), or indeterminate (9%) in color.

The cytoplasm usually is homogeneous (60%) or contains small vacuoles (40%).

The cytoplasmic membrane is indistinct (75%) or well defined (25%).

The nuclei are usually single, oval, and enlarged, occupying about 50% $\pm$ 10% of the cellular area.

The nuclear membrane is thick, sharp, and even in more than 95% of the cases.

The chromatin is finely granular in 95% of the cases with moderate hyperchromatism.

The nucleoli are prominent, round, and multiple in more than 50% of the cases.

The squamous cells, benign in nature, usually vary in size and shape according to their differentiation. The abundant cytoplasm is deep orange stained with a glassy appearance. The nuclei can be pyknotic or vesicular with marked hyperchromatism, and they are often seen in sheets or pearl formation.

The background of the smears shows an increased number of inflammatory cells, fibrin, cellular debris, and granular protein deposits.

It is difficult to differentiate cytologically the adenoacanthoma from an endometrial adenocarcinoma coexisting with separate cervical squamous metaplasia or dysplasia.

MIXED ENDOMETRIAL ADENOSQUAMOUS CARCINOMA (FIG. 84)

This endometrial cancer has both glandular and squamous malignant components and either may predominate. Seen in increasing frequency, it has a poor (less than 20% 5-year survival) prognosis. Numerous diag-

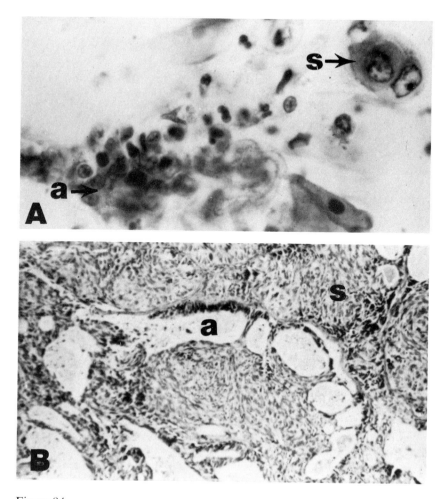

Figure 84
Cytology (A) and histology (B) of endometrial adenosquamous carcinoma
(a = adeno component; s = squamous component). (× 450)

nostic cells are exfoliated from both components. The glandular cells are
usually poorly differentiated in type, whereas the squamous cells are nonkeratinizing in 80% of the cases, keratinizing in 10%, and small-cell,
poorly differentiated in another 10%. In smears prepared from endometrial
washings, both components may often be identified in the same cluster of
cells; however, in the vaginocervical smear, they are dispersed throughout
the slide.

Because of the grave prognosis attached to this tumor, the screener
should always be very careful to continue to examine the entire smear, even
after finding diagnostic malignant cells of squamous or endometrial origin
at the beginning of the screening.

Endocervical Adenocarcinoma
(Fig. 85)

The endocervical adenocarcinoma is a relatively rare tumor, constituting about 4 to 5% of cervical cancers. It grows slowly. The cellular sample is best obtained by direct scraping of the papillary or ulcerated neoplasm.

Compared to endometrial adenocarcinoma, this tumor is likely to be better differentiated.

The size of the diagnostic exfoliated cell varies from 1 to 3 times larger than the normal endocervical cell.

The cells are shed singly or in groups.

They show very little degeneration, contrary to what is found in the cells from an endometrial adenocarcinoma.

The cytoplasm is abundant, thick, and dense, often granular and acidophilic (in 60% of the cells) or finely vacuolated (in 40%).

An occasional single, large, secretory vacuole may be found in low-grade adenocarcinoma.

The cellular outline, indistinct in 65% of the cells, is often distorted by the presence of the vacuoles, which usually contain very few leukocytes, contrary to what is found in the cells from an endometrial adenocarcinoma.

The enlarged nuclei are often eccentric and are round or oval. They are usually single with multinucleation occurring in 10% of the cells.

The chromatin is slightly increased and is finely granular and uniformly scattered throughout the nucleus (60%) or clumped in irregular angular masses with a tendency to marginal condensation (40%).

The nuclear membrane is thin, smooth, and often indistinct.

The nucleolus is prominent, round, and slightly enlarged in 40% of the cells or irregular and multiple (2–3) in 60%.

The background shows very little debris, few inflammatory cells, or little blood in early cancer (stage 0–1). It appears dirty in more advanced stages. This tumor diathesis is still less pronounced than in endometrial adenocarcinoma of a similar stage.

Fallopian Tube Adenocarcinoma

This is a rare tumor, and the exfoliated cells can be detected in a vaginal pool smear in about 30% of the cases. The cells resemble the ones exfoliating from an endometrial adenocarcinoma, and the cytologic differentiation is difficult, if not impossible. In the presence of neoplastic cells of adenocarcinomatous type in a smear of a patient who has had negative findings on dilation and curettage, a fallopian tube adenocarcinoma should be suspected.

Adenocystic Carcinoma
(Fig. 86C, D)

This is a slow-growing tumor of the vulva, vagina, or cervix that is capable of local tissue invasion and distant metastasis. The exfoliated cells are easy

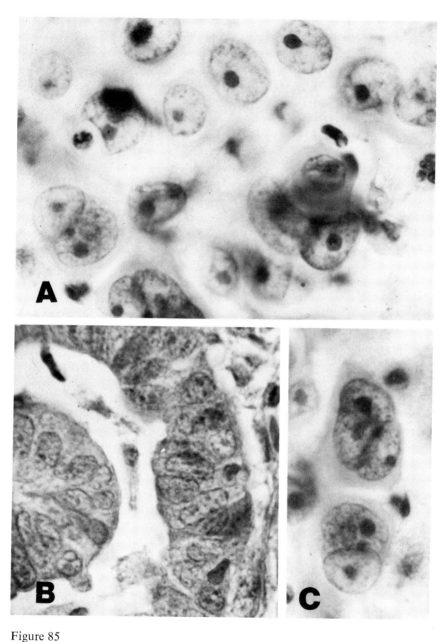

Figure 85
(A) Endocervical adenocarcinoma cells in a vaginal pool smear. Note the prominence of the nucleoli and regularity of the chromatin. (× 1300) (B) Histologic section of the tumor from the same patient. (× 750) (C) Endocervical adenocarcinoma cells showing occasional multinucleation. (× 750)

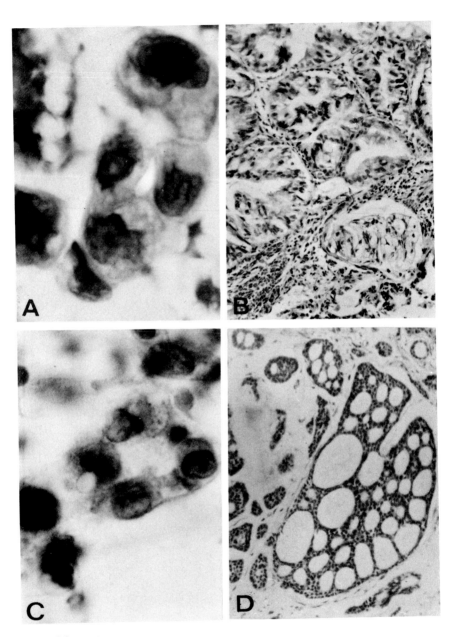

Figure 86
(A) Bartholin's gland adenocarcinoma cells in a routine vaginal smear. (× 750)
(B) Section of the Bartholin's gland carcinoma that was ulcerating in the lower
part of the vagina. (× 150) (C) Adenocystic carcinoma cells of the cervix in a
routine vaginal smear. Note the regularity of the size of the nuclei and the
chromatin pattern. (× 750) (D) Histologic section of cervical adenocystic
carcinoma. (× 150)

to overlook. Their size is small, 10 to 15 μ in diameter. The amount of cytoplasm is generally scanty. The single nuclei are round or oval, uniform in size. The chromatin is finely granular. The nucleoli are seldom seen. These cells are generally confused with benign endocervical cells, except for the regularity of their size.

Clear-Cell Adenocarcinoma (Mesonephroma) of the Vagina

This rare tumor is being reported with increasing frequency in adolescent women with vaginal bleeding whose mothers received synthetic estrogen during gestation. Located in the vagina, the tumor readily sheds numerous diagnostic cells. They are found singly, in clusters, or in papillary formation. They have generally scanty, semitransparent cytoplasm and large, single, hyperchromatic, irregular nuclei. The tumor cells vary in size (10–20 μ) and shape (oval to irregular). Nucleoli are occasionally seen but they do not predominate. The background of the smear shows protein deposits, debris, blood, and a large number of neutrophils.

In *vaginal adenosis,* clusters of benign columnar cells diagnostic of the lesion are indistinguishable from normal reactive endocervical cells; thus, the smear should originate only from the vagina.

Gartner Duct Adenocarcinoma

This rare tumor arises from the lower portion of the vagina. The exfoliated large malignant cells, usually found in sheets or clusters, are almost impossible to differentiate from endocervical adenocarcinoma cells.

Bartholin's Gland Adenocarcinoma (Fig. 86A, B)

This is a rare primary adenocarcinoma of the vulva with an incidence of approximately 1 in 1000 carcinomas of the female genital tract.

When vaginal ulceration exists, a large number of well-preserved abnormal cells with sharp, more or less clearly defined cytoplasmic boundaries may exfoliate in the vaginal pool.

The cells occur singly or in loose clusters with marked cellular molding.

The cell size varies, measuring 30 to 50 μ in diameter.

The cytoplasm is relatively abundant and acidophilic, and occasionally it is vacuolated (15% of the cells).

The nuclei are eccentric, enlarged, hyperchromatic, and irregular in shape.

Binucleation and normal or abnormal mitotic activity are common.

The chromatin pattern is irregular and coarsely granular in about 40% of the malignant cells, finely granular in 40%, and translucent in 20%.

Small, round, single, or multiple nucleoli are present in the majority of the cells, but they are not enlarged nor are they as prominent as those seen in other genital adenocarcinomas.

Table 6. Differentiation Between Endometrial and Endocervical Adenocarcinoma

Characteristic	Endocervical		Endometrial	
Number of cells	Abundant		Less abundant	
Degeneration	Mild		Marked	
Size	$190 \pm 60 \; \mu^2$		$160 \pm 40 \; \mu^2$	
Stain	Acidophilic	$80 \pm 15\%$	Acidophilic	$20 \pm 5\%$
	Basophilic	$10 \pm 10\%$	Basophilic	$60 \pm 5\%$
	Indeterminate	$10 \pm 10\%$	Indeterminate	$20 \pm 5\%$
Cytoplasmic texture	Granular	60%	Granular	40%
	Vacuolated	40%	Vacuolated	60%
Ingestion of leukocytes	Rare		Abundant	
Nuclear size	$80 \pm 20 \; \mu^2$		$60 \pm 10 \; \mu^2$	
Chromatin	Finely granular		Moderately coarse	
Hyperchromatism	Less common		Common	
Nucleoli	Multiple	70%	Multiple	20%
	Macronucleus	40%	Macronucleus	20%
Background	Relatively clean		Very dirty	

The maturation index is normal for the age of the patient, and the background of the smear usually shows considerable amounts of cellular debris and fresh and old blood.

The possibility of a Bartholin's gland adenocarcinoma should be considered in the presence of numerous well-preserved, large malignant columnar cells in a vaginal smear with no demonstrable endocervical or endometrial lesion.

Extrauterine Cancer

Various extrauterine malignant tumors may involve the uterus and exfoliate cells directly into the vagina or by passage through the tubal and endometrial lumens. The metastatic malignant cells usually are glandular in type and resemble the exfoliated cells of primary uterine adenocarcinoma. The most common primary site is the ovary (60%), followed by the gastrointestinal tract (20%).

The possibility of metastasis should be considered when:

1. No tumor diathesis is present (less than 20% of the smears have fibrin and cellular debris in the background);
2. Tumor cells are in groups or spherical clusters;
3. Tumor cells are arranged in papillary structure;
4. Psammoma bodies are present in a case of primary ovarian cancer;
5. Ascites is known to exist.

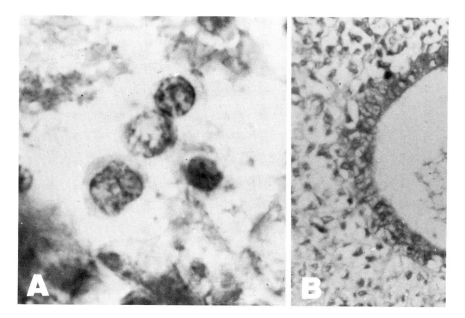

Figure 87
Cytology (A) and histology (B) of endometrial stromal cell sarcoma.

Differential Diagnosis

The differentiation between endometrial and endocervical adenocarcinoma is shown in Table 6. The adenocarcinomatous cells should be differentiated from other benign or malignant cells commonly found in the vaginal pool smear.

BENIGN REACTIVE ENDOCERVICAL CELLS. The adenocarcinomatous cell cytoplasm is more irregular in amount and shape, its nucleus and nucleoli are larger and more irregular, and its chromatin granularity is increased.

SQUAMOUS IN SITU CARCINOMA. In adenocarcinomatous cells, the cytoplasm is vacuolated and abundant, rather than thick and scanty. The nucleus is eccentric rather than central, and the nucleolus is prominent with coarser chromatin.

HISTIOCYTES. The presence of cellular molding and the absence of bean-shaped nuclei and of phagocytic vacuoles in adenocarcinomatous cells help differentiate them from histiocytes.

Genital Sarcoma

ENDOMETRIAL STROMAL SARCOMA (FIG. 87)
This is a rare polypoid tumor, characterized by an abnormal increase in uniform, small, immature stromal cells in a scanty connective tissue, which

183

invades the myometrium. Except for large ulcerated tumors, they do not exfoliate diagnostic cells, especially if the tumor is of the endolymphatic stromal meiosis type.

The exfoliated cells, when present, are numerous, usually single, and show moderate variation in size and shape (round or oval).

The pale eosinophilic cytoplasm of these cells is very scanty and often difficult to distinguish, its borders being delicate and lacy and often indistinct.

The nuclei are enlarged, about twice the size of the normal stromal cells. They are usually round with some irregularity of shape.

The chromatin is coarsely clumped in 90% of the cells and irregularly scattered throughout the nuclei. Moderate to mild hyperchromatism usually exists.

The nuclear membrane is well defined and smooth but shows irregularity of thickness by the adhesion of chromatin clumps to the inner surface.

The nucleoli are usually absent or cannnot be distinguished from the coarse chromatin clumping.

The background of the smear shows large amounts of debris, leukocytes, fibrin, fresh and old red blood cells, and phagocytes (foreign-body giant cells).

These stromal cells are usually readily recognized as malignant but often are interpreted as originating from a poorly differentiated cervical carcinoma.

MIXED MESODERMAL SARCOMA–MIXED MÜLLERIAN
TUMOR, CARCINOSARCOMA (FIGS. 88, 89A, B)
This neoplasm is often seen as a polypoid structure protruding through the cervical os in a postmenopausal woman. The histology of this tumor shows a preponderance of the epithelial components, which resemble the endometrial or endocervical adenocarcinoma. Occasional foci are seen of fibrosarcoma (from fibrous tissue), chondrosarcoma (from cartilaginous tissue), liposarcoma (from fatty tissue), osteosarcoma (from the bone), rhabdomyosarcoma (from the striated muscle), and leiomyosarcoma (from smooth muscles).

The exfoliated cells vary in size and shape. They can shed singly (40%) or in tight clusters (40%) or in small sheets (20%).

Very few changes are specific for the cytologic diagnosis of this lesion. Although the detection rate is high (90%), a majority of the anaplastic cells seen in the vaginal smear, being from the epithelial adenocarcinomatous components, are often mistaken by the cytopathologist as originating from an endometrial or an endocervical adenocarcinoma.

Changes Helpful in Diagnosis
The exfoliated cells are generally well preserved and are very anaplastic. Their size (15–90 μ) and shape are extremely variable, with pseudopod-like projections in about 30% of the cells.

The nuclei are hypertrophic (30% of the cells) and show a bland chromatin pattern that should be differentiated from degeneration by the absence of the other cellular degenerative changes.

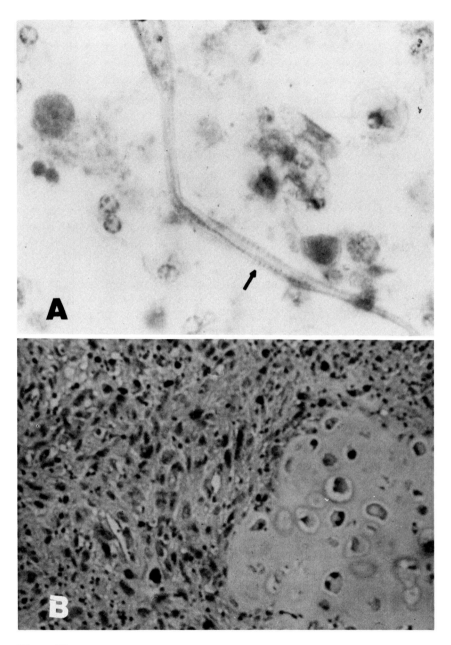

Figure 88
Cytology (A) and histology (B) of mixed mesodermal sarcoma. Note the cross striations (*arrow*) in the exfoliated cells and the cartilaginous tissue in the section.

185

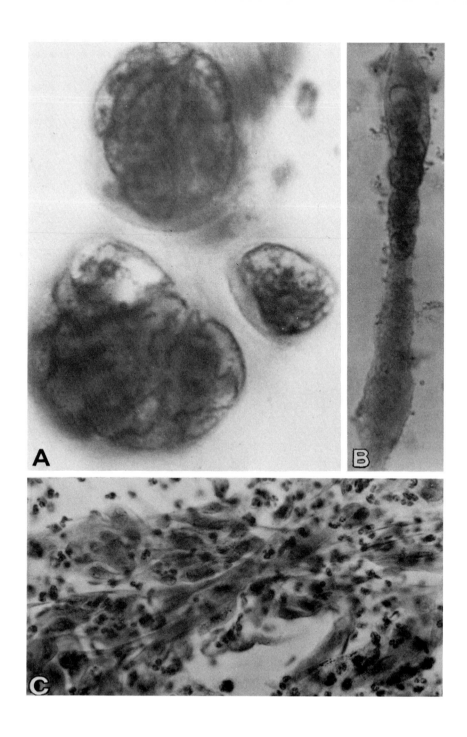

Occasionally, mixed diagnostic elements are found, especially when the neoplasm is ulcerated and directly scraped. The muscular origin of some neoplastic cells is indicated by the presence of cytoplasmic cross striations in large ovoid or elongated cells with abundant eosinophilic cytoplasm and large eccentric nuclei with prominent red nucleoli.

Cartilaginous origin is indicated by the presence of large round cells in groups of 2 or 3 with well-defined, semitransparent cytoplasm and round, prominent nuclei with regular, fine chromatin clumping.

The background of the smear is usually clouded with increased neutrophils, fibrin, and fresh and old red blood cells.

LEIOMYOSARCOMA

This neoplasm arises from the smooth muscular elements. Generally, it does not exfoliate, except when it is ulcerated or necrotized.

Single spindle cells, or whorls of them, with multiple cytoplasmic projecting structures showing variable amounts of degenerative changes, can be seen in the routine vaginal smear. This smear is usually bloody and has a dirty background with an increased amount of inflammatory cells (leukocytes).

The malignant cells have variable amounts of elongated eosinophilic cytoplasm and large elongated nuclei with sharp nuclear borders and bland or irregularly clumped chromatin pattern.

In the multinucleated giant cells (15%), the nuclei are clumped against one another with frequent atypical mitoses. Nucleoli are seldom seen.

Numerous histiocytes and foreign-body giant cells containing cellular debris are also present.

Anaplasia of the diagnostic cells with bland irregular nuclei and marked variation of size and shape should make one consider the possibility of the presence of this sarcomatous lesion.

MELANOSARCOMA

These relatively rare tumors can arise as primary neoplasms from the cervix, the vaginal wall, or the vulva, or they may be metastatic from an adjoining or distant site. Most of the specimens are obtained by direct scraping of a fungating or ulcerated lesion.

Numerous scattered, well-preserved, occasionally spindle-shaped malignant cells, single or in loose clusters, can be found in the vaginal smear. This is probably due to the poor cohesion of the melanoma cells. The cells are often multinucleated and gigantic.

Their amorphous cytoplasm, which often stains irregularly, is frequently

---

Figure 89

(A) Mixed mesodermal sarcoma cells. Note their bizarre shape and large nuclei with a relatively bland chromatin pattern. (× 2100) (B) Elongated mixed mesodermal sarcoma cells showing the diagnostic multinucleation and traces of cross striations in their granular cytoplasm. (× 1300) (C) Botryoid sarcoma cells in an 18-year-old woman with a large ulcerated lesion. (× 450)

abundant and contains, in 60% of the cells, variable amounts of brown to black irregular melanin granules that can be differentiated from similar-looking intracytoplasmic blood pigments found in some phagocytes by an iron stain of the smear.

Their nuclei are large, the number varying from 1 to 3. Occasionally, multiple small vacuoles are seen.

The nuclear border is sharp, the chromatin irregularly clumped, and the cells usually have prominent nucleoli (80%).

The nuclear contour is smooth in 50% of the cases and irregular with indentation in the remainder.

LYMPHOMAS

The genital organs are secondarily involved in 20% of all cases of a generalized lymphomatous neoplasm.

A large number of immature lymphocytes in the smear is sufficient to make one suspicious that a lesion may be present. These cells should be differentiated from the pool of benign lymphocytes seen in the routine vaginal smear during the scraping of a cervix containing lymphoid follicles resulting from a chronic inflammation.

The cytologic differentiation is not always easy. The smears from both malignant and nonmalignant lesions may show large pools of mature and immature lymphocytes.

Prominent nucleoli are more commonly found with malignancy, and the chromatin distribution is also more irregular and coarser than in the benign lymphocytes.

Similarly, an occasional single large lymphoblast should be differentiated from the small round cell of an epidermoid in situ carcinoma. The smaller size of the cell, the scantiness of its cytoplasm, and the prominence of the nucleoli help in the diagnosis of the lymphocytic cell.

Secondary involvement of the female genital tract with *reticulum cell sarcoma* results in the exfoliation of large numbers of diagnostic cells. The cytologic features are similar to the ones described for the reticulum cell sarcoma of other sites. Most of the cells are single, round to oval, vary in size, and have scanty, semitransparent cytoplasm. The usually single, large, irregularly shaped nuclei have irregular, coarse chromatin clumping and prominent nucleoli that often vary in size, shape, and number.

SARCOMA BOTRYOIDES (FIG. 89C)

Seen more commonly in infancy and preadolescence, this lesion is compatible with prolonged survival, although it may give lymph node metastasis. It may originate in the vagina, cervix, or bladder. Grossly it resembles a cluster of grapes (myxoid tissue), and it often protrudes into the vaginal orifice or the vulva. Most of the smears taken are direct scraping of the neoplasm, and the cells show only a moderate amount of degenerative change.

The exfoliated cells are found singly or, more commonly, in sheets or tight clusters. There is a marked variation of size and shape (most commonly, spindle shaped).

The cytoplasm is abundant and eosinophilic with occasional cross striations (8%).

The nuclei, often multiple (60%), are clustered together, overlapping and molding against one another.

The nuclear membrane is usually smooth, and the chromatin clumping is variable—from indistinct to coarsely granular.

The nucleoli can be prominent (35%) and vary in size, shape, and number.

The presence of a large number of anaplastic cells in the smear of a young woman should make one suspect a sarcoma botryoides.

GRANULAR CELL MYOBLASTOMA

This is a rare tumor that can occur in the vaginal wall, the cervix, or the myometrium. Its histogenesis is obscure. Probably the tumor originates from fibroblastic elements. Polypoid in shape, it readily exfoliates cells when ulcerated or subjected to friction.

The diagnostic exfoliated cells are usually large (150–250 μ) and elongated, with sharp, well-defined cytoplasmic borders.

Their cytoplasm is thick, abundant, and deeply eosinophilic, and it contains numerous fine, dark, eosinophilic granules.

In contrast, the eccentrically placed, round, vesicular nucleus is comparatively small, being about the size of the nucleus of the normal intermediate squamous cell (7–10 μ).

The nuclear membrane is sharp, regular, and well defined.

The chromatin is granular, uniformly distributed, and benign in appearance.

The round nucleolus is not prominent.

Ovarian Cytology

Type of Specimen

VAGINAL POOL SMEAR AND ENDOMETRIAL ASPIRATION. When the diagnostic exfoliated ovarian cancer cells in vaginal smears are degenerate, it is an indication that they have probably traveled slowly through the patent fallopian tube and endometrial cavity to reach the vagina. When the cells are well preserved, it is an indication that probably they exfoliated from an invasion of the uterine wall by the ovarian neoplasm.

CUL-DE-SAC ASPIRATION. Aspiration with a fine needle through the vaginal vault is one of the best methods for the cytologic detection of an ovarian lesion, provided the tumor has ulcerated through its capsule. Besides the usual inflammatory cells and fresh and old red blood cells, an increase in reactive mesothelial cells is always found mixed with clusters or sheets of neoplastic cells.

ASCITIC FLUID. The presence of tumor cells in the effusion indicates the extension of the neoplasm to the peritoneal surface (see Chap. 12).

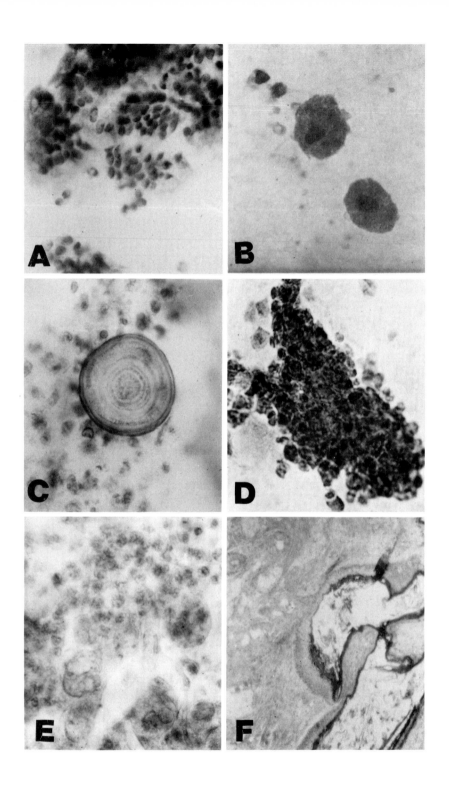

OVARIAN CYST ASPIRATION. A fine needle should be employed to prevent the cystic fluid and the live cells that it contains from spilling into the abdominal cavity. This method is best used when biopsy is contraindicated or when the ovary needs to be preserved despite the presence of multiple cystic structures.

CONTACT SMEAR. This can take the place of a frozen section when the facilities for frozen section are not available. The routine scraping of the surface of the ovaries in patients undergoing laparotomy for whatever reason has been shown to be very useful.

PERITONEAL LAVAGE. In cases where there is no natural accumulation of fluid in the abdominal cavity, 50 to 100 cc of sterile saline solution can be injected intraperitoneally and reaspirated for the detection of a primary or metastatic neoplasm.

WOUND WASHING. After the removal of an ovarian neoplasm, the presence or absence of tumor cells in the washing of the operative site with saline solution will determine, for example, whether a cystic tumor has ruptured during its removal and whether additional chemotherapy or radiotherapy is needed.

Generalities
1. The normal and neoplastic tissues of the ovary are extremely varied, and the exfoliated cells vary accordingly in their morphology;
2. The benign ovarian tumor yields fewer cells than the malignant one;
3. Most of the cystic fluid contains variable amounts of macrophages;
4. Most of the cells in benign cystic fluid are degenerate, making their differentiation difficult;
5. The cancer cells are often extremely atypical, making their diagnosis easy;
6. Some of the ovarian tumors may have a mixture of benign and malignant components;
7. The cytologic specimen obtained from one cyst is not always representative (some cysts are multilobular).

Aspiration of Functioning Ovarian Cysts

FOLLICLE CYSTS (FIG. 90A). The aspirated smears show variable amounts of well-preserved granulosa cells, which are shed singly (40%) or in sheets (60%) and show moderate size variation (10–15 μ).

The granulosa cells are round or cuboidal with cyanophilic granular

Figure 90
Ovarian cyst aspiration. (A) Granulosa cells in a follicle cyst. ($\times$ 450) (B) Luteinic cyst. ($\times$ 450) (C) Psammoma body in a serous cyst. ($\times$ 450) (D) Granulosa cells in a granulosa tumor. ($\times$ 450) (E) Dermoid cyst aspirate. ($\times$ 450) (F) Section of the dermoid cyst. ($\times$ 150)

cytoplasm and central, round nuclei with coarse chromatin clumping and prominent nucleoli.

Occasionally, mitosis is present (in about 2% of the cells).

Some of the exfoliated benign granulosa cells are degenerate. They have a shrunken, almost pyknotic, centrally located, dark nucleus with a round, transparent remnant of amorphous cytoplasm. When bleeding has occurred, hemosiderin-containing phagocytes are occasionally present.

The background of the smear is usually dirty with an abundant basophilic granular deposit.

In some cases, when the granulosa cells are numerous, it is difficult to differentiate by cytology the benign cysts from a malignant granulosa cell tumor. The vaginal maturation index determination helps in differentiation (shift to the right in the case of a malignant tumor).

LUTEIN CYSTS (FIG. 90B). These are cystic transformations of the corpus luteum. The size of the cyst varies. It can be large enough to produce symptoms in the patient by compressing the surrounding structures.

The aspirated smears show an increased number of lutein cells that are single (80%) or in sheets (20%). The lutein cells are characterized by their basophilic cytoplasm, which is large in size (15–20 μ), oval, and well defined. They may contain multiple granular basophilic coarse inclusions.

The cells are often degenerate, showing large numbers of vacuoles in their cytoplasm with variable amounts of ingested leukocytes.

The nuclei are central (40%) or eccentric (60%), round in shape with uniform chromatin clumping. The nucleoli are minute and barely visible.

Very little cellular debris and only moderate amounts of granular basophilic deposits are seen in the background, which differentiates the aspiration of lutein cysts from the follicle cyst aspiration.

ENDOMETRIOSIS (CHOCOLATE CYST). Although these cysts occur most frequently in the ovary, they can be found in almost any site. The aspiration shows large amounts of fresh and old red blood cells, giving a dark brown (chocolate color) appearance to the accumulated fluid.

A variable number of macrophages containing hemosiderin is found in 90% of the aspirated fluids.

The specimen is scanty in intact epithelial cells, but it shows all types of cellular debris.

Occasionally, groups or sheets of small endometrial cells can be found, with well-preserved, large, regular, round nuclei and scanty cytoplasm, sometimes in honeycomb formation. It is often difficult to differentiate these endometrial cells from the serous ones.

Fresh and old red blood cells are always present in abundance, and they obscure the background of the smear.

Aspiration of Nonfunctioning Ovarian Cysts

SIMPLE SEROUS CYSTS. The smears show only few epithelial cells. They are usually found singly. They have a flattened, basophilic, dense but occasionally finely vacuolated cytoplasm, and an eccentric small, round nucleus. The background of the smear is clean and shows some fine granular deposits with almost no red blood cells. Occasional mature lymphocytes and leukocytes may be found scattered in the smear.

PAPILLARY SEROUS CYSTS (SEROUS CYSTADENOMA) (FIG. 90C). The smears show variable amounts of papillary projections made of small cells with scanty cytoplasm and regular round or oval dark nuclei with fine regular chromatin pattern showing intercellular molding.

Occasionally, these cells are ciliated. Another type of nonciliated cells with more abundant cytoplasm and elongated nuclei may also be found.

The background of the smears shows a moderate amount of protein deposits with an increased number of inflammatory cells (leukocytes and lymphocytes).

Occasional blood pigment and debris-containing phagocytes are found.

Psammoma bodies (calcified granules) surrounded by a layer of epithelial cells may be present in about 10% of the specimens.

MUCINOUS CYSTS. The smears show only a few tall, columnar epithelial cells (40–80 μ), singly or in loose clusters resembling hypersecretory endocervical cells.

They have an abundant cytoplasm, overdistended by single or multiple secretory vacuoles (goblet cells) containing leukocytes and material that is positive for periodic acid–Schiff stain.

The nuclei are eccentric, often crescent shaped, with dark, coarse chromatin and occasional small, round, red nucleoli (25% of the cells).

The background of the smear is usually very dirty with large amounts of cyanophilic granular deposits in a pool of mucus.

The presence of mucus in an ovarian aspiration cyst may indicate:

1. Ovarian mucus-secreting adenocarcinoma;
2. Metastatic mucus-secreting adenocarcinoma;
3. Mucinous cystadenoma;
4. Pseudomyxoma peritonei;
5. Contamination of the specimen by endocervical mucus.

MYXOMA PERITONEI. The rupture of a benign mucinous cyst often releases actively growing cells, which may implant on the peritoneal serosa and continue to secrete mucus in abundance. Large amounts of gelatinous material may accumulate and produce complications by obstructing the bowels.

DERMOID CYSTS (BENIGN CYSTIC TERATOMA) (FIG. 90E, F). The aspirated smears show large numbers of squamous cells, some anucleated, some superficial or intermediate in type.

There is also an abundance of histiocytes, inflammatory cells, and various tissue particles (hair, glands) and debris.

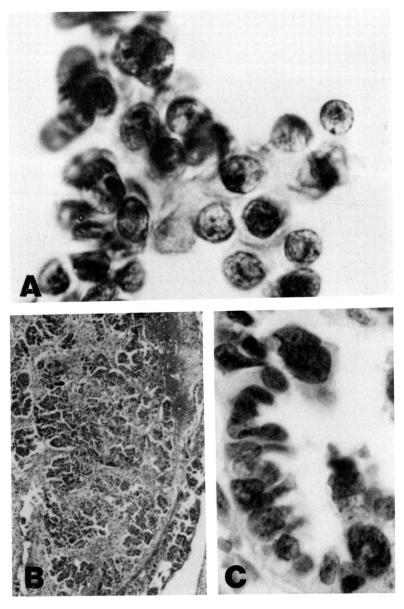

Figure 91
Ovarian papillary adenocarcinoma. (A) Aspirated cystic fluid. (× 750) (B, C)
Histology of the tumor. (B: × 150; C: × 750)

The cells are usually poorly stained because of the sebum (oily material) covering their surface.

The background of the smear is very dirty, containing such material as basophilic granular deposits, oil droplets, or debris.

Aspiration of Solid or Cystic Tumor—
Serous or Mucoid Adenocarcinoma (Fig. 91)

The aspiration of cystic tumors may produce a large amount of blood, cellular debris, and neutrophils mixed with the tumor cells.

The malignant cells show characteristic malignant features that are easy to recognize. These cells are found singly or in papillary projections.

When the cytoplasm is very abundant and hypervacuolated, this is an indication that the tumor is mucoid in type.

The pseudomucinous adenocarcinoma almost never exfoliates cells into the vagina. Serous adenocarcinoma and papillary carcinoma cells may occasionally be found in the routine vaginal smear.

FEMINIZING TUMOR (GRANULOSA CELL OR THECOMA) (FIG. 92A). The vaginal smears show a hyperestrogenic effect with a marked shift to the right of the maturation index (0/0/100). Occasional clusters of small, uniform malignant granulosa tumor cells may be seen.

A large amount of fresh and old red blood cells is usually found as the result of frequent secondary menometrorrhagia.

The needle aspiration of a cystic tumor or contact smear of a solid one will produce numerous granulosa cells resembling the normal cells found in the graafian follicles and characterized by their regular, hyperplastic, round, central nuclei, prominent nucleoli, and adequate cyanophilic cytoplasm.

VIRILIZING TUMOR—ARRHENOBLASTOMA, ADRENAL REST TUMOR, HILAR CELL TUMOR, AND GYNANDROBLASTOMA. The vaginal smears show an androgenic effect (shift to the middle or to the left of the maturation index) with a preponderance of small intermediate cells with irregular, round, cyanophilic cytoplasm and a vesicular nucleus. Except for this abnormal maturation index, exfoliative cytology is not of great help in diagnosing these tumors.

The atypical, constant maturation index in these tumors is about 30/65/5. The index reverts rapidly to normal when the tumor is removed.

POORLY DIFFERENTIATED ADENOCARCINOMA. The undifferentiated cancer cells often cannot be classified. Their marked polymorphism and scanty cytoplasm are common to all tumors. They can even be hard to differentiate from embryonal sarcoma cells.

Vaginal Pool and Ovarian Carcinoma

In about 20% of all the malignant ovarian neoplasms, abnormal neoplastic cells have been found in the vaginal pool smear, and 35% of the patients had an abnormal maturation index. No abnormality, either in the matura-

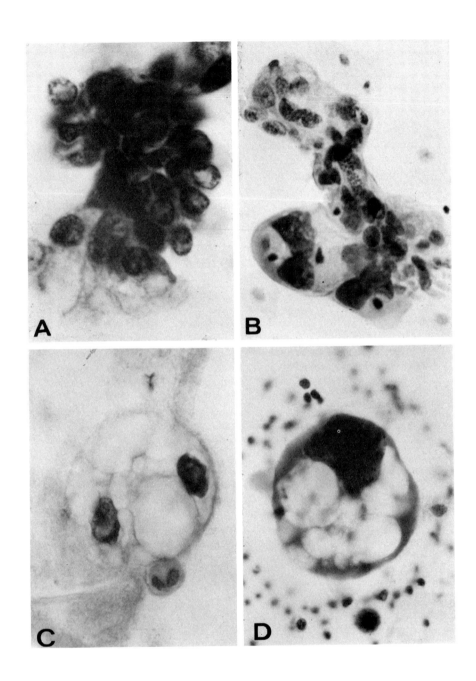

tion index or the presence of malignant cells (false-negative), was found in 60% of the cases.

The vaginal smear with malignant cells usually has a clean background with very few leukocytes, except when the wall of the uterus has been invaded by the neoplasm or when secondary infection exists.

The few malignant cells are degenerate and usually shed in tight clusters (75%).

Their cytoplasm is thin-walled, hypervacuolated if mucus-secreting, or scanty and basophilic.

Their nuclei are large, showing the classic malignant features, with abnormal chromatin clumping, prominent nucleoli, and irregular size and shape.

Occasional psammoma bodies are found in 10% of the malignant serous ovarian neoplasms. These are semicalcified with a central core surrounded by amorphous structures, like an onion ring. They are usually surrounded by attached intact malignant cells. Although primarily diagnostic of an ovarian adenocarcinoma, the psammoma bodies may also be found in rare cases of endometrial adenocarcinoma.

References and Supplementary Reading

Allen, H. H., and Fraleigh, D. M.   Carcinoma of endometrium. *Am. J. Obstet. Gynecol.* 85:302–306, 1963.

Augstrom, T., Kzellgren, O., and Bergman, F.   The cytologic diagnosis of ovarian tumors by means of aspiration biopsy. *Acta Cytol.* 16:326, 1972.

Bamford, S. B., et al.   Vaginal cytology in polycystic ovarian disease. *Acta Cytol.* 9:322, 1965.

Boutselis, J. G., et al.   Carcinoma of the uterine corpus. *Am. J. Obstet. Gynecol.* 85:994–1001, 1963.

Brux, J. A. de   The value of exfoliative cytology in the diagnosis of hormone-producing tumors. *Acta Cytol.* 4:106–108, 1960.

Brux, J. A. de, and Bret, J.   Cells originating in ulcerated submucous myomas of the uterus: The so-called "vermiform bodies." *Am. J. Clin. Pathol.* 32:442, 1959.

Buschmann, C., Hergeurader, M., and Porter, D.   Keratin bodies, a clue in the cytological detection of endometrial adenoacanthoma. *Acta Cytol.* 18:297, 1974.

Figure 92

Ovarian neoplasm. (A) Malignant granulosa tumor cells obtained by needle aspiration. Note the uniformity of the nuclei and relative scantiness of their cytoplasm. ($\times$ 750) (B) Papillary serous adenocarcinoma cells seen in the vaginal smear of a woman with an unsuspected ovarian carcinoma. Note the partial smooth border of the papillary structures and marked polymorphism of the component cells. ($\times$ 750) (C) Mucinous adenocarcinoma cells seen in the vaginal smear of a woman with an extensive lesion.   Note the large mucus-containing vacuoles overdistending their cytoplasm. ($\times$ 750) (D) Malignant cell seen in the cul-de-sac aspirate from a woman with an anaplastic ovarian adenocarcinoma. ($\times$ 750)

Ceelen, G. H., and Sakurai, M.   Vaginal cytology in leukemia. *Acta Cytol.* 6: 370, 1962.

Coleman, S. A., Rube, I. F., and Erickson, C. C.   Cytologic detection of adenocarcinoma of the uterus in a mass screening project. *Am. J. Obstet. Gynecol.* 92:472, 1965.

Dennis, E. J., Hester, L. H., Jr., and Willson, L. A.   Primary carcinoma of Bartholin's glands. *Obstet. Gynecol.* 6:291, 1955.

Differding, J. T.   Psammoma bodies in a vaginal smear. *Acta Cytol.* 11:199, 1967.

Goldwin, J.   Cytologic diagnosis of aspiration biopsy of solid or cystic tumors. *Acta Cytol.* 8:206, 1964.

Graham, J. B., Graham, R. M., and Schuler, V. F.   Preclinical detection of ovarian cancer. *Cancer* 17:1414, 1964.

Graham, R. M., Schueller, E. F., and Graham, J. B.   Detection of ovarian cancer at an early stage. *Obstet. Gynecol.* 26:151, 1965.

Grillo, D., Stienmier, R. H., and Lowell, D. M.   Early diagnosis of ovarian carcinoma by culdocentesis. *Obstet. Gynecol.* 28:346, 1966.

Hecht, E. L., and Oppenheim, A.   The cytology of endometrial cancer. *Surg. Gynecol. Obstet.* 122:1025, 1966.

Helwig, F. C.   Changing ratio of cervical to corpus carcinoma. *Am. J. Obstet. Gynecol.* 81:277–280, 1961.

Herbst, A. L., and Scully, R. E.   Adenocarcinoma of the vagina in adolescence. *Cancer* 25:745, 1970.

Hertig, A. T., and Gore, H.   Precancerous lesions of endometrium. *Z. Krebsforsch.* 65:201, 1963.

Holmquist, N. D.   The exfoliative cytology of mixed mesodermal tumors of the uterus. *Acta Cytol.* 6:373, 1962.

Howdon, W. M., et al.   Cyto- and histo-pathologic correlation in mixed mesenchymal tumors of the uterus. *Am. J. Obstet. Gynecol.* 89:670, 1964.

Javert, C. T., and Renning, E. L.   Endometrial cancer. *Cancer* 16:1057–1064, 1963.

Kanbour, A., Klionsky, B., and Cooper, R.   Cytohistologic diagnosis of uterine jet wash preparations. *Acta Cytol.* 18:51, 1974.

Katagama, I., Hajiau, C., and Eujz, J.   Cytologic diagnosis of reticulum cell sarcoma of the uterine cervix. *Acta Cytol.* 17:498, 1973.

Krupp, P. J., Jr., et al.   Malignant mixed Müllerian neoplasms (mixed mesodermal tumors). *Am. J. Obstet. Gynecol.* 81:959, 1961.

Laguna, J. C.   The diagnostic accuracy of cervical smears for detection of endometrial carcinoma. *Acta Cytol.* 2:585–586, 1958.

Lauchau, S. C., and Penner, D. W.   Simultaneous adenocarcinoma in situ epidermoid carcinoma in situ. *Cancer* 20:2250, 1967.

Liu, W.   Hypoestrogenism and endometrial carcinoma. *Acta Cytol.* 14:583, 1970.

Liu, W., et al.   Normal exfoliation of endometrial cells in premenopausal women. *Acta Cytol.* 7:211–214, 1963.

Marcus, C. C.   Cytology of the pelvic peritoneal cavity in benign and malignant disease. *Obstet. Gynecol.* 20:701, 1962.

Marcus, S. L., and Marcus, C. C.   Primary adenocarcinoma of the cervix uteri. *Am. J. Obstet. Gynecol.* 86:384, 1963.

Masukawa, T., et al.   Cytologic diagnosis of minute ovarian endometrial and breast carcinomas. *Acta Cytol.* 17:316, 1973.

Megoway, L., Bunnag, B., and Arias, L.   Peritoneal fluid cytology associated with benign neoplastic ovarian tumors in women. *Am. J. Obstet. Gynecol.* 113:961, 1972.

Montanari, C. D., et al.   Endometrial lavage as an aid in the cytochemical detection of adenocarcinoma. *Cancer* 19:1578, 1966.

Morton, D. G., Moore, J. C., and Chang, N.   Clinical value of peritoneal lavage for cytologic examination. *Am. J. Obstet. Gynecol.* 81:1115, 1961.

Moss, L. D., and Collins, D. N.   Squamous and adenoid cystic basal cell carcinoma of the cervix uteri. *Am. J. Obstet. Gynecol.* 88:86, 1964.

Nasiell, M.   Hodgkin's disease limited to the uterine cervix. *Acta Cytol.* 8:16, 1964.

Ng, A. B., Reagan, J., and Cechner, R.   The precursors of endometrial cancer: A study of their cellular manifestations. *Acta Cytol.* 17:439, 1973.

Ng, A. B., et al.   Mixed adenosquamous carcinoma of the endometrium. *Am. J. Clin. Pathol.* 59:765, 1973.

Ng, A. B., et al.   The cellular manifestation of extrauterine cancer. *Acta Cytol.* 18:108, 1974.

Parker, J. E.   Cytologic findings associated with primary uterine malignancies of mixed cell types (malignant mixed Müllerian tumor). *Acta Cytol.* 8:316, 1964.

Prudan, R., Radujkov, Z., and Rogulja, P.   Contribution au diagnostic cytolgique des tumeurs malignes de l'ovaire. *Arch. Anat. Pathol.* 15:106, 1967.

Reagan, J. W.   *The Cells of Uterine Adenocarcinoma.* Baltimore: Williams & Wilkins, 1965.

Rosati, L., and Jarzynski, D. J.   Clear cell (mesonephric) adenocarcinoma of the vagina. *Acta Cytol.* 17:493, 1973.

Roscoe, R. R.   Endometrial aspiration smear in diagnosis of malignancy of the uterine corpus. *Am. J. Obstet. Gynecol.* 87:921–925, 1963.

Rubin, D. K., and Frost, J. K.   The cytologic detection of ovarian cancer. *Acta Cytol.* 7:191, 1963.

Spjut, H. J., Kaufman, R. H., and Carrig, S. S.   Psammoma bodies in the cervico-vaginal smear. *Acta Cytol.* 8:352–355, 1964.

Taft, P., et al.   Cytology of clear cell adenocarcinoma of the genital tract in young females. *Acta Cytol.* 18:279, 1974.

Valdecasas, R. C., et al.   Malignant melanoma of the vagina. *Acta Cytol.* 18:535, 1974.

Wachtel, E.   The cytology of tumors of the ovary and fallopian tubes. *Clin. Obstet. Gynecol.* 4:1159, 1961.

Wagman, H., Brown, C. L.   Ovarian cytology. *Br. J. Cancer* 15:81, 1971.

Irradiation Effect

9

The study of the effect of therapeutic irradiation on benign and malignant cells is important because:

1. It may have a prognostic significance;
2. Irradiation changes should be recognized and differentiated from the neoplastic changes which they resemble.

Radiosensitivities

The extent of the changes varies with the type of cell irradiated. Important factors are the age of the cell, the amount of cytoplasm, the thickness of the cytoplasm, and the size of the nucleus. The amount of radiation needed varies with the type of irradiation (radium, x-ray, cobalt). The difference in the action of radium or x-ray therapy is quantitative rather than qualitative. There are also physical factors, such as cell temperature or cellular pH, that may influence this sensitivity. Other factors are still unknown. No two germinal cells will react in the same way to the same amount of irradiation. The basal squamous cells begin to show some changes during the first or second day following radiation, while it takes five to six days for the superficial squamous cells to show any changes.

Irradiation may result in the death of the cell, in an inhibition of mitosis, or in chromosomal or genetic changes that can be transmitted to its descendants. This explains why some irradiation cellular changes may persist in certain smears for the rest of the patient's life.

Cellular Changes from Irradiation
(Fig. 93)

Cellular changes resulting from irradiation are known as radiation response (RR). Most appear first in the nuclei, then in the cytoplasm.

CYTOPLASMIC CHANGES
Edema and hypertrophy of the benign and malignant cells result in a threefold to fourfold increase of cell size (Fig. 94A). The nuclei-cytoplasm ratio remains relatively unchanged because both the nucleus and cytoplasm increase proportionately in size.

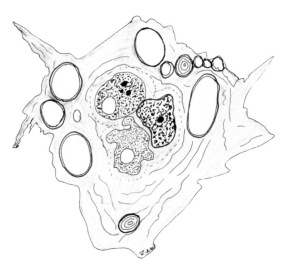

Figure 93
Irradiation changes in a squamous basal cell. Note the cellular hypertrophy, cytoplasmic vacuoles, fibrils, pseudopodlike projections, multinucleation, and nuclear vacuoles.

The cytoplasm loses its normal fine granularity and may become amorphous, homogeneous, dense, and eosinophilic or polychromatic.

The cellular outline becomes indistinct.

Vacuolization can occur, appearing first in the deep basal cells. The vacuoles have a thick border and contain no inclusions. They are multiple and surround the nucleus without distorting it. The amount of vacuolization varies and can be very extensive (Fig. 94C). With electron microscopic studies, variable degrees of destruction of the mitochondria, Golgi apparatus, and endoplasmic reticulum can be seen.

Cytoplasmic fibrils (hairlike linear structures resulting from the folding and shrinkage of the cytoplasmic membrane) are often seen in concentric or irregular arrangement around the nucleus (Fig. 94B).

The cytoplasmic membrane can rupture and fragment, indicating a late stage of irradiation change or death of the cell. This is partially responsible for the large amount of cellular debris seen in the background of a smear following recent irradiation.

These cytoplasmic changes are not always specific to irradiation injury. Some of these changes can be seen, at least partially, in other cellular degenerative processes, such as those that develop after chemotherapy, ultraviolet irradiation, or cautery.

Similar changes can also be seen in the cells of patients with folic acid deficiency. This source should especially be considered whenever irradiationlike changes are seen in the vaginocervical cells of young pregnant women with no history of irradiation.

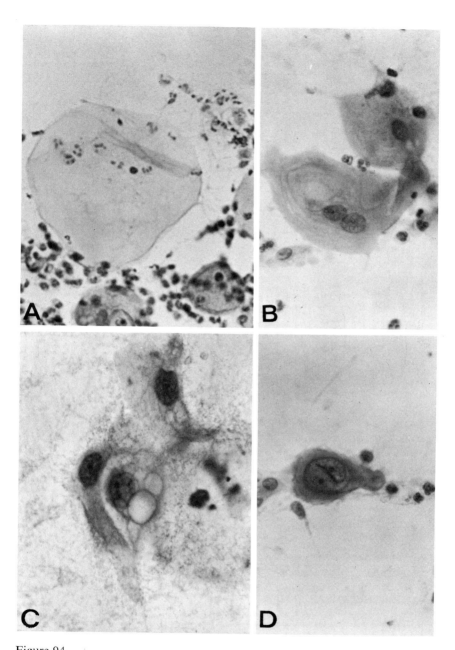

Figure 94

(A) Irradiation changes in a benign superficial cell. Compare its size to the normal superficial cell in the right lower portion. (× 450) (B) Irradiation changes in 2 benign intermediate cells. Note the cytoplasmic fibrils. (× 450) (C, D) Irradiation changes in benign basal cells. Note the cytoplasmic and nuclear vacuolization. (× 450)

CHANGES IN THE NUCLEI

One of the most consistent and common changes is nuclear hypertrophy. The size can increase, especially in the benign cells, by 2 to 10 times without losing its normal density.

The nuclei-cytoplasm ratio remains within the normal limits because of the simultaneous cytoplasmic enlargement.

The nuclei often become surrounded by a well-defined perinuclear halo.

On occasion, the nucleus will show some vacuolization, which is usually diagnostic of irradiation effect and often indicates the death of the cell.

The chromatin remains finely granular and uniformly distributed. The amount of DNA is occasionally decreased.

Multinucleation (2–6 nuclei) may occur in 30% of the cells. The different nuclei vary in size and shape and can mold or overlap.

Nuclear distortion (60%) is frequently the result of wrinkling of the nuclear membrane, which gives it the appearance of a multilobular nucleus.

The nucleoli, when present, are hypertrophic. Their size and number increase, but they remain round and regular in shape.

Often indicating cell death, condensation of chromatin at the periphery, nuclear pyknosis, and fragmentation or karyorrhexis can be seen in the benign or malignant degenerate cells.

OTHER CHANGES (FIG. 95)

Soon after irradiation, the smear usually shows a large number of inflammatory cells (leukocytes) and an increased number of repair epithelial cells with large nuclei, prominent nucleoli, and irregular, abundant cytoplasm with pseudopodlike structures (spider cells).

The histiocytes increase in number. Their size and structure vary according to their activity. Some of the histiocytes contain much ingested cytoplasmic debris. This increase in the number of histiocytes is reported as being a basis for good prognosis for the survival of the irradiated patient without a recurrence of the cancer.

Similarly, there is an increase in the number of foreign-body giant cells, which must be differentiated from the multinucleated syncytial giant irradiated epithelial cells (RR). The nuclei in the foreign-body giant cells are all alike (sister image).

There is an increase in the amount of endocervical mucous secretion and necrotic cellular debris deposited in the background of the smear.

The benign superficial cells often show an increased penetration of their cytoplasm by leukocytes, especially in the early stage of irradiation. This is a sign of the degeneration and death of these cells. The inflammatory cells and histiocytes themselves seldom show any irradiation changes.

Cellular Irradiation Changes and Cancer Treatment Prognosis

The efficacy of radiotherapy depends on:

1. Extension and size of the tumor (stage and location);
2. Resistance or sensitivity of the tumor to irradiation;
3. Amount (dosage) of irradiation given.

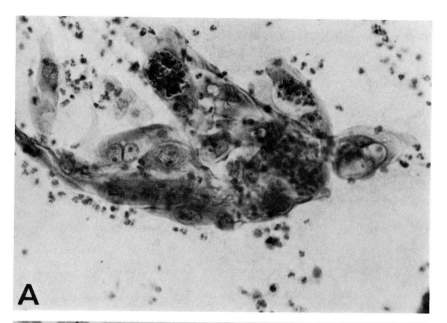

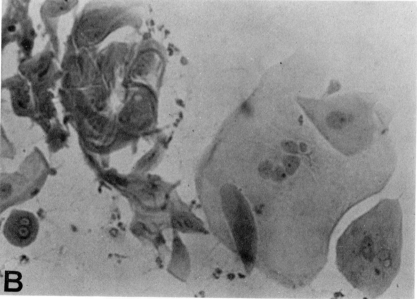

Figure 95
(A) Irradiation changes in a sheet of squamous basal repair cells seen in the smear of a woman on the fourteenth day after the first radium insertion. Note the cytoplasmic vacuoles, the cluster of phagocytized polymorphonuclear leukocytes, and the large nuclei with the prominent nucleoli. (× 450) (B) Vaginal smear of a woman irradiated 20 years previously for an early cervical carcinoma showing persistent marked irradiation effect. (× 450)

Determination of Cell Sensitization Response (SR Cell Count) (Fig. 96A)
SR cells are well-preserved basal cells that stain lavender-blue and often have an abundant dense cytoplasm. They are found singly or in clusters, are round or oval in shape, and have a sharp, well-defined cytoplasmic membrane.

To be diagnostic, the presence of uniformly distributed fine and delicate intracytoplasmic vacuoles should be obvious. Similarly, there should be no cytoplasmic inclusions or leukocytic invasion.

An incidence of 10% or more is considered a good response, and it has been reported that at least 70% of such patients will respond well to radiotherapy. The significance of the presence of these SR cells is still controversial. Several workers, including the author, have found that the determination of the sensitization response is of very little value and that they were unable to duplicate the results obtained by the promoters of this method. The accuracy of this method has been questioned and adverse results reported. Similarly, its application for the prediction of the outcome of oral squamous carcinoma treated with irradiation has not been successful.

Basal Cell Count
This is less specific but easier to master than the SR cell test. The percentage of any type of basal cell in the vaginal smear of the *premenopausal* woman is determined.

The specimen to be examined for this test is a vaginal pool secretion smear, which should not be obtained by a cervical scraping or from an eroded area of the vaginal mucosa. When the percentage of basal cells is above 45, the prognosis following radiotherapy is thought to be good; the prognosis is thought to be bad if the percentage is below 30.

Buccal Irradiation Test (Fig. 96B, C)
Before irradiating the genital tumor, a small area of the buccal mucosa is irradiated superficially. Then a smear is taken from this area for any evidence of irradiation changes, especially multinucleation. Because both buccal and vaginocervical mucosa are of the same nature, the theory is that if one is sensitive to irradiation the other will be also. It is considered a good response if more than 25% of the buccal cells show irradiation changes; if fewer than 20% of the cells show these changes, the response is considered poor.

AFTER RADIOTHERAPY

Radiation Response (RR)
The percentage of cells showing previously described irradiation changes is determined in the smear taken two weeks after the beginning of therapy in cases of radium implantation or at 3000 rads tumor dose in cases of external radiation. Several specimens, taken daily for two or three days, should be studied and the results averaged. It is considered a good re-

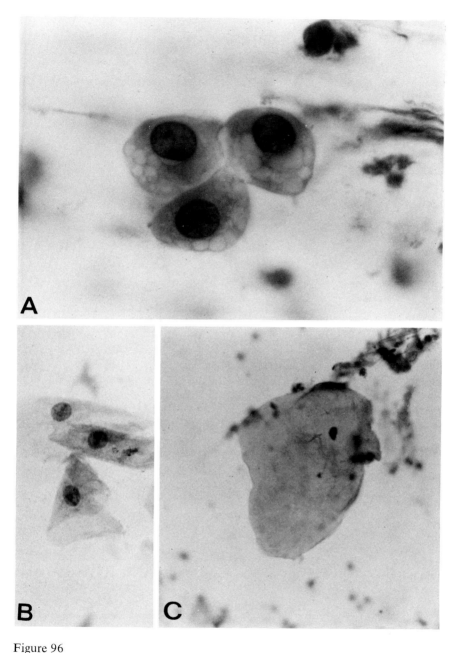

Figure 96
(A) Group of vaginal basal cells showing sensitivity response changes. (× 1300)
(B) Buccal smear at the twelfth day after irradiation with 1500 R, showing few irradiation changes, bad prognosis. (× 450) (C) Buccal smear in another patient at the twelfth day after irradiation with 1500 R, showing marked irradiation changes, good prognosis. (× 450)

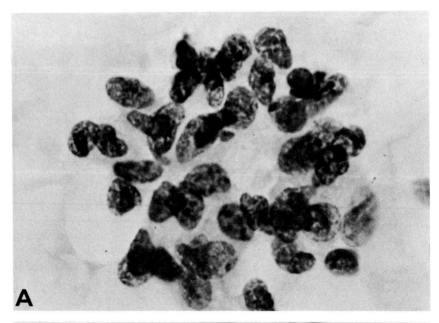

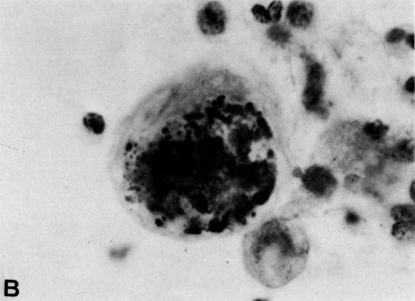

Figure 97
(A) Cluster of nonviable stripped cancer cells on the tenth day after first radium insertion in a woman with a poorly differentiated cervical carcinoma. The absence of cytoplasm differentiated them from a recurrent carcinoma. ($\times$ 1300)
(B) Irradiation changes in a malignant cell found in the vaginal smear of a woman on the seventh day after radium insertion. The nuclear degeneration differentiates these cells from persistent cells of a viable tumor. ($\times$ 1300)

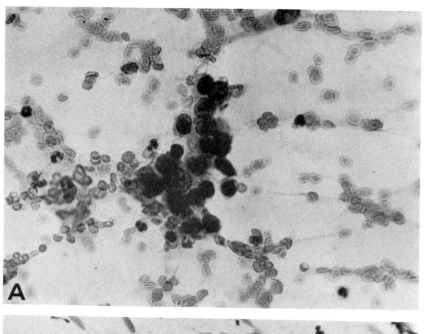

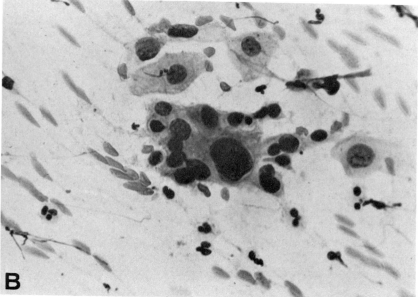

Figure 98

(A) Recurrent carcinoma cells in a woman who had been irradiated four years previously for a well-differentiated squamous carcinoma, stage 2, of the cervix. Note the scantiness of the cytoplasm, the hyperchromatic small nuclei, and the absence of any irradiation change. (× 450) (B) Persistent carcinoma cells in a woman irradiated 8 weeks previously for a well-differentiated squamous carcinoma, stage 2, of the cervix. Note the large size of the cells, abundant keratinized cytoplasm, large pyknotic nuclei, and the absence of irradiation change. (× 450)

209

sponse if about 70% of the cells show RR. It is considered a poor response if less than 60% of the cells show RR.

Recent studies again show a lack of agreement on the value of this method, and a conflict exists on the results obtained. The author has found that this prognostic method was in error in about 30% of the cases.

Superficial Cell Count

This test is valid only when the ovaries are not functioning or have been removed or destroyed, and when no external hormone therapy has been given to the patient. Only the well-preserved superficial cells are counted, those that are single or in loose clusters and have pyknotic nuclei with or without any irradiation changes. The response is considered good if the superficial cell count is below 10% (7% local occurrence). The response is considered poor if the superficial cell count is above 10% (30% local occurrence). The validity of this test has also been questioned.

Cytology of Irradiated Malignant Lesions

During radiotherapy and in the two weeks following it, as a direct result of the irradiation, cancer cells are physiologically shed in large numbers, singly, in clusters, or in large sheets. Their presence in the smear during this time does not indicate a recurrence of the neoplasm nor does their persistence indicate failure of the therapy. On the contrary, their absence indicates a poor prognosis. These malignant cells often show marked irradiation changes similar to the ones described for the benign cells. Cellular enlargement and irregularity of shape become extreme. Also commonly seen are cytoplasmic and nuclear vacuolization of the type shown in irradiation, multinucleation, and changes in the staining of the cells. The diagnosis of persistent carcinoma should not be made at this time because it is difficult to determine whether the exfoliated cells, with these changes, are dead or are still alive and capable of reproduction. The presence of vacuoles in the nucleus or its fragmentations indicates the death of the malignant cell (Fig. 97).

If, four weeks after irradiation, malignant cells showing a minimum of irradiation changes are found, they should be interpreted as indicative of a persisting neoplasm (Fig. 98B). These cells will usually have the same features as the original nonirradiated cells. If, on the other hand, only small, poorly differentiated malignant cells with scanty transparent cytoplasm and enlarged, irregular, hyperchromatic nuclei are found in the smear (cells that often do not resemble the original tumor cells), they are diagnostic of a recurrence of the original cancer (Fig. 98A).

The difference between recurrent and residual carcinoma depends, then, on the maturation of the cells, the amount of cytoplasm, and the staining qualities. It is not unusual (10%) to see irradiation changes in the benign cells in smears from certain irradiated individuals as long as 40 years after the initial treatment. Their significance is unknown. The patient with an irradiated genital tumor should be closely followed by repeated vaginal smears, taken every 3 to 6 months, to detect the possible occurrence of a second neoplasm or atypia in the irradiated site.

Table 7. Differential Characteristics of Various Cytoplasmic Vacuoles

Vacuoles	Degeneration	Viral	Phagocytic	Secretory	Irradiation
Location	Surround nucleus	Surround nucleus	Surround nucleus	One side of nucleus	Scattered in cytoplasm
Number	Many	Multiple	Many	Single or few	Multiple
Content	Healthy polymorpho-nuclear cells	Empty	Debris	Healthy leukocytes	Empty
Margin	Delicate	Thick	Delicate	Variable	Thick
Nuclear-vacuolar molding	No	Yes	No	Yes	No

Morphological Differentiation of Cellular Vacuolization

Five types of cytoplasmic vacuoles can be recognized (Fig. 99). Their differentiation can be helpful in reaching a correct interpretation of the nature of a cell (Table 7).

DEGENERATIVE VACUOLES

These vacuoles are multiple and vary in size and shape. They surround the nucleus without displacing it and contain well-preserved polymorphonuclear cells inside delicate, well-limited vacuolar margins. The cell membrane is frayed or ruptured. No vacuolar molding is present.

VIRAL (BALLOONING) VACUOLES

These vacuoles are multiple, vary in size and shape, and surround the nucleus. The thick, well-defined vacuolar margins seldom contain cellular debris or leukocytes. The nucleus is often molded and compressed by the vacuoles. The cell membrane is well defined and intact.

SECRETORY VACUOLES

This type of large vacuole is usually single, and round or oval in shape. Secretory vacuoles may contain, besides the secretion, well-preserved polymorphonuclear cells. These vacuoles may be seen pushing the nucleus to one side into a crescent shape. The cell has a well-defined thin cytoplasmic membrane and thick vacuolar margins. There is marked nuclear-vacuolar molding.

IRRADIATION VACUOLES

These multiple vacuoles surrounding the nucleus remaining in its normal position vary markedly in size, but are round and regular in shape. The vacuolar margins are thick, dark, and well defined and contain no polymorphonuclear cells. The cytoplasmic membrane, however, is usually frayed. No nuclear-vacuolar molding is present.

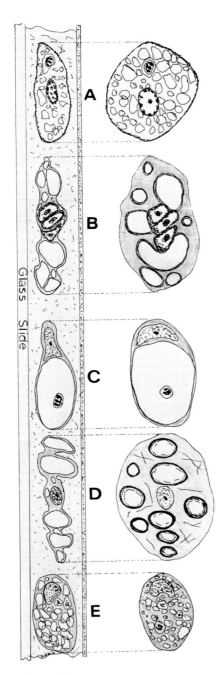

Figure 99
Different types of cytoplasmic vacuoles. (A) Degenerative. (B) Viral. (C)
Secretory. (D) Irradiation. (E) Phagocytic.

PHAGOCYTIC VACUOLES

These vacuoles are multiple, vary in size and shape, and surround the nucleus, which normally is eccentric. They contain various inclusions, cellular debris, pigment, or parts of polymorphonuclear cells. The vacuolar margin, as well as the cytoplasmic membrane, is thin and fragile but well defined. No nuclear-vacuolar molding is present.

References and Supplementary Reading

Agnew, A. M., Fidler, H. K., and Boyes, D. A.   Evaluation of radiation response. *Am. J. Obstet. Gynecol.* 79:698, 1960.

Ceelen, G. H.   Persistent radiation changes in vaginal smears and their meaning for the prognosis of squamous cell carcinoma of the cervix. *Acta Cytol.* 10:350, 1966.

Compos, J.   Persistent tumor cells in the vaginal smears and prognosis of cancer of the radiated cervix. *Acta Cytol.* 14:519, 1970.

Davis, H. J., Jones, H. W., and Dickson, R. J.   Bioassay of host radiosensitivity; an index of radiocurability applied to cervical carcinoma. *Cancer* 13:358–361, 1960.

Donlan, C., and Platt, L. I.   Evaluation of the SR and RR smear in irradiation of carcinoma of the cervix. *J. Am. Med. Wom. Assoc.* 15:576, 1960.

Feiner, L. L., and Garin, A.   Cytological response to radiation in noncervical cancers of the female genital tract. *Cancer* 16:166, 1963.

Fennell, R. H., Jr., and Vazquez, J. J.   Immunocytochemical study of the sensitization response (SR) in vaginal epithelium. *Cancer* 13:555–558, 1960.

Graham, J. B., and Graham, R. M.   The sensitization response in patients with cancer of the uterine cervix. *Cancer* 13:5, 1960.

Graham, R. M.   Cytologic prognosis in cancer of the cervix. *Am. J. Obstet. Gynecol.* 79:4, 1960.

Graham, R. M.   Accuracy of cytologic diagnosis in the treated cancer patient. *Acta Cytol.* 8:3, 1964.

Graham, R. M., and Graham, J. B.   A cellular index of sensitivity to ionizing radiation. The sensitization response. *Cancer* 6:215, 1953.

Graham, R. M., and Graham, J. B.   Factors in prognosis in cancer of the uterine cervix. *Cancer* 13:15, 1960.

Hynes, J. F.   Cancer of cervix uteri. Late local recurrence or late radiation cancer. *Del. Med. J.* 35:1, 1963.

Kangas, S.   Radiation response in patients with cervical cancer. *Acta Pathol. Microbiol. Scand.* [Suppl.] 179:151–156, 1961/1962.

Kaufmann, W., and Khan, B.   Evaluation of cytology as an aid in the prognosis of the treatment of cancer of the cervix uteri. *Am. J. Obstet. Gynecol.* 79:470, 1960.

Kjellgren, O.   The radiation reaction in the vaginal smear and its prognostic significance. Studies on radiologically treated cases of cancer of the uterine cervix. *Acta Radiol.* [Suppl.] 54:168, 1958.

Kjellgren, O.   Enhanced radiation reaction in the vaginal smear by humoral agents and surgical spaying. *Acta Cytol.* 6:455, 1962.

Kjellgren, O.   Radiation dose and radiation reaction in the vaginal smear in carcinoma of the uterine cervix. *Acta Radiol.* 58:435, 1962.

Marcial, V. A., and Boch, A.   Radiation-induced tumor regression in carcinoma of the uterine cervix; prognostic significance. *Am. J. Roentgenol. Radium Ther. Nucl. Med.* 108:114, 1970.

Mengert, W. In discussion: Cellular changes in vaginal and buccal smears after radiation. An index of the radiocurability of carcinoma of the cervix. *Am. J. Obstet. Gynecol.* 78:1099, 1959.

Moore, J. G., et al. The early assessment of irradiation therapy in cervical cancer. *Am. J. Obstet. Gynecol.* 86:677, 1963.

Neumon, M. H., and Jafarey, N. A. Cytologic study of radiation changes in carcinoma of the oral cavity: Prognosis value of various observations. *Acta Cytol.* 14:22, 1970.

Nielsen, A. M. Cytological changes in vaginal smears in radium and roentgen irradiation of uterine carcinoma and their prognostic significance. *Acta Radiol.* 37:479, 1952.

Patten, S. F., et al. Postirradiation dysplasia of uterine cervix and vagina: An analytical study of the cells. *Cancer* 16:173, 1963.

Pelzer, A. Relation entre la radiation response et la presence de metastases ganglionnaires. *Arch. Anat. Pathol.* 15:70, 1967.

Rubio, C. A., et al. Sensitization and radiation response in cases with carcinoma of the uterine cervix. Investigations in 720 cases treated at Radiumhemmet 1954–1961. *Acta Radiol.* 3:241, 1965.

Rubio, C. A., et al. Radiation response and sensitization response studies in 720 cases from Radiumhemmet, Stockholm. *Acta Cytol.* 10:191, 1966.

Shier, C. B. Prognosis in carcinoma of the cervix as determined by vaginal smear. *Am. J. Obstet. Gynecol.* 82:37, 1961.

Sugimori, H., and Taki, I. Radiosensitivity test for cervical cancer. *Acta Cytol.* 16:133, 1972.

Zerne, S. R. M., and Morris, J. M. L. Prognostic significance of cytologic response in radiation of gynecologic cancer. *Obstet. Gynecol.* 19:145, 1962.

Zimmer, T. S. Late irradiation changes; cytological study of cervical and vaginal smears. *Cancer* 12:193–196, 1959.

Cytogenetics

10

The chromosomes, except for parts of one (X) female chromosome, are scattered throughout the nucleus as finely granular chromatin in an extended state during the cellular resting stage and then appear as threads of variable size and shape during cellular mitosis (metaphase). Made of desoxyribonucleic acid (DNA), histones, and other complex acidic residual protein, they are seen as identical pairs, each composed of two threads joined together at a single point (centromere) (Fig. 100). The morphological differentiation of these chromosomes is based on their size, relative length of their arm, location of the centromere, banding characteristic, and the presence of other, often less consistent structures, such as the satellites or secondary constrictions. The great majority of the chromosomes still are hard to identify specifically and are placed in seven different main groups according to their size. The analysis of the structure of the chromosomes is called a karyotype, and the evaluation of its number is the chromosomal count. The total number of chromosomes in the normal cell is 23 pairs or 46 (44 are autosomes, chromosomes without any known link to sex, and two are sex chromatids—XX in females and XY in males). The sex chromosome is directly related to the sexual development and differentiation of the embryo. In the absence of a Y chromosome the differentiation is always female.

Clinical Application of Karyotyping

1. Congenital abnormalities. To determine:
    a. Nature of abnormality (mongolism);
    b. Degree of abnormality (in mosaicism, the greater the percentage of normal cells, the better the prognosis);
    c. Whether the abnormality is familial (chromosomal defect) or sporadic (abnormal number);
2. Sexual abnormalities (intersex, amenorrhea, sterility problems);
3. Chronic granulocytic leukemia (presence of Philadelphia [Ph] chromosome in bone marrow aspirate karyotype);
4. Amniotic fluid (to determine the possibility of erythroblastosis, familial genital abnormalities);
5. Diagnosis of malignancy from single tumor cells in effusion, which often

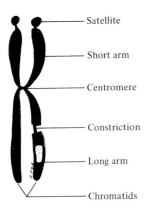

Satellite

Short arm

Centromere

Constriction

Long arm

Chromatids

Figure 100
Structure of a chromosome.

are impossible to differentiate from reactive mesothelial cells (see Chap. 12);

6. Research to determine the nature of various diseases (etiology of spontaneous abortion, cancer) or of the effect of various pollutants on the cells (drugs, tobacco, irradiation, etc.);

7. To detect the carrier of certain disease, as in sex-linked recessive inheritance of hemophilia.

Technique for Determining Karyotype

Any dividing cells can be used for karyotyping.

1. If the cells already have a high mitotic activity (e.g., anaplastic tumor, embryonal tissue, bone marrow, reactive mesothelial cells), they can be placed in a hypotonic solution, fixed, and stained immediately (in this manner, cultural artifacts may be avoided);

2. If the cells have a moderate mitotic activity, they can be incubated in colchicine solution for 4 to 5 hours, then placed in a hypotonic solution, fixed, and stained;

3. If the cells have a mild mitotic activity, they must be cultured from three days (blood) to several weeks (solid tissues).

The most commonly cultured cells are the lymphocytes that are contained in the peripheral blood. The indications for the use of other tissues are summarized in Table 8.

Of the various methods reported, the following procedure is recommended because of its simplicity and reliability.

PROCEDURE FOR DETERMINING THE KARYOTYPE
The use of commercially available kits is advocated if fewer than 120 tests a year are done.

Table 8. Chromosomal Analysis of Various Tissues

Tissue	Cultured Cells	Time Needed in Days	Best Indication
Peripheral blood	Lymphocytes	3–4	Routine
Bone marrow	Erythromyeloid series of cells	0–4	Leukemia, macroglobulinemias
Skin	Fibroblast	14	Mosaicism
Lymph nodes	Lymphocytes	3–4	Postpartum evaluation
Effusion	Various	0–4	Cancer diagnosis
Amniotic fluid	Fetal cells	3–4	Fetal abnormalities
Tumor	Tumor cells	0–4	Research
Testicle	Spermatocytes	0	Meiotic defect

1. Using sterile techniques, collect 5 ml of venous blood in a heparinized tube from the fasting patient (early in the week, if one wishes to avoid weekend work);
2. Allow the blood to sediment for 1 hour at room temperature;
3. Mix the supernatant plasma with the prepared TC–199 medium (1 part of plasma to 4 parts of medium);
4. Incubate the culture at 37° C for 72 hours and check the pH daily to rule out contamination. The medium should remain pink-red (phenol red is already present in the culture medium). If the culture becomes too acid (orange-yellow color of the indicator; the cap of the culture vessel may be opened for 30 minutes or a few drops of sterile 1% sodium bicarbonate solution added), bacterial contamination should be suspected;
5. Add 0.1 mg per ml of colchicine solution to arrest the mitosis in metaphase and vastly increase the number of cells available for analysis;
6. Incubate the solution for 5 hours at 37° C;
7. Centrifuge the culture at 800 revolutions per minute (rpm) for 8 minutes and aspirate all supernatant liquid;
8. Resuspend the cells in 5 ml of Hank's solution, centrifuge again at 800 rpm for 10 minutes, and decant all the supernatant except 1 ml;
9. Add 4 ml of distilled water, resuspend the cells, and incubate at 37° C for 10 minutes. Prepare the fixative by mixing 1 ml of glacial acetic acid to 9 ml of methanol;
10. Centrifuge the cells at 800 rpm for 5 minutes, and decant the supernatant;
11. Slowly add 5 ml of fixative and let stand at room temperature for 15 minutes;
12. Gently resuspend the cells, then centrifuge the solution at 600 rpm for 5 minutes, aspirate all liquid, add 5 ml of fresh fixative, and resuspend the cells gently and let stand for 10 minutes;

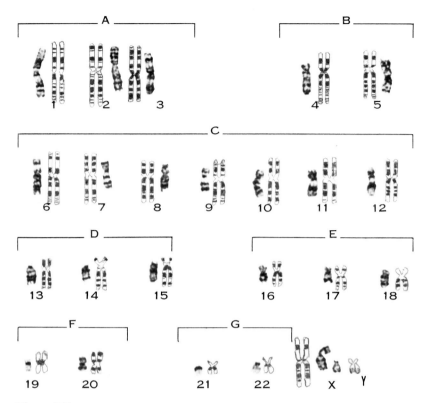

Figure 101
Structure of the different chromosome groups.

13. Centrifuge for 5 minutes at 800 rpms, then aspirate all the supernatant except 0.2 ml. Gently homogenize and place 2 to 3 drops of suspension on clean slides;
14. Heat the slides and air dry them. Dip the slides in distilled water and add 5 drops of Giemsa stock solution on each slide. Cover the cells by tilting the slide and let it stain for 10 minutes. Rinse with tap water. After placing the smear in xylol for 15 minutes, coverslip, using mounting media;
15. Examine the slides after placing a green filter in front of the light source;
16. Locate and photograph the cells in mitosis that have well-scattered pairs of chromosomes. Then, after enlarging the photographs, cut out the individual chromosomes. Several photographic prints are needed when the chromosomes overlap;
17. Arrange the groups according to their size and structure and determine the karyotype (Table 9, Fig. 101). The longest chromosome gets the lowest number (Fig. 102).

Table 9. Morphological Identification of the Human Chromosome Groups

Group	Chromosome Number	Description
A	1, 2, 3	Largest of all the chromosomes (7–8 μ), they are enumerated in relation to their decreasing size. All have median centromere (number 2 being slightly less metacentric).
B	4, 5	Next largest. Difficult to differentiate from each other. Both are submetacentric and almost of identical size.
C	6–12 + X chromosome	Next largest. Difficult to differentiate. They are all submetacentric.
D	13, 14, 15	Next largest. Acrocentric. Number 13 is the largest and has satellites; number 14 is the next largest; number 15 is the shortest of the group and has secondary constrictions.
E	16, 17, 18	Next largest. They are relatively easy to differentiate from each other. Number 16 is largest and is metacentric with long arm constriction; number 17 is submetacentric; number 18 is shortest and is submetacentric.
F	19, 20	Next largest. Difficult to identify. Both are metacentric.
G	21, 22 + Y chromosome	Shortest chromosome (1–2 μ). Acrocentric. Number 21 often has a satellite on the short arm. The Y chromosome found in the male is the largest of this group and has no satellite.

BANDING (FIG. 101)

Fluorochrome stain (quinacrine mustard or quinacrine dehydrochloride) has a variable affinity for certain zones of the arms of the chromosomes, producing the distinctive dark and bright banding appearance.

Each chromosome seems to have a specific banding pattern, which allows their precise recognition. Unfortunately, these fluorescent stains fade rapidly, and repeat examination of the cells is not always possible.

Recently it was found that if the chromosomes are denatured, they can be stained with Giemsa stain and identical banding patterns can be seen in a more permanent way.

Because the tips of the long arms of certain chromosomes are not stained and are invisible with this method, a so-called reversed banding technique has been developed that stains only the previously unstained parts.

The bands are easier to see in the chromosomes of early metaphase. The secondary constrictions are usually unbanded. The banding of the Y chromosome is not consistent. Often its long arm stains dark, and no band may be discerned.

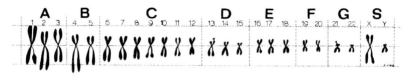

Figure 102
Chromosomal banding in a normal male karyotype.

Sex Chromosome Determination

The sex chromosome mass, or Barr body (found in 1949), is the small basophilic body peculiar to the nuclei of female cells in interphase. It is located next to the inner surface of the nuclear membrane of the cell in the resting stage. Ideally, in a woman it should be found in almost all her cells during their interphase stage. But owing to the generally poor preservation of some of the cells, the numerous artifacts, the poor staining, and other factors, the sex chromosome mass is usually found in only 30 to 60% of the female cells examined. The genetic sex is determined by the presence of the sex chromosome mass (genetic female) or its absence (genetic male). The newly discovered method of demonstrating the presence of the Y chromosome is becoming as important as the demonstration of the presence of the Barr body. Although any tissue can be studied for sex chromosome determination, the most popular source is the blood smear (for the presence of so-called drumsticks in the nuclei of the leukocytes), skin (in biopsies), and exfoliated squamous cells from the vaginal or buccal mucosas.

To be accepted for evaluation, the cells should be intact and have non-pyknotic vesicular nuclei, as seen in the columnar cells and intermediate and basal squamous cells. Also, the cells must have been handled carefully to prevent mechanical distortion. The specimen obtained by a gentle scraping of the inner surface of the cheek (from which intermediate squamous cells mainly are collected) is preferred for sex chromosome determination.

MATERIAL AND METHOD

Technique
To obtain a satisfactory specimen for the scraping of the inner surface of the cheek, the following procedure must be observed:

1. Before scraping, it is very important to remove all cellular debris and other foreign material from the surface of the mucosa. Rinsing the mouth and brushing the teeth prior to the scraping is recommended.
2. Immediate fixation of the cells is critical for two reasons. If the cells are even slightly air dried, the sex chromosome mass will not stain well and this can produce an error in the interpretation (false-negative, pseudo-male). If the cells are degenerate, the chromatin tends to clump and migrate toward the periphery, resembling the Barr bodies (false-positive,

pseudofemale). The bottle of fixative should be opened and held by the patient near the mouth ready to receive the slides.

3. With the edge of a tongue depressor or Ayre spatula, the inner surface of the cheek along the buccal reflex is energetically scraped. The material obtained is spread in a moderately thick smear on a clean glass slide, which is dropped immediately into the fixative solution. Several smears should be done at one time, and at least four slides should be submitted.

Staining for the X Sex Chromatin Mass (Barr Body)

Two of the slides are stained according to the Papanicolaou technique to keep a permanent record of the case. The remaining slides are stained as follows:

1. The appropriately labeled smear of the patient and a control smear taken from the buccal mucosa of a female technician are left for at least 2 hours in the mixed fixative of glacial acetic acid and alcohol.
2. During this time orcein stain is prepared by adding 0.1 gm of orcein to 10 ml of warm 50% solution of glacial acetic acid. The solution is cooled and mixed by shaking the bottle under running water. The mixture is then filtered several times through a #1 filter paper to eliminate debris and precipitate.
3. The slides are covered with the staining solution for at least 10 minutes, and coverslips are applied very lightly to allow oil immersion studies.

This rather simple method has the advantage of staining only the chromatin components of the nucleus, leaving the remainder of the cells transparent. Since the staining solution is slightly hypotonic, the cells and the nuclei become edematous and increase in size, allowing the chromatin clumps to stand out. To obtain the best results and sharpest contrast, a green filter should be placed in front of the light source during the examination.

EXAMINATION

The control smears (taken from the female technician and stained together with the diagnostic slides) are examined first. If the cells show well-stained, dark chromosome masses, the patient's smear is then examined. If the patient is a normal female, there should be present the same percentage of chromatin-positive cells as is found in the control smear. An average of 200 to 300 cells is usually studied, of which 100 to 200 are rejected as unsatisfactory because of the presence of artifacts (microorganisms), distortion, or diffuse degenerative chromatin clumping.

MORPHOLOGY OF SEX CHROMOSOME MASS IN THE EPITHELIAL CELLS (FIG. 103A)

Found only in the cells with more than one X chromosome, the sex chromosome mass should be seen adherent to the slightly indented inner sur-

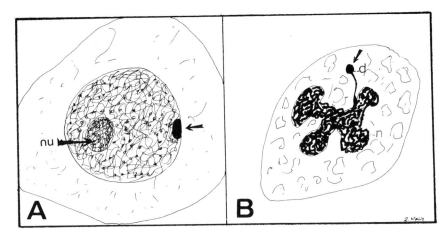

Figure 103
(A) Chromosomal mass (Barr body) in a basal squamous cell (nu = nucleolus).
(B) Drumsticklike appendage in a polymorphonuclear leukocyte.

face of the nuclear membrane. This is an important condition for its acceptance as a true sex chromosome mass.

The size of sex chromosome masses, although slightly variable, averages about 1 μ in diameter.

The shape of the masses can vary—they can be round, biconvex, triangular, or rectangular.

The borders of the mass should be well-defined and sharp.

The mass should be at the same focusing level as is the nucleus.

INTERPRETATION OF THE SEX CHROMOSOME MASS IN THE EPITHELIAL CELLS

As illustrated in Figure 104, to be acceptable, the close relation of the chromosome mass to the inner surface of the nuclear membrane should be evident under oil immersion examination.

As shown in Figure 104, the length of the surface in comparison to the axis of observation (Zone WX and ZY) is greater when the nucleus is not compressed and is allowed to remain round. Zones WZ and XY are considered blind because the relation of sex chromatin to the nuclear membrane cannot be demonstrated. In these zones, the female chromosome mass, even if adherent to the nuclear membrane, will appear as though it is situated in the middle of the nucleus. This chromatin mass should not be accepted as a sex chromosome because of the danger of confusing it with other nuclear masses, such as degenerative chromatin clumping, prominent nucleoli, viral inclusion bodies, or superimposed bacteria.

ABNORMALITIES OF THE X SEX CHROMOSOME MASS (BARR BODY)

VARIATION IN NUMBER. Various sex-related congenital abnormalities may be indicated when no Barr body or more than one Barr body is present.

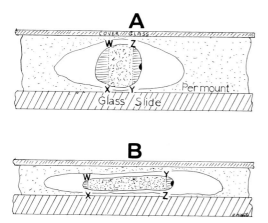

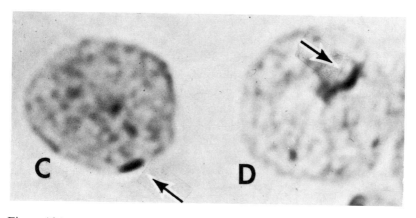

Figure 104

Length of the portion of the inner surface of the nuclear membrane where its relation to the sex chromatin mass (*arrows*) with respect to the axis of observation is visible. (A) The surface WX and ZY is long in the W to X direction when the nucleus is not compressed. (B) The surface WX and ZY is short in the W to X direction when the nucleus is compressed. In C and D are shown sex chromosome masses (Barr Bodies) in the nuclei of two basal squamous cells stained with orcein. (C) The chromosomal mass is acceptable, its relation to the inner surface of the nuclear membrane being plainly visible. (D) The chromosomal mass is not acceptable. Its relation to the inner surface of the nuclear membrane is not visible, and it appears as if situated in the middle of the nucleus. It is hard to differentiate it from other nuclear masses, such as nucleoli, inclusion body, or bacteria. (× 1800)

The number of Barr bodies equals the number of X chromosomes minus one.

VARIATION IN SIZE. Often the result of artifact (staining, drug ingestion, etc.), it may also indicate abnormality of the length of the long arm of the X chromosome.

VARIATION IN INCIDENCE may indicate a mosaicism (incidence lower than 40% of the cells).

MORPHOLOGY OF SEX CHROMOSOME MASSES IN POLYMORPHONUCLEAR LEUKOCYTES (FIGS. 103B, 105A)

The study requires well-preserved leukocytes, as seen in an inflammatory vaginal smear or in the peripheral blood.

In the female only, a drumsticklike appendage is attached to one lobe of the nucleus. It is a spherical mass that measures 1 to 2 μ and is attached by a very slender long stalk.

The size of the drumstick appendage can vary in accordance with the size of the X chromosome.

The drumsticks are present only in the mature polymorphonuclear leukocytes.

The drumsticks should be differentiated from other small nuclear projections normally seen in both male and female. Their larger size, darker appearance, and slender stalk help in their recognition.

The incidence of this drumstick in the neutrophils is lower than the incidence of the chromosome mass found in the epithelial cells. (Only 5 per 100 of the leukocytes examined in a normal female will contain it.) This number decreases to less than 1% during pregnancy.

The drumstick appendages are rarely found in the leukocytes of a normal male (an average of about 0.1%).

With few exceptions, the number of drumsticks does not increase with the abnormal increase of the number of X chromosomes. Therefore the number of drumsticks cannot be used for the diagnosis of a superfemale with 3 or 4 X chromosomes, for example.

A very low incidence of this drumstick in a female may suggest a mosaicism.

FACTORS INFLUENCING THE PERCENTAGE OF CELLS THAT ARE POSITIVE FOR SEX CHROMOSOME MASSES

STAINING TECHNIQUE USED. A poorly stained nucleus or overstained cytoplasm will decrease the number of Barr bodies counted in a smear.

QUALITY OF THE SMEAR. The Barr body can be confused with overlapping bacteria, and it can be difficult to see or absent in air dried or degenerate cells.

CHROMOSOME ABNORMALITY. A decreased count often indicates a mosaicism.

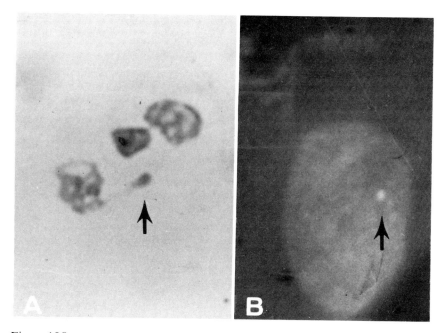

Figure 105

(A) Drumstick (*arrow*) in a "female" leukocyte. (× 1300) (B) Y chromo-some (*arrow*) fluorescing in a "male" epithelial buccal cell. (× 1300)

SMEARS TAKEN AT THE POSTPARTUM PERIOD. The incidence of chro-matin-positive cells from the buccal or vaginal mucosas seems to decrease in the first week after delivery (lower than 15% of the cells).

NEWBORN. The incidence of chromatin-positive cells often decreases in the first week after birth (lower than 10%). The sex chromosome determina-tion should be done at least two weeks after birth; otherwise it is not always valid. Just after birth the baby's buccal smear may still be contaminated by cells from the mother's blood, genitalia, cervical mucus, or debris.

HORMONE THERAPY. An excess of estrogen or corticosteroid hormone administration to a woman will decrease the incidence of cells that are positive for sex chromosome mass. This emphasizes the importance of hav-ing a complete history of the patient before the determination in order to rule out a mosaicism.

MENSTRUAL CYCLE. The frequency of sex chromatin was found to be significantly lower in the late proliferative phase and the midcycle period. The late luteal phase and during menstruation seem to be the best time to take a smear from a young woman for sex chromosome mass determination.

DRUGS. Some of the sulfanilamides and some antibiotics may decrease the incidence and size of the buccal chromatin masses. The mechanism of this action is unknown.

HYPOPARATHYROIDISM. This can produce an increase of equivocal appendages in the drumstick count of the leukocytes.

IRRADIATION. Irradiation may produce an abnormality in the incidence, number, and size of the chromosomal masses.

DEMONSTRATION OF Y CHROMOSOME (FIG. 105B)
All cells with no Barr bodies are not necessarily "male" cells. A cell with an XO sex chromosomal pattern, for example, has no Barr body but is still a "female" cell.

The male cell must have a Y chromosome that can be specifically stained with a quinocrine stain. The distal segment of the long arm of the Y chromosome fluoresces brilliantly in cells in the interphase stage. This brightly fluorescing long arm can be seen in more than 60% of the cells of normal male patients with normal Y chromosomes. This test will also show whether more than one Y chromosome is present.

Technique for Demonstrating the Fluorescing Y Chromosome
The buccal smear is placed in:

95% ethyl alcohol	
Acid alcohol (1% of HCl in 70% ethyl alcohol):	10 minutes
Distilled water:	3 changes
0.5% aqueous quinocrine hydrochloride:	10 minutes
Distilled water:	3 changes
Acid alcohol (0.25 ml of 1% acetic acid in 100 ml 70% ethyl alcohol):	10 minutes
0.01 M acetic acid–phosphate buffer (pH 5.5):	6 minutes
Distilled water:	3 changes
0.01 M phosphate buffer (pH 7.4):	2 changes
Mount and coverslip	

CHROMOSOMAL NONDISJUNCTION (FIG. 106)

BY MEIOSIS. The chromosomal abnormalities are uniform and are found in every cell of the body (XXX type, producing two Barr bodies in all the cells, for example).

BY MITOSIS. Mosaicism: The sex chromosomes vary from one cell to another in the body, or in the same organ (XO or XXX type, producing either no Barr bodies or two of them).

Remark: An individual can live with only one X chromosome (XO) but not with one Y chromosome (YO).

An individual cannot live if any of the autosomes is missing, but a surplus of autosomes may be compatible with life.

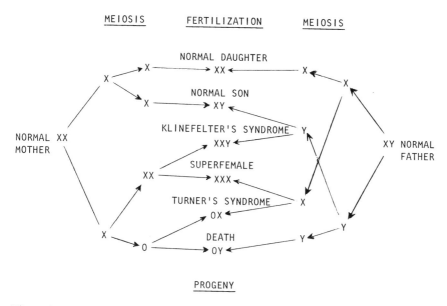

Figure 106
Fertilization with normal and abnormal gametes.

SEX CHROMATIN AND BARR BODIES

The relation between sex chromatin and Barr bodies in various syndromes is shown in Table 10.

Remark: The number of Barr bodies is usually 1 less than the number of X chromosomes.

NORMAL AND ABNORMAL SEX CHROMOSOME DETERMINATION

Normal Male

44 autosomes

1 X chromosome

1 Y chromosome

0 Barr bodies

Normal external male genitalia. Normal male urethral cytohormone (shift to the middle or left).

Normal Female

44 autosomes

2 X chromosomes

0 Y chromosomes

1 Barr body

Normal external female genitalia. Variable maturation index.

Table 10. Sex Chromatin and Barr Bodies in Various Syndromes

Chromosomes	Syndrome	Barr Bodies
X0	Turner's syndrome	0
XY	Normal male	0
XX	Normal female	1
XXX	Superfemale	2
0Y–00	Death	

Female Pseudohermaphroditism

44 autosomes

2 X chromosomes

1 Barr body

0 Y chromosomes

The external genitalia are ambiguous or male. The condition may be an in utero result of male hormone therapy of the mother or of an adrenal cortical hyperplasia (androgenital syndrome). The maturation index varies $(25/50/25 \pm 20)$.

Male Hermaphroditism (Testicular Feminization)

44 autosomes

1 X chromosome

1 Y chromosome

0 Barr bodies

The external genitalia are female but infantile. There is a small vagina but no uterus. Immature testicles (but with no spermatogenesis) are present. The maturation index may vary from 0/10/90 to 90/10/0, according to the amount of estrogen produced.

Superfemale

44 autosomes

3 or 4 X chromosomes

0 Y chromosomes

2 to 3 Barr bodies

This condition may be the cause of sterility and low normal or subnormal intelligence. The maturation index is variable, but is normal female in type.

Turner's Syndrome (Gonadal Agenesis)

44 autosomes

1 X chromosome

0 Y chromosomes

0 Barr bodies

This syndrome is characterized by such conditions as sexual infantility, amenorrhea, and neck webbing. The maturation index is variable (60/30/10).

Klinefelter's Syndrome (Testis Dysgenesis)

    44 autosomes

2, 3, or 4 X chromosomes

  1 or 2 Y chromosomes

1, 2, or 3 Barr bodies

    Klinefelter's syndrome produces a eunuchoid male with gynecomastia.

Mongolism (Down's Syndrome)

    45 autosomes (no. 21 has an extra autosome)

1 or 2 X chromosomes

0 or 1 Y chromosomes

0 or 1 Barr body (normal)

    In mongolism the maturation index is variable. Mental retardation and mongoloid features are characteristic.

Edward's Syndrome

    45 autosomes (no. 18 has an extra autosome)

1 or 2 X chromosomes

0 or 1 Y chromosome

0 or 1 Barr body

    In trisomy 18, the child fails to thrive, is often mentally retarded, and has congenital malformations.

References and Supplementary Reading

Bamford, S., et al. Neutrophil appendages as indicators of sex chromosome aberrations. *Acta Cytol.* 8:323, 1964.

Bamford, S., Cassin, C. M., and Mitchell, B. S. Sex chromatin determinations in selected cases of developmental sex abnormalities with an assessment of results. *Acta Cytol.* 7:151, 1963.

Bamford, S., et al. Further observations on neutrophil appendages in sex chromosome aberrations. *Acta Cytol.* 10:323, 1966.

Barr, M. L., and Bertram, E. G. A morphological distinction between neurones of the male and female, and the behavior of the nucleolar satellite during accelerated nucleoprotein synthesis. *Nature* 163:676–677, 1949.

Blanco de del Campo, M. S., and Ramirez, O. E. G. Fluctuations of the sex chromatin during the menstrual cycle. *Acta Cytol.* 9:251, 1965.

Chu, E., Malmgren, R., and Kazam, E. Variability of sex chromatin counts. *Acta Cytol.* 13:72, 1969.

Colby, E. B., and Calhoun, L. Accessory nuclear lobule on the polymorphonuclear neutrophil leukocyte of domestic animals. *Acta Cytol.* 7:346, 1963.

Comings, D. E. Sex chromatin, nuclear size and the cell cycle. *Cytogenet. Cell Genet.* 6:120–144, 1967.

Dougherty, C., and McCormick, M. X chromosomal anomalies in the newborn. *Acta Cytol.* 17:423, 1973.

Forssman, H., Lehmann, O., and Thysell, T. Reproduction in mongolism. Chromosomal studies and re-examination of child. *Am. J. Ment. Defic.* 65:495, 1961.

Grob, H. S., and Kupperman, H. S. Experiences with technics of chromatin sex determination. *Am. J. Clin. Pathol.* 36:132–138, 1961.

Hagy, C., and Brodrick, M. M. Variation of sex chromatin in human oral mucosa during the menstrual cycle. *Acta Cytol.* 16:314, 1972.

Iliya, F., Meisner, L., and Copenhaver, E. H. Testicular feminizing syndrome. *Obstet. Gynecol.* 25:451, 1965.

Jones, H. W., Jr., Ferguson-Smith, M. A., and Heller, R. H. Pathologic and cytogenetic findings in true hermaphroditism. *Obstet. Gynecol.* 25:435, 1965.

Jones, H. W., Jr., et al. Chromosomes of cervical atypia, carcinoma in situ, and epidermoid carcinoma of the cervix. *Obstet. Gynecol.* 30:790, 1967.

Jones, H. W., Jr., and Zourlas, P. A. Clinical, histologic, and cytogenetic findings in male hermaphroditism. *Obstet. Gynecol.* 25:597, 1965.

Kailin, E. W., and Platt, L. Comparison of X chromatin frequency in buccal and vaginal cells from ectopic women. *Acta Cytol.* 15:294, 1971.

Kirkland, J. A., Stanley, M. A., and Cellier, K. M. Comparative study of histologic and chromosomal abnormalities in cervical neoplasia. *Cancer* 20:1934, 1967.

Klinger, H. P. Morphological Characteristics of the Sex Chromatin. In K. L. Moore (Ed.), *The Sex Chromatin.* Philadelphia: Saunders, 1966. Pp. 76–90.

Leviz, I. S., and Carel, R. Variation in the incidence of sex chromatin—A reappraisal. *Acta Cytol.* 12:352, 1968.

Miles, C. P. Chromatin elements, nuclear morphology and midbody in human mitosis. *Acta Cytol.* 8:356, 1964.

Milunsky, A., et al. Prenatal genetic diagnosis. *N. Engl. J. Med.* 283:1498, 1970.

Moore, K. L., and Barr, M. L. Smears from the oral mucosa in the detection of chromosomal sex. *Lancet* 2:57–58, 1955.

Naib, Z. M. Nuclear chromatin sex determination in patients with genital abnormalities. *Obstet. Gynecol.* 18:64, 1961.

Naujoks, H. Culture of tissues from spontaneous human abortions (preliminary work for chromosome analysis). *Acta Cytol.* 7:300, 1963.

Pansegrau, D. G., and Peterson, R. E. Improved staining of sex chromatin. *Am. J. Clin. Pathol.* 41:266–272, 1964.

Pearson, P., Bobrow, M., and Vosa, C. Technique for identifying Y chromosomes in human interphase nuclei. *Nature* 226:78, 1970.

Piver, M. S., et al. Testicular feminization. *Obstet. Gynecol.* 28:397, 1966.

Platt, L., and Kailim, E. Buccal X-chromatin frequency in numerous diseases. *Acta Cytol.* 13:700, 1969.

Siracky, J. Sex chromatin in gynecologic cancer. Incidence and limitations of its clinical interpretation. *Acta Cytol.* 16:105, 1972.

Sohval, A. R., and Casselman, W. G. B. Alterations in size of nuclear sex chromatin mass (Barr body) induced by antibiotics. *Lancet* 2:1386–1388, 1961.

Tagher, P., and Reisman, L. E. Reproduction in Down's syndrome (mongolism): Chromosomal study of mother and normal child. *Obstet. Gynecol.* 27:182, 1966.

Taylor, A. I. Sex chromatin in the newborn. *Lancet* 1:912–914, 1963.

Thuline, H. Y-specific fluorescence in peripheral blood leukocytes. *J. Pediatr.* 78:875, 1971.

Vakil, D., Lewin, P., and Cohen, P. Value of fluorescent Y chromosome and sex chromatin test. *Acta Cytol.* 17:220, 1973.

Zourlas, P. A., and Jones, H. W. Clinical, histologic and cytogenetic findings in male hermaphroditism. *Obstet. Gynecol.* 25:768, 1965.

The Respiratory Tract

Anatomy and Histology

The lung is essentially a system of airways and blood vessels constructed as interlacing, ramifying ducts that make possible an efficient gas exchange between air and blood. The proximal, or upper, airways, the larynx, trachea, bronchi, and bronchioles, are thick-walled passageways through which air is conducted into the thin-walled distal air spaces, the alveoli, where gas exchange occurs. The right lung is divided into three lobes, the left lung into two.

The lower part of the larynx (below the vocal cords), the trachea, and the bronchi, are lined by ciliated pseudostratified columnar epithelium containing occasional goblet cells. The surrounding submucosa contains mucus-producing glands that become less numerous in the more distal bronchi. The outer wall consists of a coat of smooth muscle and cartilage rings, which lend elasticity and rigidity to the conducting passageways.

The smaller bronchi are lined by a tall columnar epithelium, devoid of goblet cells, resting on a single layer of basal cells. This epithelium gradually flattens in smaller bronchi into a layer of cuboidal cells resting on an occasional basal cell. The epithelium is supported by a basement membrane, subadjacent to which is a thin submucosa devoid of mucus-producing glands. The surrounding wall consists of a thin coat of smooth muscle. Cartilage is absent. Peribronchial fibrous tissue surrounds the trachea and bronchi.

Finer airways devoid of surrounding fibrous tissue, known as terminal bronchioles, arise from the smallest bronchi and, accompanied by a small companion pulmonary artery, enter into and lie in direct contact with the adjacent alveolar tissue. The wall of these passageways consists of a thin layer of smooth muscle, supporting elastic fibers, a basement membrane, and overlapping cuboidal epithelium.

Arising from the bronchioles are the alveolar ducts, which form the final conducting passageway from which the alveoli arise. The walls of the alveolar ducts are thin and are composed of smooth-muscle elements, supporting elastic and collagen fibers, a basement membrane, and an exceedingly thin epithelial covering that is apparent only in the electron microscope. This epithelial lining is continuous with that of the alveolar spaces.

The alveolar wall, or interalveolar septum, consists of opposed thin epithelial lining cells and their underlying basement membranes, between which the pulmonary arterioles have divided into a dense network of capillary loops. These are supported by a network of fine elastic and collagen fibers containing occasional fibroblastic cells and leukocytes situated between pulmonary capillary loops.

Within the alveolar wall, ameboid phagocytic cells (histiocytes) may remain for a time in the alveolar septum or migrate between alveolar lining cells to enter the alveolar space. By virtue of their ameboid and phagocytic activity, these cells ingest debris and foreign material in the alveoli and alveolar ducts and may eventually make their way to the smallest bronchi, where ciliary motion transports them toward the trachea.

Routine Sputum Collection and Processing (Modified Saccomano Technique)

In the morning upon awakening, the sitting patient should first brush his teeth or rinse his mouth and gargle with saline solution to eliminate the accumulated exfoliated oral cells that otherwise contaminate the sputum.

He should then inhale repeatedly to the full capacity of his lungs and exhale the air with an expulsive cough. The expectoration produced should be collected in a wide-mouthed bottle, preferably made of transparent glass, or in a disposable plastic container about half full with 50% ethyl alcohol containing 2% polyethylene glycol (2 ml of 1% Carbowax 1540 added to 50 ml of 50% ethyl alcohol). The container is sealed and shaken briskly for a few seconds to disperse the mucus threads and cells, allowing adequate fixation. The specimen may remain in the fixative for many days, if necessary, before further processing. Alcohol disinfection of the outside of the container at the bedside of the patient is advised because it is estimated that about 10% of the patients whose sputum is examined for malignant cells have an unsuspected active tuberculous infection.

A sufficient quantity of 50% ethyl alcohol is added to the specimen to make 100 ml total. If the sputum is unusually abundant, the specimen may be diluted to 200 ml volume with additional 50% ethyl alcohol.

The specimen is then blended in a high-speed (21,000 rpm) blender* for 3 to 5 seconds. If the specimen does not become uniformly blended, the blending may be continued for another 5 seconds. We have not found blending to be injurious to cells if it is not done excessively.

The blended material is centrifuged in round-bottom test tubes (15 by 50 mm size), for about 10 minutes at 1000 to 1500 rpm.

Decant the supernatant fluid, leaving very little fluid to admix with the granular, pale centrifugate. This is best accomplished in a vibrator; a thick, granular, almost pasty mixture is formed.

Six slides are prepared from each centrifuged tube. To make smears,

* Waring Timer-Light Blendor, Scientific Products (catalog No. 8350–1), with semi-micro, stainless steel Eberbach container (catalog No. S8395–1). Scientific Products, 1430 Waukegan Road, McGaw Park, Ill. 60085.

place centrifugate (1 to 2 drops if it is thick, 2 to 4 drops if it is thin) on slides. Gently spread the material as evenly as possible by superimposing a clean slide. Manipulate gently until the material is dispersed between the two slides. Then pull them apart in a sliding motion.

Allow the slides to dry completely, and then hydrate four slides by placing them in tap water for 5 minutes to remove the Carbowax before beginning the staining with hematoxylin. Subsequent staining is done by the modified Papanicolaou method. The use of 2% polyethylene glycol (Carbowax 1540) prevents shrinking of the cells during drying. Since the slides are allowed to air dry before staining, it is not necessary to use albumin. The finished stained slides have a fine, uniform, granular surface. The two extra air dried slides are stored for possible special stain (silver stain for fungi, for example).

To obtain the best results in cancer detection, this procedure should be repeated on three consecutive morning sputums with a total of 12 smears stained and examined.

The smears are considered to be satisfactory only when an adequate number of columnar cells and carbon-bearing histiocytes is present.

The sputum that is obtained after bronchoscopy, or on the day following it, is potentially the richest in cancer cells.

If too many inflammatory cells are found, antibiotic and expectorant therapy should be administered for three to five days, then the sputum series repeated.

AEROSOL INHALATION

This is indicated only for the patient who has slight or no expectoration. A solution of 150 gm of sodium chloride (NaCl) dissolved in 200 ml of propylene glycol and mixed with 800 ml of distilled water is filtered before using. This solution is heated to 115° F in a nebulizer and vaporized by means of an air pump. The patient breathes in and out through the mouthpiece for 5 to 10 minutes. The resulting cough-produced sputum is collected fresh, labeled, and sent to the laboratory and processed as a routine specimen.

DIRECT COLLECTION OF BRONCHIAL SECRETIONS

This is indicated when the patient is unable to produce any sputum.

TRACHEAL ASPIRATION. A French catheter, No. 14 to 16, is moistened in saline solution and introduced through the mouth or the nares into the trachea of the sitting patient until it reaches the carina. The secretion is continuously aspirated with a 20-ml syringe and smeared on two or three slides and fixed immediately.

BRONCHIAL ASPIRATION AND WASHINGS. After local anesthesia of the upper respiratory tract, a French catheter, moistened with saline solution, is introduced through the nares of the sitting patient.

The catheter is first advanced to the carina, after a solution of Pontocaine (tetracaine) has been instilled into the trachea, then introduced into

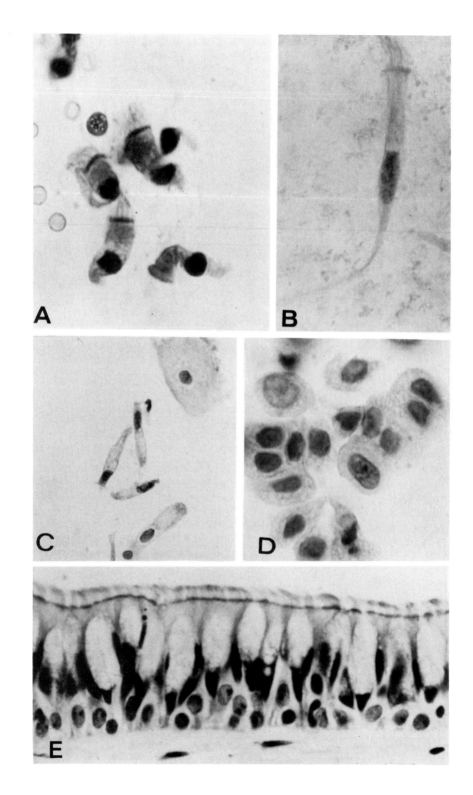

the main suspected bronchus by sharply rotating the patient's head. Any secretion is aspirated and smeared as in a tracheal aspiration.

Pontocaine and normal saline solution are then instilled and allowed to fill up and remain in the bronchus for about 5 to 10 minutes, with the patient in a recumbent position. This liquid is reaspirated with the catheter in various positions. The catheter is slowly withdrawn, flushed with saline solution, and the collected material, labeled *right* or *left* main bronchial specimen, is sent to the laboratory for processing.

At the laboratory, the specimen is diluted with 20 ml of saline solution and passed through one to three membrane filters, according to the amount of cells and mucus contained, then fixed with 95% alcohol.

BRONCHIAL BRUSHING.   The use of the flexible bronchofiberscope permits the scraping of an increased number of bronchial lesions under direct observation. The few cells obtained are often highly diagnostic and well preserved. To prevent the cells from air drying, the smearing should be done directly on an alcohol-moistened slide.

NEEDLE ASPIRATION.   Accurate needle aspiration of small peripheral tumors (coin lesions) located radiologically can be of great help in determining the nature of lesions that cannot, or do not, exfoliate in one of the main bronchi. This method is especially helpful in determining the nature of suspected metastatic tumors to the lung and in the diagnosis of *Pneumocystis carinii* infection.

Cytology of the Upper Respiratory Tract— Trachea, Larynx, and Nasopharynx

NORMAL CYTOLOGY

Ciliated Columnar Cells (Fig. 107A)

Ciliated columnar cells are shed singly, in tight clusters, or in sheets, the last mainly as the result of traumatic exfoliation (postbronchoscopy) or in certain inflammations (influenza virus).

Their size varies (10–50 μ) and their shape, although mainly pyramidal (70%), can be irregular (10%), round (10%), or oval (10%).

The cytoplasm is adequate to abundant, angular, vesicular, or coarsely granular in texture with sharp, smooth, distinct borders.

Figure 107

Normal columnar cells in sputum. (A) Ciliated upper respiratory tract columnar cells. Note the prominent terminal bar, the dense cytoplasm, and the hyperchromatic nuclei. ($\times$ 450) (B) Ciliated cell from the lower respiratory tract in a bronchial washing. Note the elongated, slender cytoplasm, a faint terminal bar, and the vesicular nucleus. ($\times$ 450) (C) Goblet columnar cell. Note the mucus-containing vacuoles distending the cytoplasm and the eccentric hyperchromatic small nucleoli. ($\times$ 350) (D) Nonciliated basal columnar cell in a bronchial washing. Note the uniform, oval nuclei, and the vesicular cytoplasm. These cells are differentiated from histiocytes by the cellular molding. ($\times$ 750) (E) Histology of normal bronchial mucosa. ($\times$ 750)

The cells stain deep purple to dark blue with the Papanicolaou stain (80% of the cells).

On the large end of the cell there is a prominent, but hazy, thick terminal bar from which emerge dense, slender eosinophilic cilia of variable length.

The relatively large nuclei (5–15 μ) are centrally or eccentrically located, round or oval, single (70%) or multiple (30%) with a finely granular uniform (75%) or coarsely clumped (25%) chromatin that is evenly distributed, except for a few karyosomes.

The nuclear membrane is even and smooth but often dark and heavy.

Occasionally (25%), prominent red, round nucleoli, varying in size and number, are present.

These cells, when found in a sputum or bronchial washing, especially if degenerate and having lost their cilia, may be a pitfall in the diagnosis of bronchogenic carcinomas.

Goblet Cells (Mucus-Secreting Columnar Cells) (Fig. 107C)

Goblet cells are shed singly (80%) or in loose clusters (20%). They vary in size (15–80 μ) and shape—oval (40%), round (20%), or irregular (40%), but usually stout.

Their lightly stained cytoplasm is abundant and vesicular, containing hypersecretory single or multiple vacuoles with numerous ingested large healthy neutrophils in 30% of the cells.

The periodic acid–Schiff (PAS) stain is positive.

No cilia or terminal bars can be seen. The cytoplasmic membrane is delicate and often indistinct.

The eccentric, large, oval nuclei, usually single, can be multiple (2–6) in 10% of the cells and may overlap without molding.

The chromatin is finely granular, uniformly distributed, and homogeneous. The nuclear membrane is smooth and delicate.

Prominent nucleoli are usually seen, varying in size and number from one cell to another.

In hypersecretory cells, the nucleus may be pyknotic and eccentric in position owing to the presence of large intracytoplasmic secretory vacuoles.

The cells, especially if degenerate, can be difficult to differentiate from a low-grade bronchogenic adenocarcinoma, when seen in a sputum specimen.

Inflammatory Cells

A variable number of neutrophils, eosinophils, and basophils, mixed with pink-stained mucus, is frequently seen, even in the absence of symptomatic inflammation.

Inflammatory histiocytes and foreign-body giant cells, lymphocytes, and other monocytes, indicating the existence of chronic tracheobronchial inflammation, are also found in variable amounts.

Squamous Cells

Anucleated, superficial, intermediate or basal oral cells are commonly seen and are not usually of diagnostic importance. Their numbers are markedly decreased if the patient brushes his teeth before obtaining the specimen.

CYTOLOGY OF ALLERGIES

In respiratory allergic conditions (rhinitis, sinusitis), a marked increase in the number of goblet cells (50% of the cells) is found in the nasopharyngeal smear, frequently combined with a moderate increase in the number of eosinophilic polymorphonuclear cells and occasionally of basophilic polymorphonuclear (mast) cells. Charcot-Leyden crystals may also be seen (p. 249).

Large numbers of stripped oval hypertrophic nuclei with prominent nucleoli, coarse chromatin, and smooth membranes, are often present. Although little degeneration is seen, these nuclei originate from the breakup of some of the goblet cells as a result of the fragility of their cytoplasm.

The background of the smear shows very few bacteria but an increased amount of mucus and cellular debris.

CYTOLOGY OF VIRAL INFECTIONS (FIGS. 108, 109)

Parainfluenza Viral Infection (Fig. 108A, 109A)

With the parainfluenza viral infection, the tracheal or nasal smear shows an unusually large number (more than 60%) of injured ciliated columnar cells with advanced degenerative changes, absence of cilia (ciliocytophoria of Papanicolaou) and multiple small intracytoplasmic eosinophilic red inclusions ($2–5$ μ in diameter).

These degenerate, nonciliated cells are mononucleated, enlarged ($10–30$ μ in diameter), and retain their normal elongated shape and their vacuolated basophilic or cyanophilic cytoplasm.

Loose fragments of detached cytoplasm with intact terminal bar and cilia (cellular tuft) are common.

Usually single (80%), they may be arranged in groups or clusters in association with long strands of mucus.

Their cytoplasmic membrane is thin, and the outline of the cells is indistinct.

The nuclei are slightly hypertrophic and show mild degenerative signs in the moderate clumping of their chromatin and its margination toward the nuclear membrane. No intranuclear inclusions can be recognized. Occasionally, karyopyknosis is found.

Large amounts of cellular debris, some resembling the remnants of ciliated tufts, are found mixed with moderate numbers of inflammatory cells (neutrophils, eosinophils, and monocytes).

No morphologic differences can be seen when the cells are infected by the parainfluenza I or III type virus.

Herpes Simplex Viral Infection (Figs. 108B, 109B)

The changes involve only the young columnar and squamous exfoliated cells, and their characteristics are similar to those described under Herpes Progenitalis.

Adenoviral Infection (Figs. 108C, 109B)

In contrast to the effect of other viruses, the cellular changes resulting from adenoviral infection seem to be compatible with cellular life.

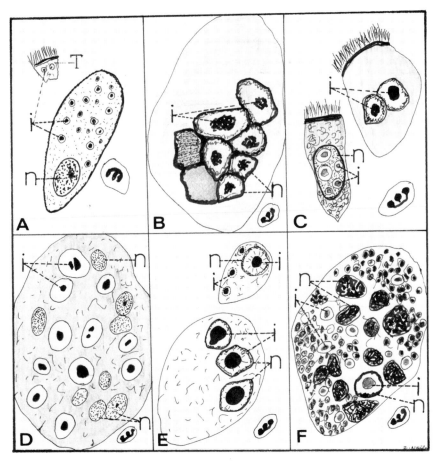

Figure 108
Structures of virally infected respiratory tract cells (i = inclusion; n = nucleus; t = tuft). (A) Parainfluenza viral infection. (B) Herpes simplex viral infection. (C) Adenoviral infection. (D) Respiratory syncytial viral infection. (E) Cytomegalic viral infection. (F) Measles viral infection.

The characteristic changes are confined to the nuclei. The cytoplasm is not affected.

Little cellular degeneration is seen, and the terminal bar and cilia are preserved in spite of the marked nuclear alteration.

The adequate cytoplasm shows no changes, and stains pink or blue, as in the noninfected nasotracheal cells.

Representing the early stage of cellular infection, small (2–4 μ) multiple, round, eosinophilic inclusions, surrounded by a more or less prominent individual halo, are first seen in the nuclei, resulting in a honeycomb appearance of the nuclei.

The inclusions are granular and seem to pack the enlarged nuclei, but with no apparent disruption of the normal chromatin pattern.

In the later stages of the infection, the inclusions seem to merge into a larger (7–10 μ), densely packed, central, dark basophilic mass. The inclusions are surrounded by a very prominent halo, occupying most of the surface of the enlarged nucleus.

At this advanced stage of cellular infection, the cytoplasm often shows some degenerative vacuolization and distortion but still has some persistent diagnostic cilia or a recognizable terminal bar.

The background of most of the smears is clear of debris, with variable amounts of inflammatory cells with monocytic and lymphocytic predominance.

Respiratory Syncytial Viral Infection (Fig. 108D)

The presence in the nasopharyngeal or tracheal smear of large multinucleated cells (100 μ^+) with an abundant, basophilic, vacuolated cytoplasm and well-defined but irregular borders is diagnostic. These irregularities are the result of the bulging of the individual component cells in syncytia.

No cilia or terminal bars are present in the cells.

Multiple (10–30) prominent, round, and extremely basophilic, intracytoplasmic inclusions are surrounded by an individual halo, the largest seen in any type of virus-infected cells.

The numerous enlarged nuclei are loosely scattered throughout the cytoplasm with no molding but with overlapping. Vesicular, round, or oval in shape, the nuclei have a uniformly distributed granular chromatin with an occasional small nucleolus. They resemble the nuclei of normal tracheal columnar cells, except for the size.

No intranuclear inclusions are seen.

The background of the smear shows bacteria, cellular debris, and variable amounts of mixed inflammatory cells.

Cytomegalic Viral Infection (Figs. 108E, 109D)

The affected nasotracheal columnar cells are usually single and nonciliated with a relatively scanty, lacy cytoplasm with no terminal bar or cilia.

The cells are moderately enlarged and vary in proportion to the size of their intranuclear inclusions.

Their stained color varies from blue to dark lavender; they appear darker than the noninfected tracheal cells.

In the cytoplasm, round or oval basophilic inclusions with moderate halos are occasionally found and are sometimes difficult to differentiate from ingested debris or red blood cells.

The cytoplasmic membrane is thick, regular, and well defined.

The round or oval, often enlarged, nuclei show coarsely granular basophilic chromatin condensed into the form of a ring along the inner surface of the nuclear membrane and around a central large, single, dark basophilic inclusion and its well-defined halo.

The cells are usually mononucleated (95%), but on occasion there are giant, inclusion-bearing, multinucleated syncytial columnar cells. Each of the 3 to 6 nuclei scattered in the abundant basophilic cytoplasm contains an

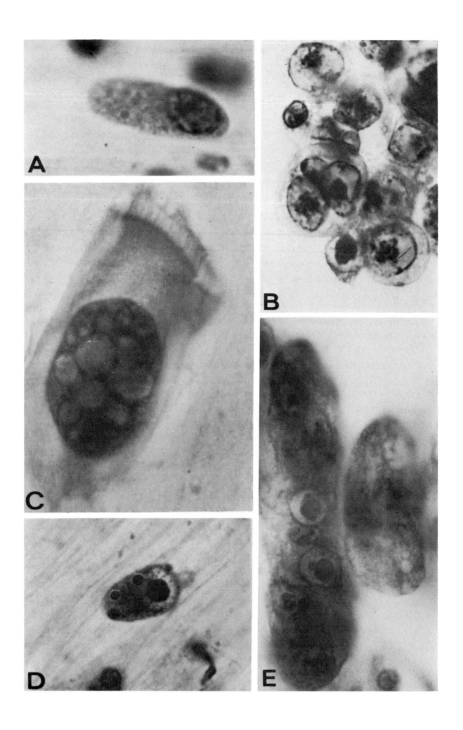

individual dark inclusion, surrounded by a halo, as seen in the mononucleated cells.

Measles Viral Infection (Figs. 108F, 109E)

The cytologic diagnosis of measles from the nasopharyngeal secretions can precede the appearance of the skin rash by two to three days.

Giant multinucleated cells (30–180 μ in diameter) with an abundant, well-defined, granular, acidophilic, and vacuolated cytoplasm are diagnostic. They contain 10 to 200 round nuclei of regular size that resemble the nuclei of large lymphocytes. These cells are easily seen even under low-power magnification examination of secretions from the coryza of measles.

A large proportion of these nuclei show evidence of degeneration (karyopyknosis and large, hazy chromatin clumping).

The numerous diagnostic multiple, small, round-to-oval, eosinophilic inclusion bodies surrounded by minute halos are generally found in the cytoplasm (85%) and occasionally in the nuclei.

The background of the smear shows no bacteria and only a moderate number of polymorphonuclear and mononuclear inflammatory cells.

CYTOLOGY OF BACTERIAL INFECTION

In acute or chronic bacterial tracheobronchitis, the smear from the tracheal aspiration usually contains a huge quantity of neutrophils, monocytes, and histiocytes and a variable number of large phagocytes.

The background, most of the time, is very dirty with blood, protein deposits, and colonies of mixed bacteria.

The epithelial cells, squamous or columnar, may show, besides the vacuolated degenerative changes, an increase in eosinophilia of their cytoplasm and variations in their shape and size, which cause them to resemble metaplastic bronchial cells. The semitransparency and granularity of their cytoplasm helps in the differential diagnosis.

A large number of stripped oval epithelial nuclei is also seen; they should not be confused with the exfoliated cells from a poorly differentiated lung carcinoma. The regularity of their nuclear membrane and of their chromatin pattern helps in their differentiation.

Figure 109

(A) Bronchial cell infected with parainfluenza virus, showing multiple intracytoplasmic inclusions. Note the absence of cilia. ($\times$ 750) (B) Bronchial cells infected with herpes simplex virus, showing occasional multinucleation and granular intranuclear inclusions. ($\times$ 750) (C) Bronchial cells infected with adenovirus in its early stage. Note the multiple intranuclear inclusions and the persisting terminal bar and cilia. ($\times$ 1800) (D) Cytomegalic viral infection of a bronchial cell showing the basophilic intracytoplasmic and intranuclear inclusions. ($\times$ 750) (E) Measles viral infection of 2 bronchial cells. Note the large size and the multiple intracytoplasmic and intranuclear eosinophilic inclusions. ($\times$ 750)

The occurrence of these tumors is rare, and most of the benign neoplasms (mesenchymal in origin) do not exfoliate recognizable diagnostic cells unless they become ulcerated or a malignant change occurs.

Well-Differentiated Squamous Cell Carcinoma
This tumor sheds keratinized cells similar to the ones described for the invasive cervical carcinoma.

Undifferentiated Squamous Cell Carcinoma
The cells of this tumor usually shed in clusters. They have very scanty basophilic cytoplasm and vesicular nuclei of variable size ($10–50\ \mu$), occupying about four-fifths of the cellular volume. The chromatin is slightly increased and coarsely granular. The nuclear membrane is sharp with some irregularity. Occasional multiple prominent nucleoli are present. Multinucleation occurs in about 20% of the cells.

Transitional Cell Carcinoma
Numerous cells may shed singly (20%), in clusters (30%), or in sheets (50%). The shape of these cells varies from polygonal to spindle. The cytoplasm is adequate to abundant, granular, and acidophilic. The cytoplasmic membrane is poorly defined. The nuclei are round to oval with moderate size variation. The chromatin is increased and coarsely granular. Multinucleation occurs in large syncytial-type cellular formations.

Lymphoepithelioma
Pools of mature lymphocytes are found mixed with the malignant transitional cells described in the preceding paragraph.

Cytology of the Lower Respiratory Tract

NORMAL CELLS

Benign Squamous Cells (Fig. 110A)
Benign squamous cells may originate from the buccal or the nasal cavities, the pharynx, or the larynx. They often indicate, especially when seen in quantity, that the sputum is mixed with saliva. The exact origin of the cells cannot be determined. Their number is drastically decreased if the patient brushes his teeth or gargles with a saline solution and cleanses his mouth before coughing to produce sputa. Their features are similar to those of the squamous cells of vaginal origin.

The cytoplasm of benign squamous cells is abundant, thin and pink, often semitransparent, which differentiates them from the metaplastic bronchial squamous cells, which have an opaque, less abundant thick cytoplasm that stains deep orange.

Their nuclei are normal in size, either pyknotic or vesicular, similar to the previously described nuclei of normal superficial and intermediate squamous vaginal cells.

When present in large amounts, benign squamous cells can become a nuisance by masking the diagnostic bronchial cells (pitfall of diagnosis).

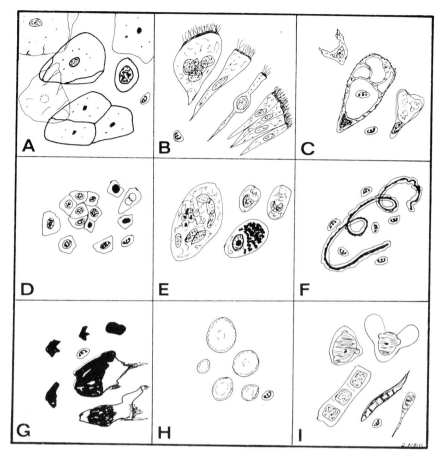

Figure 110
Normal structures seen in a sputum. (A) Benign squamous cell. (B) Ciliated bronchial cells. (C) Nonciliated goblet cells. (D) Basal bronchial cells. (E) histiocytes. (F) Curschmann's spiral. (G) Inspissated mucus and calcified concretions. (H) Corpora amylacea. (I) Contaminants.

They may show pronounced degenerative changes (cytoplasmic vacuolization or pigment deposits), which can make them difficult to recognize.

Occasionally, smaller cells, structurally similar to the parabasal squamous cells of the vaginal smear, are found singly or in sheets. Their presence often indicates an erosion or ulceration of the bronchial or buccal epithelium.

Ciliated Bronchial Cells (Fig. 107B, 110B)
Intact ciliated bronchial cells are more common in the bronchial brush or washing specimen than in the sputum.

Elongated and pyramidal in shape, they are shed singly, in tight groups, or in sheets, often in a palisade formation.

Their cytoplasm is delicate, pale pink or blue, elongated, with a distinct terminal bar at the broad end, from which emerge moderately short cilia, stained pink to red. The lateral borders are often indistinct.

Occasional granular pigments (lipofuscin) may be found in cytoplasmic vacuoles adjacent to the nuclei. They may be related to abnormal lipid and protein metabolism and are found mainly after irradiation therapy.

Their eccentric nuclei are uniformly round or oval, single (85% of the cells) or multiple (2–6). Their size varies from 7 to 15 μ.

The chromatin is uniformly finely granular and regularly distributed throughout the nuclei.

On occasion (25%), small, central, single or multiple, round, red nucleoli can be seen.

The traumatically exfoliated bronchial cells often degenerate quickly. They first lose their cilia, then the cytoplasm, which explains the presence of a variable number of stripped but relatively well-preserved nuclei, seen in some bronchial washing specimens or in the sputum that follows them.

The presence of ciliated multinucleated cells in small numbers has no significance. If found in excessive amounts in a patient with known lung tuberculosis, they may indicate the possible coexistence of a pulmonary carcinoma.

The exfoliated bronchogenic neoplastic cells are not ciliated, and the presence of cilia in a cell is sufficient to interpret it as being a benign one, even if it possesses several good features for malignancy.

Goblet Cells (Figs. 107C, 110C)

Single, in clusters, or in sheets, these nonciliated mucus-secreting cells are seldom seen in a routine smear, except in certain diseases (asthma, chronic bronchitis, or bronchiectasis).

Goblet cells are relatively stout, large cells (15–30 μ) with marked variation in their shape—round (20%), elongated (40%), and irregular (40%).

Their cytoplasm is seldom found intact because of their delicate texture. Vesicular and abundant, the cytoplasm is occasionally distended by hypersecretory single or multiple mucus-containing vacuoles, often with ingested well-preserved neutrophils. These secretions are periodic acid–Schiff (PAS) stain positive. The chromatin is evenly distributed. The cytoplasmic border is thin and poorly defined with no discernible terminal bar.

The nucleus, usually single and peripherally located, is large (10–18 μ), round (30%), oval (40%), or crescent-shaped (30%), and occasionally almost pyknotic.

The chromatin is finely granular and regularly distributed.

The nuclear membrane is thin, smooth, and well defined. The nucleoli are usually single (90%), small, and red.

Basal Columnar Cells (Figs. 107D, 110D)

Basal columnar cells, which originate in the deep layer of the bronchial epithelium, are nonciliated cells and are rarely shed physiologically. Most of the time they are the result of traumatic exfoliation of proliferated bronchial epithelium produced by repeated trauma, mucosal ulceration, chronic irritation (heavy smoking), or other inflammatory conditions.

They may shed singly (20%), in clusters (20%), or in small sheets in the form of papillary projections (60%). They are mainly seen in bronchial washings.

Their size varies moderately (8–12 μ); their shape is usually round or triangular.

Their dense cytoplasm is adequate to scanty and can be blue, pink, or dark orange. No debris, secretory vacuoles, or inclusions are present.

Their cytoplasmic borders are well defined when single but indistinct in a sheet.

Their single nuclei are small, uniform in size, and round or oval in shape with a dense granular, hyperchromatic or occasionally pyknotic appearance.

The basal columnar cells should be differentiated from squamous metaplastic cells because of the similar density and occasional tendency of the cytoplasm to stain orange. They should also be differentiated from the oat-cell carcinoma because of the scantiness of their cytoplasm and the hyperchromatism of their nuclei.

The regularity of the chromatin and its uniform distribution, the smooth nuclear membrane, and their small uniform size aid in the recognition of these cells as benign basal cells.

Pulmonary Histiocytes (Figs. 110E, 111A, B)

The presence of numerous phagocytes (macrophages) in a sputum is not associated with any specific pathologic lesion. The source of these cells is still conjectural, but they probably arise, in part, from the alveolar cells and the reticuloendothelial system. They all have a phagocytic power, and their mission is to eliminate or neutralize unwanted particles from the pulmonary air passages.

Pulmonary histiocytes are always shed singly, sometimes in clusters, but never in sheets.

The absence of cellular molding is a very important criterion in differentiating them from some tumor cells, which they may closely resemble.

Their size varies (10–100 μ), and their shape can be round (40%), oval (20%), or irregular with pseudopodlike projections (40%).

The cytoplasm is usually vacuolated and varies from very scanty to abundant. The cytoplasm is sometimes free of ingested material, but more often it contains carbon particles (dust cells), blood (hemosiderin), debris (heart-failure cells), or lipids (lipid pneumonia).

The cytoplasm stains irregularly and varies from pink to dark orange and from green to dark blue.

The cytoplasmic border is well defined and sharp (80%), or indistinct, often as the result of degeneration (20%).

The nuclei are eccentric in 80% and central in 20%. Normally single (in 85% of the cells), the nuclei may be multiple (2–20 nuclei), particularly in the sputum of a patient with pneumonia or other irritative condition.

The nuclei are round, oval (60% of the time), or indented in the shape of a bean (40%). The nuclear border is distinct, well defined, and regular.

The amount of chromatin is variable and can be finely granular to coarsely clumped.

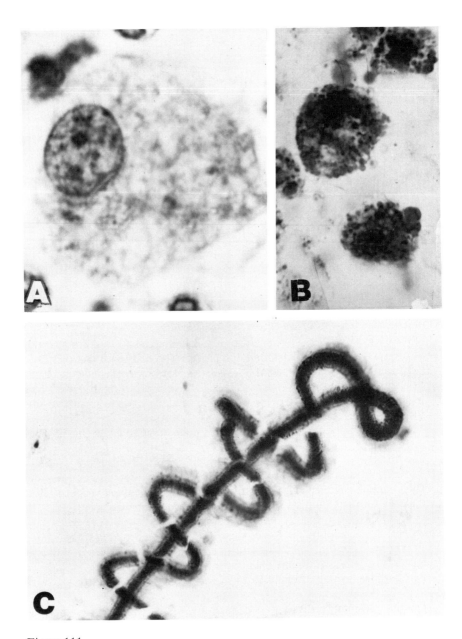

Figure 111
(A) Histiocytes in a sputum with little ingested material. Note the vacuolated cytoplasm, the eccentric nucleus, and the prominent nucleolus. ($\times$ 2100) (B) Histiocyte containing dust and debris. ($\times$ 750) (C) Curschmann's spiral in a sputum specimen. Note the central dark core surrounded by the semitransparent inspissated mucus. ($\times$ 1300)

On occasion (15%), 1 or 2 small, prominent, round, red nucleoli can be found; they should not be mistaken for a malignant feature.

It is sometimes difficult to differentiate these phagocytic cells from the alveolar adenocarcinoma, giant cell, or even poorly differentiated bronchial cell carcinoma, especially when degenerative changes have occurred. The nuclei of these phagocytes can be as dark, prominent, and irregular as are the malignant nuclei, and their cytoplasm can vary from semitransparent to deep orange, as in the malignant cells. However, the irregularity of the staining of the cytoplasm, the presence of ingested debris, the occasional bean-shaped appearance of their nuclei, and the lack of cellular molding help in their identification.

The presence of dust cells indicates that at least part of the sputum originated from the lower respiratory tract.

Smooth-Muscle Cells

In sputum following traumatic bronchoscopy or penetrating wound injuries and in necrotizing ulcerative tracheobronchitis (Wegener's granulomatosis), bizarre smooth-muscle cells may become a pitfall in diagnosis. Their size and shape vary, often producing a caudate or other extremely bizarre shape, with glassy eosinophilic cytoplasm and lobular irregular nuclei. Their fuzzy cytoplasmic boundaries and the regularity of their fine nuclear chromatin aid in the differentiation of these cells from those of squamous cancer or metaplasia.

Inflammatory Cells

An increase in the number of polymorphonuclear leukocytes in a sputum specimen is found in pulmonary malignancy, bronchitis, pulmonary abscess, or any other irritative buccal, tracheal, or bronchial lesion. Most of the time, their absence indicates that the smear contains no cancer cells or that the specimen is unsatisfactory.

A moderate increase in the number of monocytes and lymphocytes is observed in chronic bronchitis, tuberculosis, and atelectasis. The presence of a large number of lymphocytic pools in strands of mucus indicates either the traumatic scraping of lymphoid tissue (for example, the tonsils) or that a lymphoid vessel has been ruptured traumatically or by an invading neoplasm, which explains the often noted coexistence of lymphocytic pools with bronchial carcinoma.

The inflammatory lymphocytes are differentiated from the malignant lymphomatous cells by the uniformity of their size and shape and the absence of a prominent nucleolus.

The number of eosinophilic or basophilic polymorphonuclear leukocytes and of plasma cells may increase in cases of respiratory allergy, fungal and parasitic diseases, or other nonspecific chronic inflammation.

ACELLULAR STRUCTURES

Curschmann's Spirals (Figs. 110F, 111C)

These exfoliated castlike formations result from the expelled precipitation of mucus and glucoprotein in the small bronchioles. They vary in length and characteristically have a spiral shape. They show a central dark core that is surrounded by a semitransparent, jellylike, amorphous, pink, mucoid material. They are found in greater abundance in patients with asthma, long-standing chronic bronchitis, incomplete bronchial obstruction, or history of heavy cigarette smoking. Their presence indicates that at least part of the sputum originated from deep in the lung. These Curshmann's spirals, when found surrounded by inflammatory cells, have been attributed to air pollution rather than to asthma.

Inspissated Mucus (Figs. 110G, 112A)
These are irregular masses of mucus and proteinaceous material that vary in size and shape and color (pink to dark purple). They are seen mainly in the sputum of patients with bronchial obstruction. They are usually mixed with a large number of bacteria and can be a handicap to diagnosis by masking the tumor cells or simulating a degenerate, stripped, pyknotic, malignant nucleus. The absence of chromatin structures helps in the differential diagnosis.

Calcified Debris (Fig. 112B)
Variable amounts of loose clusters of granules of irregularly crystalline shaped concretions, dark purple in color, can occasionally be found in the sputum. Some may have lamellar structures. They have no particular significance, but it is thought that they may indicate the presence of a healed, calcified, chronic tuberculous lesion, pulmonary fibrosis, or alveolar cell carcinoma.

Corpora Amylacea (Figs. 110H, 112C)
They are small (5–40 μ) waxy-looking round bodies, pink-to-purple stained. They are found singly or in clusters. They overlap but do not mold. Formed by the condensation of precipitated protein, they usually originate from the alveolar space and are found in increased numbers in the sputum of patients with long-standing pulmonary edema.

Charcot-Leyden Crystals (Fig. 112D)
Elongated, octohedral in shape, pink to orange, these crystals may be found in sputum of patients with allergic or asthmatic pulmonary lesions and in the healing stage of Löffler's pneumonia. Various amounts of eosinophilic leukocytes are also often present in the smears. These crystals are thought to originate from the condensation of degenerate eosinophilic granules found in the cytoplasm of the eosinophils.

Contaminants (Figs. 110I, 113)
Contaminants can be food particles from the buccal cavity, inhaled pollen, or debris contained in the staining solution or the mounting medium. They can be mineral, vegetable, or animal in origin. Their size, shape, and color vary enormously according to their nature. Some of the orange, well-preserved vegetable cells may imitate the appearance of malignant cells by their size, dark nuclei, and staining (diagnostic pitfall), but usually they

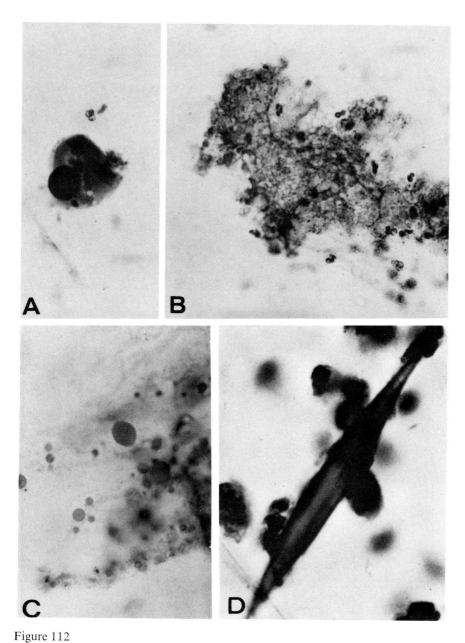

Figure 112

(A) Inspissated mucus, densely basophilic but amorphous. May be mistaken for a degenerate pyknotic nucleus. ($\times$ 450) (B) Calcified debris in a sputum, which may become a handicap for diagnosis by masking the tumor cells. ($\times$ 450) (C) Corpora amylacea in the sputum of a male with a long-standing pulmonary edema: Round and stained pink to yellow, they vary in size. ($\times$ 450) (D) Charcot-Leyden crystals. ($\times$ 450)

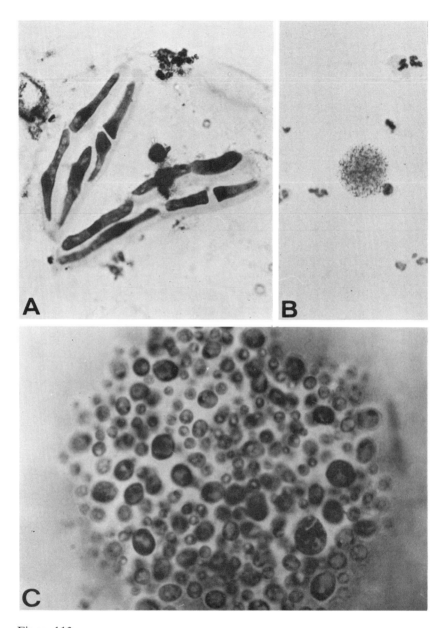

Figure 113
(A) Vegetable fibers as a contaminant in sputum. Note the elongated dark nucleus and transparent cytoplasm, not to be confused with malignant squamous spindle cells. (× 450) (B) Droplet of oil in the sputum of a man who was using nose drops. (× 450) (C) Colony of yeast in a sputum kept overnight before processing. (× 750)

are easy to recognize by their transparent cellulose capsule, amorphous nuclei, uniform, rectangular (vegetable cell), or butterfly shape (seed and pollen).

Benign Pulmonary Lesions

DISEASES OF VASCULAR ORIGIN

Pulmonary Congestion and Edema

With pulmonary congestion and edema, the number of ciliated epithelial cells is scantier than normal, especially in the sputum specimen. The ciliated epithelial cells are hypertrophic and have larger round or oval nuclei, coarser chromatin and more prominent nucleoli than the normal cells, which explains why they are sometimes mistaken for alveolar neoplastic cells. They shed singly (95%), as opposed to the alveolar carcinoma cells, which shed in clusters. A moderate increase in the number of hemosiderin-laden macrophages (heart-failure cells) is usually present. The amount and tenacity of the strands of mucus are often decreased. The background of the smear shows a mild degree of finely granular pink deposits. The number of monocytic inflammatory cells varies. It is generally increased. In long-standing edema, numerous round pink-orange corpora amylacea may be present. A few have several small black particles in continuous shaking movement (Brownian movement) when examined under oil immersion magnification.

Pulmonary Infarction and Hemorrhage

With infarction and hemorrhage, the number of epithelial cells in a sputum is normal, and the cells show very few characteristic changes. Large numbers of heart-failure cells (hemosiderin-containing histiocytes) and multinucleated foreign-body giant cells containing hemosiderin debris may be seen. The background of the smear is often obscured by large amounts of coarsely clumped cellular debris, basophilic protein deposits, and abundant fresh and old red blood cells. The added increase of the number of leukocytes and the presence of colonies of microorganisms should suggest the possibility of a septic infarction.

DISEASES OF BRONCHIOALVEOLAR EXPANSION

Atelectasis

With atelectasis, there is a prevalence of mononucleated small macrophages and multinucleated foreign-body giant cells containing a minimum of phagocytized debris. Various numbers of orange-stained squamous cells, metaplastic, dyskaryotic, or dyskeratotic in type, are often seen and may become pitfalls in the diagnosis of bronchogenic squamous cell carcinoma. The background of the smear is usually clean, with almost no deposits. There is no increase in inflammatory cells.

Löffler's Pneumonia

An unusual increase of eosinophils in a sputum should suggest Löffler's pneumonia. A moderate number of other inflammatory cells may also be seen. Eosinophilic granules, resulting from ruptured eosinophils, may be seen in the background. Occasional Charcot-Leyden crystals may also be seen. The epithelial cells are usually not remarkable.

Asthma

With asthma there is often a prevalence of well-preserved, hypersecretory goblet cells, found singly or in clusters, with large, round, or triangular nuclei and prominent nucleoli. In the background, there is an increase in the amount of inflammatory cells (leukocytes and eosinophils) and in coarsely granular basophilic protein deposits. The presence of Curschmann's spirals and inspissated mucus trapped in the usually abundant, thick, tenacious, purple-stained strands of mucus is helpful in the diagnosis. On rare occasions, fusiform, orange Charcot-Leyden crystals may be seen. The common occurrence of metaplastic, dyskaryotic, and dyskeratotic squamous cells in chronic asthma can also be a pitfall in the diagnosis of bronchogenic squamous cell carcinoma.

Bronchiectasis

There is a prevalence of degenerate hypersecretory goblet cells with large numbers of polymorphonuclear neutrophils, cellular debris, phagocytes, bacteria, and protein deposits in the background. Multinucleated giant cells and squamous metaplastic cells are frequently seen.

INFLAMMATORY DISEASES

Viral Diseases

In viral diseases, the diagnostic cells are similar to the ones originating from the upper respiratory tract. Exfoliative cytology is a very useful, simple, and rapid diagnostic tool in all kinds of viral pneumonias.

Bacterial Diseases

Large numbers of leukocytes, cellular debris, and colonies of various kinds of microorganisms are present in the abundant, thick mucus found in the sputum smears during a bacterial bronchitis, pneumonia, or pulmonary abscess. Occasionally, hyperplastic, small, triangular, and nonciliated bronchial cells can be seen singly or in groups (Papanicolaou cells) in chronic inflammation. Fresh and old red blood cells indicate a bronchitis, rather than pneumonia, and an abscess formation is often indicated by the presence of round metaplastic squamous cells with bland or pyknotic nuclei and finely vacuolated orange cytoplasm. Similarly, the presence of very large numbers of degenerate epithelial and inflammatory cells usually indicates an acute or massive inflammatory lesion. An abundance of colonies of microorganisms without inflammatory or degenerative cellular reaction probably indicates that the microorganisms have multiplied after the specimen has been obtained and are contaminants rather than the cause of the infection (Fig. 113C). These colonies may be so abundant that they can prevent the detection of tumor cells by obscuring them in a sputum that has not been fixed or refrigerated and when processing has been delayed.

Allergic Diseases

The sputum smear from a patient with respiratory allergy often shows an increase in the number of hypersecretory goblet cells, shed in clusters or tight sheets (honeycomb appearance). Occasionally a large number of eosinophilic and basophilic (mast cells) polymorphonuclear leukocytes can be seen. A large number of well-preserved, stripped, hypertrophic, columnar cell nuclei, round or oval in shape, with smooth nuclear membranes, coarse chromatin, and prominent nucleoli, are often found and should be distinguished from oat-cell carcinoma. There is also an increase in the number of histiocytes and foreign-body giant cells having no apparent phagocytized debris. The background shows a moderate increase in the number of strands of mucus with very few bacteria.

Tuberculosis (Fig. 114A)

Pulmonary tuberculosis is difficult to diagnose by cytology. Occasionally, Langhans' cells are recognized by the peripheral location of their multiple nuclei. It is necessary to differentiate these Langhans' cells from similar appearing foreign-body giant cells whose nuclei are more central and whose cytoplasm may show clusters of ingested cytoplasmic debris. The abundant pink cytoplasm of Langhans' cells may contain more than 30 elongated or round-to-oval, often overlapping nuclei with fine granular chromatin and no nucleoli. The cytoplasmic membrane is usually well defined.

On occasion, epithelioid cells resembling large, elongated phagocytes but with centrally located elongated nuclei with tapered ends can be found trapped in the mucus. The cytoplasm of these cells is dense, basophilic, amorphous, and free of inclusions. The cytoplasmic margins are hazy.

The frequent presence in the smear of metaplastic squamous cells, elongated, spindle, tadpole, or very irregular in shape, can be a pitfall in the diagnosis of well-differentiated squamous carcinoma. An acid-fast stain of the air-dried, but unfixed, smears may show the elongated red diagnostic microorganisms in the midst of pale blue epithelial and inflammatory cells.

Fungal and Parasitic Diseases

Candida albicans (Fig. 115A) is diagnosed by the presence of yellow-to-brown stained long bodies (hyphae) and budding round bodies (conidia) similar to the ones infecting the genital tract. Often these bodies indicate a secondary infection in a debilitated patient or in one who has received large quantities of antibiotics or extensive chemotherapeutic treatment. The fungi can also be a contaminant, originating from an oral lesion, especially if they are seen in very large colonies. Their unexpected presence in a routine smear may indicate the possibility of a coexisting or advanced pulmonary neoplasm, and this smear should be screened more carefully than usual. On occasion, in a poorly dehydrated sputum smear, the water trapped in the Candida hyphae (long bodies) may evaporate, producing a shining linear image that should not be confused with crystals. The organism grows well on Sabouraud's or blood agar culture medium.

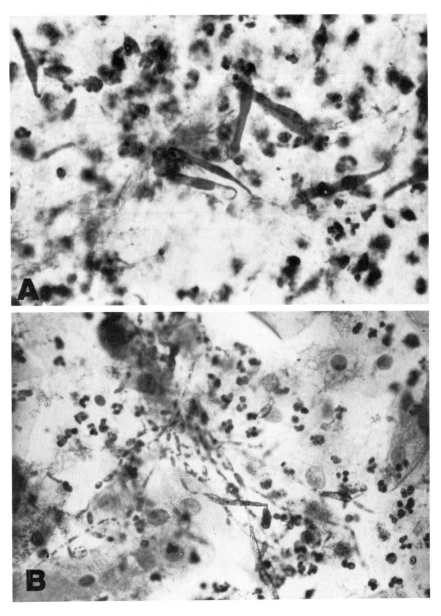

Figure 114
(A) Elongated epithelioid cells in a bronchial washing of a patient with active pulmonary tuberculosis. (× 450) (B) *Geotrichum* fungus in a sputum specimen. Note the irregular branching hyphae. (× 450)

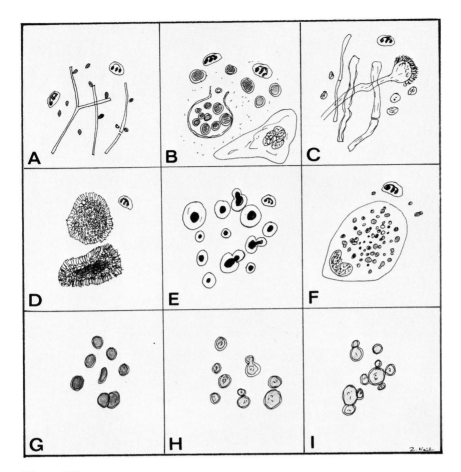

Figure 115
Different fungi that may be seen in a sputum specimen. (A) *Candida albicans.*
(B) *Coccidioides.* (C) *Aspergillus.* (D) *Actinomyces.* (E) *Cryptococcus.* (F)
Histioplasma. (G) *Chromoblastomyces.* (H) *Blastomyces dermatitidis,* which
causes North American blastomycosis. (I) *Blastomyces brasiliensis,* which
causes South American blastomycosis.

COCCIDIOIDOMYCOSIS (*Coccidioides immitis*) (FIGS. 115B, 116A, B) is indicated by large (50–20 μ), free, cystic structures (spherules or sporangia)
that contain multiple small, round endospores, and they are often overlooked because of their resemblance to the pollen seen frequently as a contaminant in certain seasons. When the cyst ruptures, a larger number of
these endospores may be seen free in the background of the smear. They
are usually poorly stained and overlooked. They can be recognized by the
presence of a double-shelled membrane, their poorly stained pink color,
and their size. Large amounts of granular deposits and necrotic debris are
also often present in the smear. When they are cultured in Sabouraud's
glucose medium only the mycelial forms are seen.

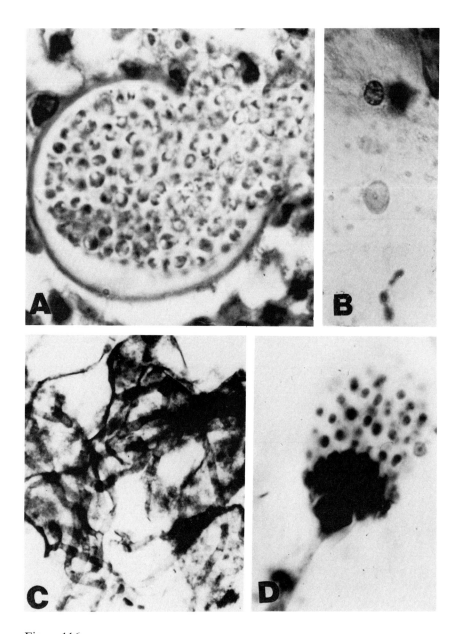

Figure 116
(A) Coccidioidomycosis in bronchial aspirate showing rupturing cystic structures (sporangia). (× 450) (B) One of the released, free endospores in the sputum. (× 750) (C) Aspergillosis. Note the thick hyphae. (× 450) (D) Conidiospore in aspergillosis. (× 450)

ASPERGILLOSIS (FIGS. 115C, 116C, D) is an infection that is often secondary to the debilitation of a patient with a malignant lesion, or one that occurs after prolonged steroid, antibiotic, or chemotherapeutic treatment. It is acquired by inhalation. As in infection with *Candida,* long and round bodies can be seen in the sputum smear. The long bodies (hyphae) are larger and more tortuous, and the segments are shorter. The round bodies (conidia) often vary in size (5–20 μ). In rare instances, when an abscess is present, a bulbar fruiting body (a complex conidiospore with tubular structures [phialides] supporting chains of greenish conida [spores]) can be found at one end of a hypha and is diagnostic (Fig. 116D). It should be differentiated from similar-looking conidiospores of *Penicillium* fungi, which are frequent contaminants of our smears.

ACTINOMYCOSIS (FIGS. 115D, 117A). The conglomeration of *Actinomyces israelii,* an anaerobic bacteria that is misrepresented as a fungus, forms yellow, sulphurlike, grossly visible (100–300 μ) granules. Their presence in the sputum should be noted by the technician preparing the smear. Microscopically, they are large structures, easily seen during a low-power magnification examination of the smear, red to purple stained, with distinct, rod-like projections on the surface. In higher-power magnification examination, the red-to-purple stained, tangled, branching filaments may be seen. The danger of missing this diagnosis is in mistaking them for contaminants. Besides the inflammatory cells, numerous phagocytes with large lipid-containing vacuoles may be present in pulmonary actinomycosis.

CRYPTOCOCCOSIS, OR TORULOSIS (FIG. 115E). The fungus *Cryptococcus* is found in the smear in small clusters of poorly stained, round, pink bodies measuring 5 to 15 μ with occasional budding. Their semitransparent mucinous capsules are thick and create a clear halo around the reddish-purple spore. The buds are attached to the mother cells by narrow necks, giving them their characteristic teardrop shape. They are often overlooked and confused with contaminants or talcum powder crystals. The smears show large amounts of inflammation, protein, and cellular debris deposits and an increase in the number of histiocytes and foreign-body giant cells. They occur more frequently in patients with preexisting lymphoma or other advanced malignant disease. The pigeon is the main source of this infection.

HISTOPLASMOSIS (FIG. 115F). This infection is rarely diagnosed by sputum examination and is often overlooked. The diagnosis is based on the presence of large mononucleated or binucleated phagocytes, whose cytoplasm is packed with numerous brown-to-black round organisms (*Histoplasma capsulatum*) (2–4 μ) showing occasional budding. These organisms should not be confused with ingested carbon particles or with hemosiderin pigment. A Gomori's methenamine silver or periodic acid–Schiff stain helps in the differentiation. They are usually mistaken for irregular granular protein deposits or degenerative debris when seen outside the phagocytes.

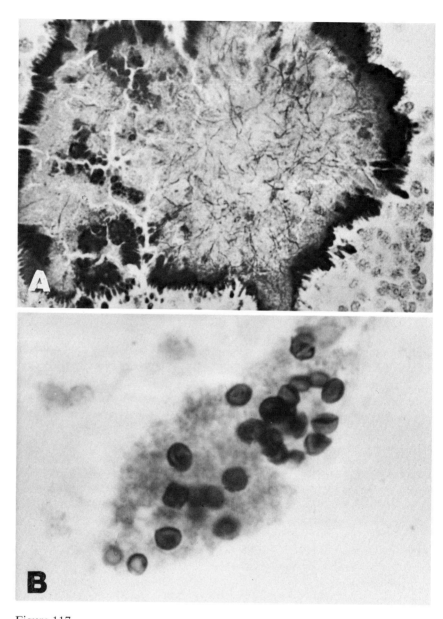

Figure 117
(A) Actinomycosis. Note the conglomeration of tangled, branching filaments. (× 450) (B) *Pneumocystis carinii* organism in needle aspirate. (×750)

NORTH AMERICAN BLASTOMYCOSIS (FIGS. 115H, 118A). The fungi (*Blastomyces dermatitidis*) are poorly stained, slightly orangeophilic, and vary in size (from the size of a red blood corpuscle to that of a large histiocyte, 15 to 20 μ). The organisms are recognized by their darker inner wall and outer fuzzy wall separated by a refractile zone. Single budding with a large base attaching them to the mother cells is characteristic. The smears also show large numbers of foreign-body giant cells and inflammatory cells. Frequently, metaplastic, dyskaryotic, and dyskeratotic squamous superficial cells are also present, which can be a pitfall in the diagnosis of squamous cell carcinoma.

SOUTH AMERICAN BLASTOMYCOSIS (FIGS. 115I AND 118B). Similar to the fungi in North American blastomycosis, *Blastomyces brasiliensis* have multiple, rather than single, teardrop-shaped buds surrounding the parent capsule, giving them the characteristic ship's-wheel appearance.

CHROMOBLASTOMYCOSIS (FIG. 118C). Rarely seen, these fungi are found in tropical regions. They are round, spherical, or moon-shaped organisms, about the size of the red blood cells. They stain dark brown to deep orange with a bland center. No budding is seen. Large amounts of inflammatory and phagocytic reaction are usually present in the smear.

GEOTRICHOSIS (GEOTRICHUM CANDIDUM) (FIG. 114B). Geotrichosis is a mild fungal disease of the mucous membrane of the mouth and lung. Sputum and bronchial washings may contain large numbers of rectangular-to-oval arthrospores (4–8 μ) with occasional septate mycelia. Chains of these arthrospores may also be found. They are thick-walled and nonbudding. The hyphae vary in size, thickness, and length, and they show occasional branching. Because of the frequency of buccal infestation, a diagnosis in sputum of pulmonary geotrichosis should be made only if the possibility of buccal contamination of the smear can be eliminated. Presence of the arthrospores in smears from direct tracheal aspirates or bronchial washings is more significant.

PNEUMOCYSTIC PNEUMONIA (FIG. 117B). The presence of clusters of *Pneumocystis carinii* organisms is diagnostic of this often fatal pneumonia of children or adults who have received frequent immunosuppressive therapy. Sputum and bronchial washing smears are poor specimens. They usually will not contain the organisms that are embedded in thick, hard-to-cough, honeycombed mucoid exudate. Only a transcutaneous needle aspiration smear or open lung impression smear will contain the diagnostic loose cluster of organisms. They are recognized by their pink-to-orange stain and encapsulated round or pear shape. They vary in size (5–10 μ). They occasionally have a small, central, blue, microdotlike structure. They are seen singly or in loose clusters in strands of foamy mucus. The sputum usually shows a moderate degree of secondary inflammatory reaction, polymorphonuclear cells, and thick, tenacious strands of mucus. A methenamine silver stain of the extra unstained slide helps in the diagnosis. They

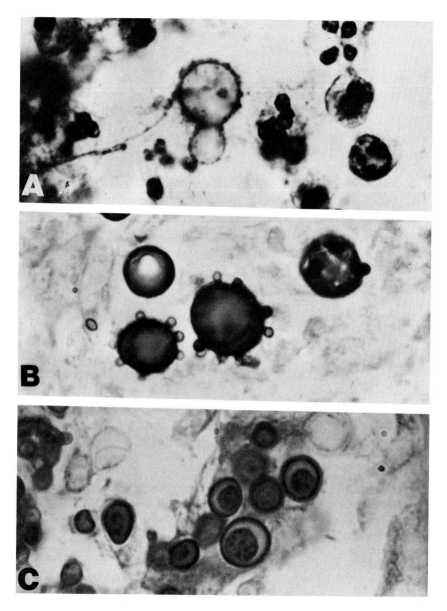

Figure 118
Sputum specimen with fungi of (A) North American blastomycosis, (B) South American blastomycosis, and (C) chromoblastomycosis. ($\times$ 1300)

stain poorly with eosin, but somewhat better with Giemsa's stain. They should be differentiated from contaminant yeast, red blood cells, or other round, yeastlike artifacts.

Diagnostic Pitfalls—Contaminant Fungi and Pseudofungi

Saprophyte fungi and other similar-looking contaminant structures may be found in poorly obtained or processed sputum and mistaken for pathogenic fungi. Most of the saprophytes grow well at room temperature and are usually pigmented. If the processing of a sputum is delayed, an abundance of fungal organisms may be present in the smear. The most common are:

PENICILLIUM (FIG. 119c). These are fast-growing fungi that often contaminate the various staining solutions or even the fixative sprays used in cytology. The diagnosis is based on the presence of chains of spores pinched off from flask-shaped sterigma on top of branched or unbranched conidiophores. These structures should be differentiated from the more regular conidiophores of pathogenic *Aspergillus* fungi.

SCOPULARIOPSIS SP (FIG. 119b). These nonpathogenic fungi are recognized by the numerous, chainlike, small, lemon-shaped conidia at the tip of many short, single, branched hyphae. Because of their larger size, they are easily differentiated from chains of cocci. The hyphae may be mistaken for *Candida* except for the absence of regular segmentation and attached grapelike spores.

HELMINTHOSPORIUM SP (FIG. 119a). These contaminants are mainly found in sputum specimens. They are recognized by their brown-to-yellow hyphae and their long, septate, simple or branching conidiophores with their characteristic knotted, twisted appearance. Their ovoid conidia are dark brown and may contain 4 or more cells.

ALTERNARIA SP (FIG. 119d). A frequent contaminant of almost all cytological specimens, these fungi belong to the group Fungi Imperfecti. They are recognized by their short, dark yellow-to-purple conidiophores and yellow-to-brown transversely and longitudinally septate conidia. Their presence in man has little significance, although they may be the cause of several diseases in plants.

POLLEN. At certain seasons of the year, sputum may be contaminated by various inhaled floating pollens. The presence of a capsule, abundant uniformly stained cytoplasm, and swollen polar structures aid in the recognition of these pollens.

CRYSTALS. Numerous crystals with various shapes that simulate fungi, parasites, or other infectious agents may be found in almost any cytologic specimen. For recognition of their true nature, they often need to be examined with a polarized light. Commonly found round structures, often stained purple and having pseudocapsules, are starch and talcum powder

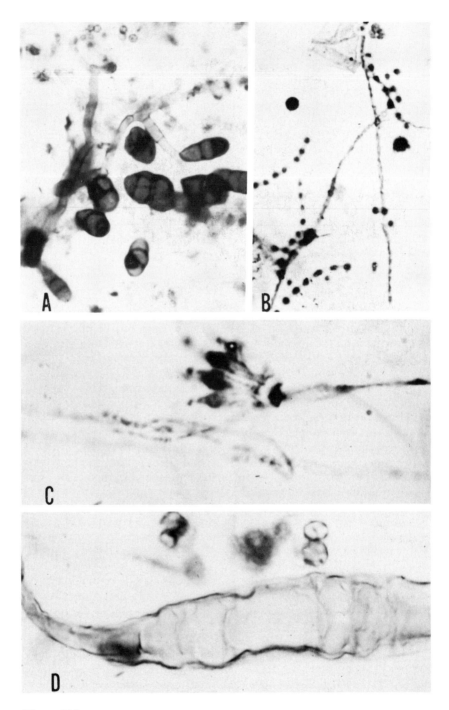

Figure 119
Contaminant fungi. (A) *Helminthosporum*. (B) *Scopulariopsis*. (× 450) (C)
Penicillium. (× 450) (D) *Alternaria*. (× 800)

crystals, which may be mistaken for pathogenic fungal spores. In the sputum of some asthmatic patients, Charcot-Leyden crystals, eosinophilic and octahedral in shape, have been mistaken for fungi or parasites.

Another structure commonly seen in sputum and bronchial washings and often mistaken for fungi is the previously described corpora amylacea (see Fig. 112C).

Parasitic Infection

AMOEBIASIS (FIG. 120A). The pulmonary involvement is usually secondary to the expansion of a diaphragmatic amebic abscess resulting from hepatic lesions. These protozoa are more commonly seen in areas to which this parasitic infection is endemic. They are often overlooked in the sputum smears and mistaken for cytoplasmic debris.

Grossly, the sputum is often abundant, doughy, and brown. The presence of numerous, poorly stained, round-to-oval, gray organisms (18–25 μ), with relatively large, pale, occasionally multiple, nuclei (1–4), is diagnostic. They resemble the vaginal *Trichomonas*, except for their refractile wall, larger nuclei, and the occasional presence of intact phagocytized, intracytoplasmic blood cells. Special iron or silver stain, of the unfixed but air-dried smear, is helpful in the diagnosis. These organisms must be differentiated from *Entamoeba gingivalis*, which is a relatively common parasite of the buccal cavity and is occasionally encountered in sputum specimens. Morphologically they are the same, except that *E. gingivalis* has a leukophagocytic capacity (leukocytes in the cytoplasm). Both *E. gingivalis* and *E. histolytica* phagocytize erythrocytes.

ACANTHAMOEBA. This ameba of the soil and water may become pathogenic to man. It is found in various cytologic specimens (sputum, spinal fluid); it may also be found in asymptomatic carriers within the nasal discharge smear. Unlike *E. histolytica*, *Acanthamoeba* has large distinct nuclei with central prominent karyosomes. The cytoplasm is usually scanty and vacuolated. The nucleus is darkly stained red to purple. Large numbers of inflammatory cells (polymorphonuclear cells) are present. The main danger lies in mistaking these parasites for isolated malignant cells.

Diseases of Unknown Origin

PLASMA CELL PNEUMONIA. The excess of single or large clusters or pools of basophilic plasma cells in a sputum smear is diagnostic of this rare disease. The individual plasma cells vary in size (8–15 μ). They have an eccentric, large nucleus and an adequate blue-stained, vesicular cytoplasm with sharp borders. These basophilic plasma cells should be differentiated from the similar-looking, small cells of oat-cell carcinoma by the absence of cellular and nuclear molding and by their relatively more abundant cytoplasm. The hyperchromatism and the chromatin clumping of the plasma cells are approximately the same as in the cells of oat-cell carcinoma.

ALVEOLAR PROTEINOSIS. In this condition, there is an increase of single or clustered, hypertrophic, and hyperplastic, nonciliated alveolar epithelial

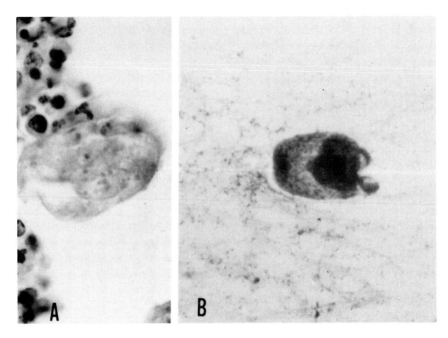

Figure 120
(A) Amebiasis. Large protozoa in a sputum specimen from a patient with ruptured diaphragmatic abscess. (× 750) (B) *Echinococcus scoleces* in a sputum
specimen. (× 125)

cells with prominent nucleoli that should be differentiated from similar-
looking cells exfoliating from alveolar cell carcinoma. The regularity of
the nuclear membrane and of the chromatin and nucleoli in the epithelial
cells helps in this diagnosis. Small or large clumps of granular protein deposits, pink or purple stained, can be seen in the background. There are
occasional pink spherical structures (lipids) or elongated, empty-looking,
transparent cleftlike structures (embedded cholesterol crystals). An increase of structures like corpora amylacea is often noted. The background
of the smear is usually very dirty, resulting from degenerate cellular debris
and protein deposits, but very few inflammatory cells are usually present.
A periodic acid–Schiff (PAS) stain of the extra unfixed slides will show an
abundance of PAS-positive material in the background.

SARCOIDOSIS.   This infection cannot be specifically diagnosed by cytology,
but it can be the cause of the presence in the smears of bright orange-
stained metaplastic cells with pyknotic nuclei, which can become a pitfall
in the diagnosis of squamous cell carcinoma.

IDIOPATHIC HEMOSIDEROSIS (GOODPASTURE'S SYNDROME).   The presence
of a large number of hyperplastic alveolar cells in the sputum of a young

adult, mixed with an abundance of large hemosiderin-laden phagocytes is suggestive. These alveolar cells are found singly, with degenerate vacuolated cytoplasm and central, large, round nuclei with prominent nucleoli. Variable amounts of cellular debris and protein deposits are also found. The refractile brown-yellow hemosiderin may be further identified by a special iron stain (Prussian blue) of the extra unstained slides.

Pneumoconiosis

SIDEROSIS.   Elementary iron and ferrous and ferric oxide may be found in the sputum of iron foundry workers. Macrophages with dark blue granules of hemosiderin pigment in their cytoplasm are found in more than 90% of workers exposed to inhalation of excessive amounts of iron oxide dust. Curschmann's spirals, eosinophilic leukocytes, and reactive hyperplastic alveolar cells are also commonly found. There is a need to screen the specimens of these workers with extra care because of the reported increase of lung cancer among iron workers.

LIPID PNEUMONIA (FIG. 121F).   The smears show a large number of macrophages, single or in clusters, with large, reactive, round-to-oval nuclei and prominent nucleoli. Their cytoplasm is packed with phagocytized spherical lipid droplets that are transparent to pink-brown and variable in size ($4–15$ μ) and number and are contained in large, prominent, distinct vacuoles. No dust particles can be seen in these cells. Numerous multinucleated foreign-body giant cells with similar cytoplasm are usually present. Occasionally, signet ring-like cells are found, with single or multiple, large cytoplasmic vacuoles containing lipid material and compressing the nuclei into a moon shape. Spherical lipid droplets, stained pink to red and often covered with microorganisms, also can be found free in the background of the smear.

These lipid-containing phagocytes may be differentiated from the cells shedding from an alveolar cell carcinoma by the absence of their cellular molding and the presence of phagocytized intracytoplasmic lipid droplets. The specific diagnosis often is confirmed only after a stain for sudanophilic materials is performed on the air-dried smear. These phagocytic cells, diagnostic of an aspirated lipid pneumonia, should also be differentiated from the smaller macrophages laden with fatty debris and cholesterol that are often seen in the sputum of patients with nonspecific chronic inflammatory process or with a large carcinoma with necrobiosis. The cytoplasm of the cholesterol-containing phagocytes is finely vacuolated and granular, and no giant multinucleated cells are seen in the smears.

BYSSINOSIS (FIGS. 121B, 122).   This is a pneumoconiosis that is seen in patients who have been exposed to the dust that arises from cotton processing. The sputum may show an abundant amount of free, brilliant, rectangular, prism-shaped, transparent crystals that shine brightly with a polarized light. Variable amounts of inflammatory cells are usually present.

SILICOSIS (FIGS. 121C, 123).   Clusters of particles of coarse dust ($3–15$ μ) are easily detected because of their higher refractive index when the

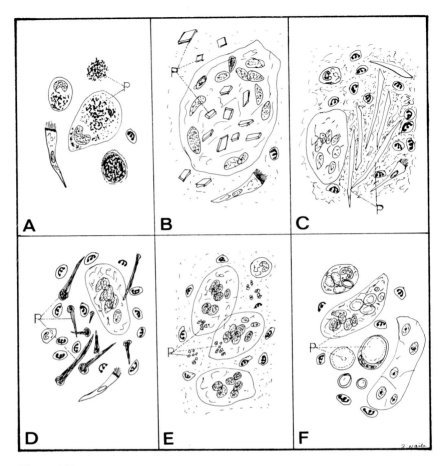

Figure 121
Diagnostic features of smears in pneumoconiosis and lipid pneumonia. (A) Anthracosis. (B) Byssinosis. (C) Silicosis. (D) Asbestosis. (E) Berylliosis. (F) Lipid pneumonia.

sputum is examined with polarized light. These sharp, fragmented, and elongated crystals are transparent and can very easily be missed with ordinary light microscopy. Besides these silica crystals, the sputum shows large amounts of inflammatory cells (polymorphonuclear cells) mixed with increased numbers of monocytes and reactive macrophages, which often have prominent nucleoli. Multinucleated foreign-body giant cells are also commonly seen.

ASBESTOSIS (FIG. 121D). The presence of clusters of elongated, slender, club-shaped or dumbbell-shaped asbestos needles in the sputum or bronchial washing, stained yellow to dark brown and measuring 10 to 80 μ is diagnostic. In addition, the smears show an increased amount of inflammation and larger numbers of histiocytes and multinucleated foreign-body giant

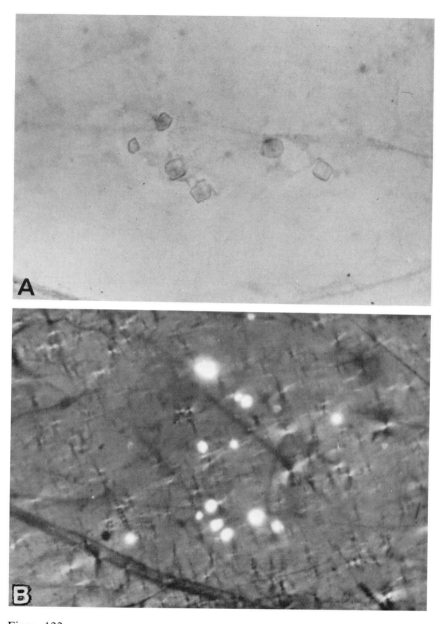

Figure 122
(A) The rectangular small crystals of byssinosis seen in the sputum specimen from a patient with a diffuse pulmonary lesion who had worked for a long time in a cotton gin. (× 1300) (B) The same crystals seen with a polarized light. (× 450)

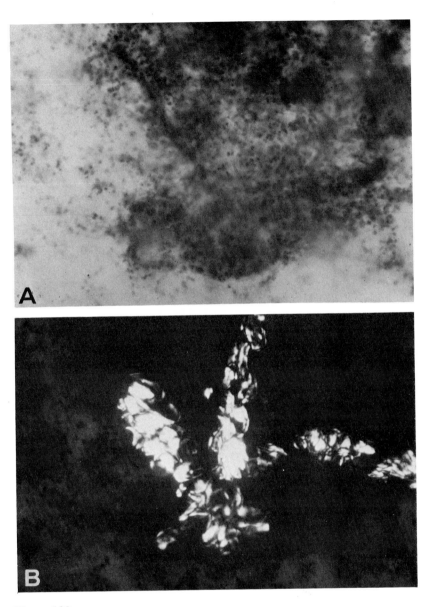

Figure 123
(A) Pulmonary silicosis. The crystals are hardly visible in an inflammatory sputum specimen and may easily be missed. (× 325) (B) The same crystals seen with a polarized light. (× 325)

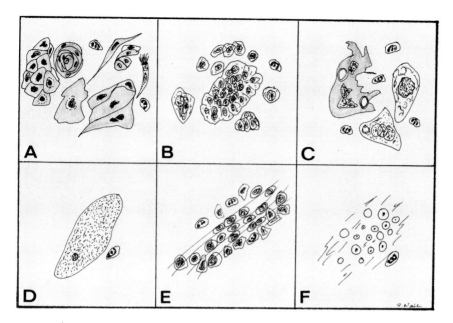

Figure 124

Smears from various benign pulmonary lesions. (A) Squamous metaplasia. (B) Bronchial cell hyperplasia. (C) Irradiation effect. (D) Granular cell myoblastoma. (E) Hamartoma. (F) Pneumocystic pneumonia.

cells. A possible association between asbestosis, mesothelioma, and bronchogenic carcinoma has been reported.

BERYLLIOSIS (FIG. 121E). It may be possible to make a diagnosis from the occasional presence of finely granular beryllium oxide crystals, which are often difficult to demonstrate, except when the slide is examined with a polarized light. The crystals can be found inside the cytoplasm of the numerous macrophages and foreign-body giant cells or extracellularly in the background. The smear usually shows an abundant basophilic protein deposit in the background, with a large number of mixed inflammatory cells.

ANTHRACOSIS (FIG. 121A). The presence of large amounts of carbon pigment deposited in the cytoplasm of the macrophage is diagnostic. Occasionally, the carbon is also found phagocytized in reactive alveolar cells with prominent nuclei, coarse chromatin, and large nucleoli. These black carbon dust particles may be seen free in the background of the smear.

Cytology of Benign Pulmonary Lesions

SQUAMOUS METAPLASIA (FIGS. 124A, 125A, B)

The replacement of pseudostratified ciliated or nonciliated columnar epithelium by stratified squamous epithelium may occur as the result of a

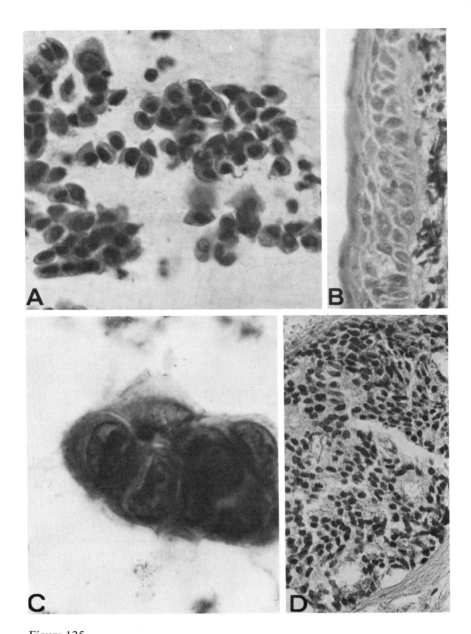

Figure 125
(A) Cluster of small orangeophilic squamous cells diagnostic of squamous metaplasia in a sputum specimen from a patient with chronic bronchitis. Note the absence of cilia, scantiness of the dense cytoplasm, and hyperchromatism of the round-to-oval nuclei. (× 450) (B) Squamous metaplasia of the bronchial mucosa, probable source of exfoliation of the cells in A. (× 450) (C) Papillary tight clusters of benign bronchial cells with occasional terminal bar but no cilia in the bronchial washing of a woman with bronchiectasis and bronchial mucosa hyperplasia. (× 1300) (D) Bronchial mucosa hyperplasia in the resected pulmonary lobe. (× 450)

chronic chemical irritation (heavy smoking), pollution, or any chronic lung disease (tuberculosis, pulmonary abscess, asthma, bronchiectasis). It is generally thought that most bronchogenic squamous carcinomas arise from squamous metaplastic epithelium. The cells are shed equally, singly, or in clusters. Their size varies from 10 to 25 μ, and their shape is irregular. Occasional pearl-like formations and isolated spindle cells are seen.

Their cytoplasm is usually adequate and dense with occasional heavy keratinization, giving them a distinctive bright orange-stained, glassy appearance. The smooth and regular cytoplasmic border can be thin and sharp (70%) or indistinct (30%).

The nuclei are centrally located and vary moderately in size from cell to cell (8–15 μ). Usually round, the nuclei show occasional irregularity of shape as the result of pyknotic degeneration. The chromatin is increased and dense, often appearing semipyknotic, with no discernible pattern. No nucleoli are present. Occasional multinucleation is seen (12%).

Besides the nuclear pyknosis, the cells may show other degenerative changes in the form of cytoplasmic vacuoles and haziness of structure.

The background of the smear usually shows only a moderate increase of inflammatory cells.

These metaplastic cells can be difficult to differentiate from the degenerate normal squamous cells originating from the buccal or pharyngeal mucosa. The increased and varying density of the cytoplasm in metaplastic cells helps in the differential diagnosis.

They may also be confused with the cells desquamated from a well-differentiated low-grade squamous carcinoma. The uniformity of the size and shape of the benign metaplastic cells is helpful in the differentiation.

In the presence of metaplastic cells, the sputum smears should be screened far more carefully than usual for the detection of coexisting malignant cells, and the patient should be closely followed until these cells have disappeared from his sputum.

BASAL CELL HYPERPLASIA (FIGS. 124B, 125C, D)

The diagnostic exfoliated basal hyperplastic cells are more commonly found in bronchial washings than in sputa. The proliferation of the subcolumnar basal cells is usually the result of chronic or acute irritation (bronchiectasis, tuberculosis, or asthma).

The hyperplastic cells, basal columnar in type, usually shed in clusters or sheets. Their size varies; their nuclei are hyperchromatic, surrounded by scanty basophilic and nonciliated cytoplasm. Their nucleoli may be prominent. Occasionally, large goblet cells with mucus-containing vacuoles are mixed with these undifferentiated cells.

The benignity of the lesion is based on the uniformity of the size and shape of the cells, their vesicular, finely vacuolated but uniformly scanty cytoplasm and hyperchromatic round-to-oval nuclei.

These benign cells are differentiated from the cells of bronchial adenoma by their smaller size and more uniform shape, and from the oat-cell carcinoma, which has scantier cytoplasm and more irregular hyperchromatic, malignant nuclei.

SQUAMOUS DYSPLASIA

When irritation or other factors that have produced squamous hyperplasia and metaplasia continue, progressively more atypical squamous cells may be found in the sputum. These are hyperkeratinized cells with an increased nuclei-cytoplasm ratio and irregular nuclei with moderately abnormal chromatin clumping. These cells can be considered as premalignant; they often precede the development of the lung cancer by several years. If the irritation is removed, these atypical cells often disappear from the sputum. The differentiation of regular metaplasia with moderate or marked atypia from in situ and invasive bronchial carcinoma may be difficult and is based mainly on excessive:

1. increase and variation of the size and shape of the malignant cells;
2. polymorphism of their nuclear size and shape;
3. increase of the nuclei-cytoplasm ratio;
4. increase of the abnormality of their chromatin clumps.

IRRADIATION AND CHEMOTHERAPY
CHANGES (FIGS. 124C, 126, 127)

The irradiated columnar cells lose their cilia and their cytoplasm often hyalinizes, acquiring a deep orange color, with occasional vacuolization. Their cytoplasmic membranes are frequently irregular with marked distortion of their shape.

The nuclei usually become larger and increase to two or three times their normal size. Binucleation and trinucleation, with or without vacuolization, and abnormal mitoses are commonly found.

Obscuring the background, there is also an increase of goblet cells, mucus, histiocytes, and inflammatory cells.

Frequently, it can be difficult to differentiate these irradiation changes from certain malignant ones. With a history of irradiation or chemotherapy, the cytopathologist should be very conservative in his interpretation.

The number of exfoliated malignant cells reaches its apex toward the middle of the irradiation treatment and disappears toward the end. Because of this, in a patient with a lung cancer, a negative sputum prior to treatment will often become positive when repeated during the radiotherapy. The sputum smears generally will contain numerous clusters of malignant cells with or without detectable irradiation changes.

The persistence of live tumor cells in the sputum three weeks after the end of the radiotherapy indicates a poor prognosis and may mean possible survival and persistence of the neoplasm.

The administration of chemotherapeutic drugs may produce similar changes, with an increased number of abnormal metaplastic cells showing large pyknotic, irregular nuclei. This, again, may be a source of error causing the incorrect diagnosis of well-differentiated squamous cell carcinoma if the history of previous chemotherapy is not known by the screener.

GRANULAR CELL MYOBLASTOMA (FIGS. 124D, 128A)

Granular cell myoblastoma can occur in the bronchi. Grossly, it appears as a fragile, fungating neoplasm obstructing their lumen that can be difficult

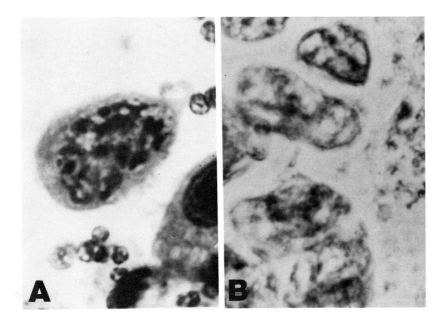

Figure 126
Nonviable tumor cells in sputum of patients receiving irradiation in the treatment of (A) bronchogenic squamous carcinoma. (B) Adenocarcinoma. (× 1300)

to differentiate grossly from a bronchogenic carcinoma. The exfoliated cells are easy to differentiate from squamous carcinoma and squamous metaplastic cells by the granularity of their abundant cytoplasm and the regularity of their usually single, small, round, and vesicular nuclei.

HAMARTOMA AND CHONDROMA (FIG. 128B)
These are peripheral, rare tumors that seldom exfoliate cells into the sputum. However, when they do, the cells are often overlooked or confused with the cells desquamated from a malignant neoplasm. The microscopic section usually shows masses of cartilagenous and fibrous stroma, separated by clefts that are lined by multilayered cuboidal epithelium that is devoid of cilia. When ulcerated, an abundance of diagnostic cells is found in the strands of mucus in the sputum and in bronchial washing smears. The cells are shed singly or in groups and are uniformly small with moderate variation of their size (9–12 μ) and shape. Usually scanty, the vesicular cytoplasm is poorly stained, gray to blue, and its borders are ill defined. The nuclei vary little in size. Round or oval, with little molding, they are centrally located and show a moderate degree of hyperchromatism. These cells should be differentiated from the cells exfoliating from the oat-cell carcinoma.

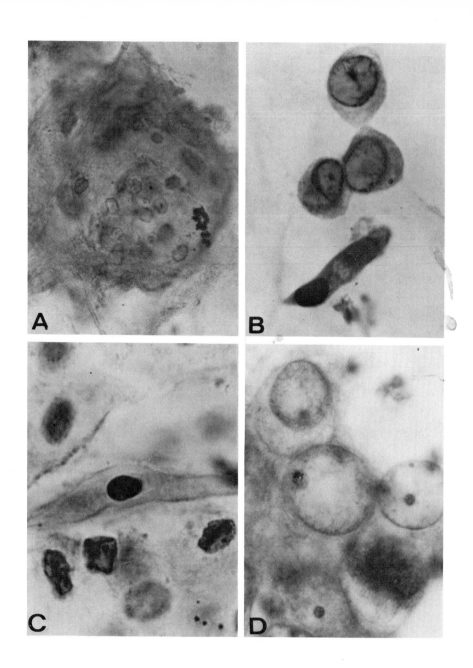

Cytology of Malignant Pulmonary Lesions (Table 11)

The recognition of the malignant exfoliated cells in a sputum specimen should not be more difficult than the recognition of malignant cells in a vaginal smear. The precancerous atypical lesions, previously described, are not as varied in the respiratory tract as in the uterine cervix. The cytologic report should be rendered as definitely positive or negative for the presence of a cancer. Interpreting a sputum as suspicious is of little help to the clinician. One should not hesitate to ask for a repeat specimen if the cells are not completely diagnostic or are too scanty and not representative.

WELL-DIFFERENTIATED SQUAMOUS CELL CARCINOMA (FIGS. 129A, 130A)

This is one of the most common bronchogenic cancers of the lung and the easiest to diagnose by exfoliative cytology. The features of well-differentiated bronchogenic squamous cancer cells are about the same as those of invasive cervical well-differentiated squamous cancer cells. Some of the cells that would be classified as dysplasia in a vaginal smear will be considered to be diagnostic of a squamous cell carcinoma when seen in the specimen originating from the lung.

General Cellular Criteria

The number of diagnostic cells in a sputum or bronchial washing depends mainly on the following conditions, one of which is the *location* of the neoplasm. The cells are more numerous when the neoplasm involves a large bronchial tree. Other factors that influence the number of cells are the *technique* used in collecting and processing the specimen and the *size* of the tumor. The bronchial mucosal congestion, edema, and inflammation produced by an advanced cancer often narrows the lumen of the distal bronchus through which the exfoliated cells must pass. This explains why an early, small carcinoma may exfoliate more and better cells than an advanced large one.

The cells exfoliate singly (60%), in small clusters (20%), or in sheets (20%).

Their size varies greatly (40 ± 30 μ), and their shape is very irregular, often showing ameboid elongation with multiple pseudopodlike projections.

Figure 127
(A) Giant multinucleated cell seen in the sputum of a woman receiving radiotherapy for a carcinoma of the breast. Note the hypertrophy of the cell, the multinucleation, and the prominence of the nucleoli. ($\times 350$) (B) Irradiation changes in normal bronchial cells in a sputum specimen from a patient the sixth day after the beginning of radiotherapy for a pulmonary carcinoma. ($\times 450$) (C) Squamous metaplastic and dysplastic cells in the sputum of a patient receiving chemotherapy for a generalized lymphosarcoma. ($\times 1300$) (D) Hypertrophy of bronchial cells seen in the bronchial washing of a woman receiving radiotherapy for breast carcinoma. These cells may be easily confused with an alveolar cell carcinoma. ($\times 1300$)

275

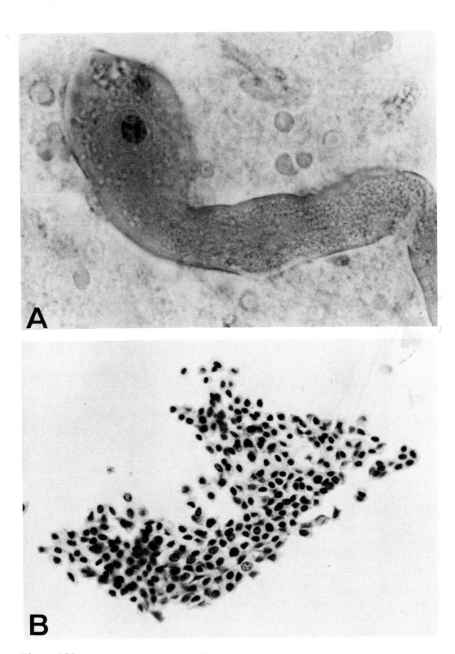

Figure 128
(A) Granular cell myoblastoma in the bronchial washing of a young woman with a fungating bronchial neoplasm. Note the size and abundance of the granular, orangeophilic cytoplasm and the small, round, single, vesicular nucleus. (× 1300) (B) Sheet of small cuboidal cells in the bronchial washing of a man with histologically proven ulcerated endobronchial hamartoma. (× 450)

Table 11. Differential Characteristics of Lung Carcinoma

Criteria	Well-Differentiated Squamous Carcinoma		Giant Cell Carcinoma		Adeno-carcinoma		Oat-Cell Carcinoma		Alveolar Carcinoma	
Size (μ)	40 ± 30		80 ± 30		40 ± 20		20 ± 8		35 ± 5	
Shape	Irregular		Round-oval		Irregular		Irregular		Round-oval	
Occurrence	Single	60%	Single	99%	Single	20%	Single	90%	Single	15%
	Sheet	40%	Sheet	1%	Sheet	80%	Sheet	10%	Sheet	85%
Cytoplasmic stain	Basophilic	20%	Basophilic	40%	Basophilic	95%	Baso- philic	100%	Basophilic	15%
	Orange	80%	Orange	60%	Orange	5%	Orange	0	Orange	85%
Amount of cytoplasm	Abundant		Adequate		Abundant		Scanty		Adequate	
Cytoplasmic secretory vacuoles	0%		0%		99%		0%		80%	
Nuclear size (μ)	15 ± 10		25 ± 5		25 ± 5		18 ± 5		20 ± 5	
Nuclear shape	Round	20%	Round to oval		Irregular		Irregular		Round to oval	
	Oval	20%								
	Irregular	60%								
Chromatin pattern	Clumped	90%	Clump	75%	Clump	60%	Clump	90%	Clump	5%
	Granular	10%	Granular	25%	Granular	40%	Granular	10%	Granular	95%
Multinucleation	10%		75%		20%		10%		20%	
Nucleoli	8%		85%		80%		2%		90%	

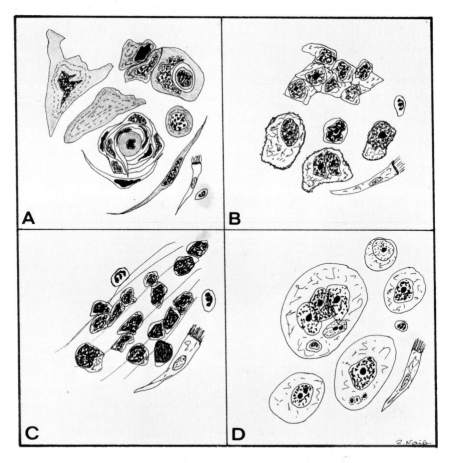

Figure 129
Exfoliated cells from various pulmonary types of squamous cell carcinoma. (A) Keratinized, well-differentiated. (B) Nonkeratinized, well-differentiated. (C) Poorly‑differentiated, small‑cell (oat-cell) carcinoma. (D) Giant‑cell carcinoma.

The spindle and tadpole cells are not as common as they are in cervical, well-differentiated, invasive squamous cell carcinoma, except when the lung cancer is necrotic and has formed a central cavitation.

An increased number of phagocytes and nonspecific inflammatory cells is always present in the sputum. The background of the smears is usually obscured by moderate amounts of fresh and old red blood cells, mucus and fibrin, as well as by abundant necrotic cytoplasmic debris (tumor diathesis).

Cytoplasmic Changes
The amount of cytoplasm varies, being rather scanty in the single round cells and abundant in the irregularly shaped, polygonal cells. It forms 20% to 80% of the total surface of the cells.

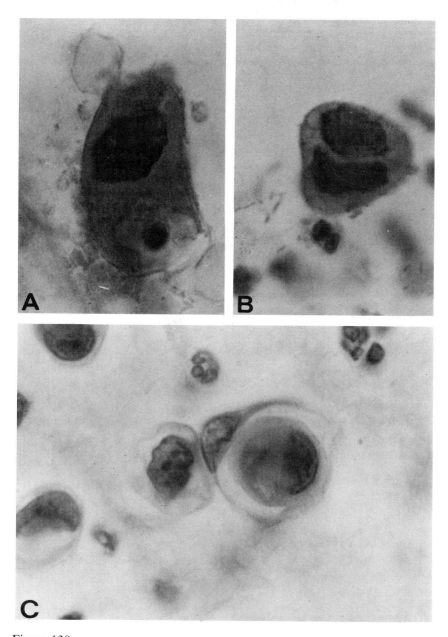

Figure 130
Bronchogenic squamous carcinoma cells in sputum and bronchial washing.
(A) Well-differentiated carcinoma. Note the abundant keratinized cytoplasm
and irregular, almost pyknotic nucleus. (× 1300) (B) Poorly differentiated
squamous carcinoma. Note the scantiness of the cytoplasm and molding of the
irregular nuclei. (× 1300) (C) Pseudocannibalism of the malignant squamous
cells. Note the prominent nucleoli in one of the cells. (× 1300)

With the Papanicolaou stain, the cytoplasm appears bright orangeophilic, dense and granular, and refractile with a glassy appearance (80%). There is an increased amount of keratin that is often deposited in concentric perinuclear rings (20% of the cells).

The endoplasm is abundant and extends almost to the borders of the cells and often shows fibrillar linear formations, resulting from the wrinkling of the cytoplasmic membrane.

The cellular borders are sharp, well limited, and clearly visible.

Often diagnostic of a long-standing, well-differentiated, degenerate squamous cell carcinoma is the occasional presence of anucleated, elongated or oval cells stained deep orange with an abundant, glassy cytoplasm (ghost cells). These cells must be differentiated from the normal anucleated buccal cells. The latter have a more transparent cytoplasm and contain less keratin.

Nuclear Changes

The sizes of the nuclei vary. The mean nuclear surface is 80 ± 60 μ^2. The nuclei-cytoplasm ratio is increased, the nuclei occupying from 20% to 80% of the volume of the cell.

The polymorphism of the nuclei becomes one of the most important diagnostic criteria when the cells are degenerate and no other structural detail can be seen. The nuclear shape is round (20%), oval (20%), or markedly irregular (60%). The nuclear variation depends on the shape of the entire cell, being elongated when the cell is spindleshaped and round or irregular when the cell is polygonal.

Because of the usual marked hyperchromatism, the nuclei often appear very dark and pyknotic with no discernible chromatin pattern (40%). In the vesicular nuclei, the chromatin is irregularly clumped, with pointed projections (40%) or coarsely granular (20%).

In approximately 8% of the malignant cells with vesicular nuclei, a slightly enlarged nucleolus, red, prominent, and irregular in shape with pointed projections, can be seen.

Malignant pseudophagocytosis (cannibalism) is commonly found (75% of the cases) and indicates the rapid growth of the tumor in a narrow space.

On occasion, pearl-like structures are present, with individual cells showing features of malignancy. Similarly, elongated spindle or tadpole cells may be seen in the sputum, but they are less common than in the vaginal smear.

IN SITU SQUAMOUS CELL CARCINOMA

The in situ carcinoma of the bronchial mucosa does occur; however, no specifically diagnostic cells are found such as those seen in cervical in situ carcinoma. The cells shedding from a bronchial in situ squamous carcinoma resemble the ones exfoliating from an invasive one. They are more usually seen in clusters and their malignant features are less extreme.

A patient should be suspected of having an in situ lesion if he has repeated cytologically positive sputum smears without any clinical evidence of the presence of a neoplasm. Such a patient should be followed very carefully until the site of the neoplasm can be detected by clinical findings, bronchoscopy, or pulmonary x-ray.

Some of these exfoliating bronchogenic squamous in situ carcinomas may precede by several years the clinically detectable invasive carcinoma.

NONKERATINIZING SQUAMOUS CELL CARCINOMA (FIGS. 129B, 130B)

An abundance of diagnostic cells is usually found in the smears. They are shed in clusters (50%), sheets (25%), or singly (25%).

The size of the cells varies moderately ($50 \pm 20 \ \mu^2$), and their shape is regular, round, or oval (75%), irregular (20%), spindle, tadpole, or polygonal (5%).

The cytoplasm is granular and usually scanty. Its often poorly defined borders are irregular and uneven. Semitransparent, usually basophilic, the cytoplasm contains multiple, small, lacy vacuoles with no keratinization in 85% of the cells.

The centrally located nuclei vary moderately in size (mean nuclear surface = $40 \pm 10 \ \mu^2$), and the nuclei-cytoplasm ratio is increased, with the nucleus occupying about four-fifths of the volume of the cell.

The nuclei show marked irregularity of shape, being round (30%), oval (30%), and irregular and angular (40%); and the nucleoli are often prominent and irregular in shape, size, and number. The chromatin is irregularly clumped and abundant.

The smears show large amounts of nonspecific inflammation but only a minimum of necrotic debris.

The differentiation of the cells from the poorly differentiated anaplastic bronchial adenocarcinoma is difficult, often impossible. It is mainly based on the occasional presence (1%) of well-differentiated, keratinized, malignant squamous cells.

In 90% of the sputum specimens, other atypical cells diagnostic of benign squamous metaplasia coexist with the malignant cells.

POORLY DIFFERENTIATED SMALL-CELL (OAT-CELL) CARCINOMA (FIGS. 129C, 131A, B)

The cells of oat-cell carcinoma are usually seen in large numbers in association with long strands of mucus. They often form dark lines of elongated, compact masses of numerous small hyperchromatic, neoplastic cells with satellites of scattered single cells trailing the cellular clusters in the strands of mucus. They usually are more frequent in sputum specimens than in bronchial washings.

The size of the cells varies moderately, some being only slightly larger than a small lymphocyte ($15 \pm 8 \ \mu$). Their shape varies from round (20%) to oval (20%) to irregular (60%).

The cytoplasm is very scanty and transparent and is often difficult to distinguish under low-power magnification. The cells should be examined with a high-power oil immersion objective, and the cytoplasm should be unequivocally identified before a diagnosis is rendered. The cytoplasm is better seen in the larger cells.

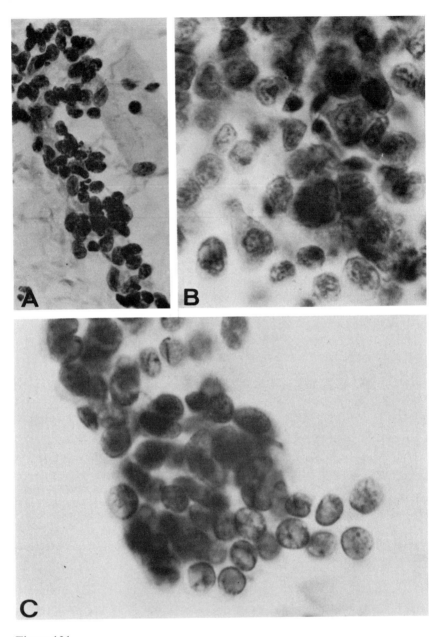

Figure 131
(A) Poorly differentiated oat-cell carcinoma. Note the association of the small, dark, neoplastic cells with a strand of mucus. (× 450) (B) Higher power magnification of the tumor cells showing the scantiness of the cytoplasm and nuclear molding. (× 1300) (C) Cluster of stripped benign columnar nuclei often confused with an oat-cell carcinoma. Note the regularity of the nuclear membrane and blandness of the chromatin. (× 1300)

The usually basophilic cytoplasm shows minute vacuolization.

The nuclei are usually single (90%) and hyperchromatic. They are uniform and small ($19 \pm 5 \mu$) and are only slightly larger than the nucleus of an immature lymphocyte. The nuclear shape is variable—round or oval (60%) or irregular with angles resulting from the marked in situ cellular molding (40%).

The chromatin is increased and clumped in 90% of the cells and coarsely granular in 10%.

Nucleoli may be present in about 2% of the tumor cells.

The oat cells should be differentiated from clusters of reserve or basal cells exfoliating from a bronchial hyperplasia. Both types of cells have scanty cytoplasm and a tendency to be distributed in linear form in the strands of mucus. The fine, uniform nuclear chromatin and the regularity of the nuclear membrane of the benign bronchial cells help in the differential diagnosis (Fig. 131C).

The oat cells can also be confused with clusters of benign stripped nuclei of degenerate bronchial columnar cells commonly seen in the sputum of patients with asthma, chronic bronchitis, or other diseases. These degenerate bronchial columnar cells differ from the malignant oat cells by their more orderly arrangement, absence of cytoplasm, and regularity of chromatin clumping. The occasional presence of well-preserved benign hypersecretory columnar cells with intact cytoplasm and identical nuclei in their midst confirms the benign nature of these clusters. Some of these degenerate cells have irregular wisps of cytoplasm still attached to their nuclei.

The oat cells are also sometimes hard to differentiate from the clusters of small histiocytes retaining only foamy remnants of degenerate cytoplasm as seen in the sputum of patients with pulmonary inflammation. The bean-shaped, eccentric nuclei, cytoplasmic inclusions, and the lack of molding of the benign cells are helpful in the differential diagnosis.

GIANT CELL CARCINOMA (FIGS. 129D, 132)

The cells shed singly (85%) or in clusters with overlapping but no molding. They show marked variation of size ($80 \pm 30 \mu$) but have a fairly constant shape (round or oval, 95%).

Mitoses are frequent (2% of the tumor cells) and atypical in 60%.

The cytoplasm is dense, adequate to scanty, amorphous, eosinophilic, orangeophilic (60%) or basophilic (40%) without keratinization but with occasional round pink-to-orange inclusions, the significance of which is not well known (possibly, abnormal keratin precursors).

The cytoplasmic membrane is smooth, distinct, and well limited.

Multinucleation is frequent, with the different nuclei showing marked variation in their size and shape. Overlapping or molding against each other differentiates them from the nuclei seen in benign foreign-body giant cells (sister image).

The nuclear membrane is usually smooth but with sharp, occasional indentations. It can be uneven in shape and thickness in individual cells.

There is a marked irregularity of the chromatin clumping with pointed

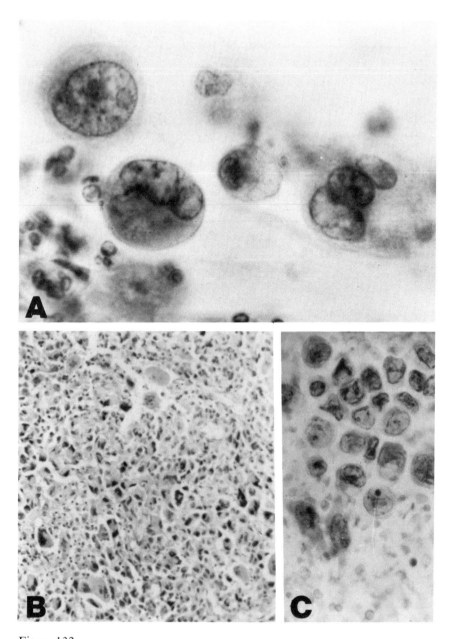

Figure 132
(A) Giant cell carcinoma. The large, single, malignant cells in the sputum might be confused with reactive histiocytes, except for the marked irregularity of the number, size, and shape of their nuclei and the prominent nucleoli. ($\times$ 1300) (B) Histologic section of the bronchial wall showing similar cells. ($\times$ 450) (C) Loose giant carcinoma cells in a sputum specimen. ($\times$750)

projections (75% of the cells), hyperchromatism, irregular distribution, and apparent empty spaces between the clumps.

Prominent nucleoli are usually present (85%) and centrally placed, varying in size, number, and shape. They constitute important clues in the diagnosis.

Giant cell carcinoma is one of the most malignant pulmonary neoplasms. Clinically, at the beginning, it is often confused with a pulmonary inflammation. The prognosis is bad; the condition often is fatal in fewer than 6 months, and leads to brain metastasis. Some of the exfoliated giant cancer cells are hard to differentiate from large histiocytes. The absence of intra-cytoplasmic inclusions (carbon particles), the size of the nuclei, the prominence of the nucleoli, and the irregularity of the chromatin pattern differentiate the malignant cells.

ADENOCARCINOMA

Bronchial Adenoma (Figs. 133A, 134A)

Found mainly in the young adult, bronchial adenoma is a rare, slow-growing, but locally invading and occasionally metastasizing neoplasm of the large bronchi. The diagnostic cells are more numerous in the bronchial washing than in sputum.

The cells shed in sheets (45%), clusters (25%), or singly (20%). Their size ($25 \pm 5 \mu$) and shape (round or oval) are uniform.

The cytoplasm is adequate to scanty, delicate, eosinophilic (60%) or basophilic (40%) with sharp, smooth, regular borders. No cilia or terminal bars are seen.

The nucleus is central, uniformly round, and slightly hyperchromatic. It covers approximately one-third of the cellular volume.

The nuclear membrane is smooth, well defined, and regular.

Occasional small, round nucleoli are present.

The cells are differentiated from the hyperplastic basal columnar cells because in bronchial adenoma there is a complete absence of cilia and terminal bars although uniformity of size and shape persists.

Primary Adenocarcinoma of the Lung (Figs. 133B, 134B, C, D)

This neoplasm occurs as frequently in men as in women and seems unrelated to cigarette smoking. The tumor cells may shed in tight clusters, sheets, acini, or palisade formation, rarely singly. The cell size varies ($40 \pm 20 \mu$). Although usually round or oval (40%), the cells are often irregular in shape, especially when seen in sheets.

In primary adenocarcinoma of the lung, the cytoplasm is lacy in appearance, blue or, occasionally, orange stained, but with no keratinization. The cytoplasm can be finely or grossly vacuolated, depending on the number and size of droplets of mucus present. Occasionally, large secretory vacuoles may compress the nuclei into an irregular moon shape, giving them the appearance of signet ring cells. In such cases, large numbers of well-preserved neutrophils can often be seen ingesting the contents of these vacuoles.

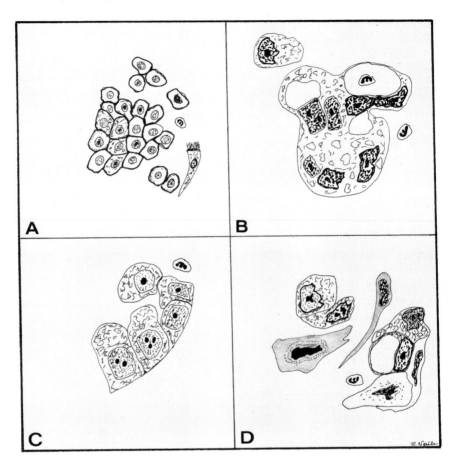

Figure 133
Exfoliative cells of the various pulmonary adenocarcinomas. (A) Adenoma.
(B) Bronchogenic adenocarcinoma. (C) Alveolar cell carcinoma. (D) Muco-
epidermoid bronchogenic carcinoma.

The nucleus is oval (20%), round (20%), or irregular (60%). Its size
is fairly constant ($25 \pm 5 \ \mu$). A moderate hyperchromatism is present,
with the chromatin coarsely clumped in 60% and evenly granular in 40%.
The nuclear border is sharp and well defined but irregular, showing multiple indentation.

Multinucleation (1–3 nuclei) occurs in 20% of the cells, and a large,
prominent, single or multiple diagnostic nucleolus is found in about 80%
of the cells.

A periodic acid–Schiff (PAS) stain of the unfixed, air dried smears helps
in the diagnosis of this neoplasm by the demonstration of a large accumulation of PAS-positive droplets of mucinlike substances in the cytoplasm.

Secondary Adenocarcinoma of the Lung

Unless a metastatic tumor involves a main bronchus, malignant cells are not found in the cytologic specimen. Because of the low incidence (35%) of this involvement, exfoliative cytology, except in the case of a direct needle aspiration of the mass, often cannot detect a metastatic neoplasm, in spite of massive radiologic evidence of its existence. The diagnosis is based on the presence of clusters of malignant cells, varying in size, shape, and structure according to the nature of the primary neoplasm. These clusters of cells have a tendency to appear in the sputum as ball-like structures. The aspect of the individual cells varies, depending on the origin and nature of the metastasis.

Mucoepidermoid Carcinoma

The diagnosis is based on the exfoliation of a mixture of cells typical of an adenocarcinoma and of a well-differentiated squamous cell carcinoma. For a definite cytologic interpretation, the two types of malignant cells must be found in continuity in at least one cluster. The PAS stain of the unfixed smear helps in the demonstration of mucous secretion in the cytoplasm of the suspected adenomatous tumor cells.

Alveolar Cell Carcinoma—Terminal Bronchiolar Carcinoma, Pulmonary Adenomatosis (Figs. 133C, 135)

Alveolar cell carcinoma, also known as terminal bronchiolar carcinoma and pulmonary adenomatosis, is a tumor that arises from the alveoli or from the terminal bronchiolar mucosa. The cells shed in clusters or sheets (85%). The cells are mononuclear, and their size is fairly uniform (35 ± 5 μ), with regular round or oval shape (90%).

The cytoplasm is adequate, vesicular, and basophilic with no cilia or terminal bars, and it often contains an abundant mucinous secretion in over-distended vacuoles.

The cells usually contain single nuclei (80%), but occasional multinucleation with giant cell formation occurs.

The nuclei are uniformly round or oval in shape and vary little in size (20 ± 5 μ).

The chromatin is not increased. Coarsely granular in 95% of the cells and clumped in 5%, the chromatin pattern differs very little from that of the reactive benign cells.

The nucleolus is prominent, single or multiple.

Occasional calcified psammoma bodies with diagnostic concentric lamination may be found in the sputum and bronchial washings of patients with papillary adenocarcinoma.

A PAS stain of the smears will show variable amounts of mucin present in the cytoplasm. These cells can easily be overlooked because of their benign appearance. The absence of cilia, the large size of the nuclei, and the prominence of the nucleoli are helpful in the detection of this neoplasm. The presence in a routine sputum of well-preserved clusters or sheets of large, nonciliated columnar cells, in spite of the benign appearance of the nuclei, should make one suspect an alveolar cell carcinoma.

287

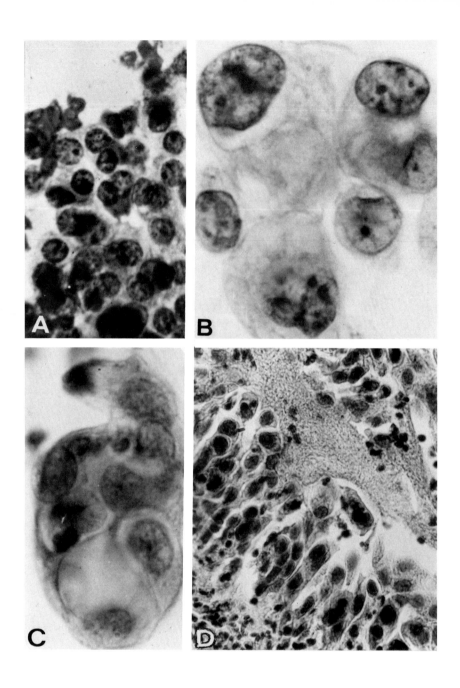

Pulmonary Sarcoma

These are rare tumors that are difficult to diagnose by cytology. The cells exfoliate only when traumatized and when a main bronchus is involved.

FIBROSARCOMA.   These are usually secondary to a pleural fibrosarcoma. On occasion, elongated spindle cells can be seen, mainly in the bronchial washing. They shed singly or in sheets. Their size varies $(60 \pm 30 \ \mu)$, with irregular shape and pseudopodlike bifurcated projections. The cytoplasm is lacy and semitransparent. The nucleus is single or multiple, often elongated, benign, and bland-looking. Their nuclear membrane is smooth and regular. A large number of inflammatory cells and protein deposits are usually found in the background. Because of the transparency of their cytoplasm and their bland nuclei, these cells are often overlooked.

MALIGNANT LYMPHOMA (Fig. 136A, B).   A large number of clusters of lymphocytes and lymphoblasts in the sputum, with marked variation in their size and shape, increased hyperchromatism, and prominent nucleoli, is diagnostic of a lymphomatous involvement of the lung. These cells exfoliate either by rupture of an involved mediastinal lymph node or as a true pulmonary metastasis and infiltration. Primary pulmonary lymphoma is rare. The most common tumor is the lymphosarcoma. The Reed-Sternberg cells, with their prominent acidophilic nucleoli in their several or multilobular nuclei, and their scanty basophilic cytoplasm indicate Hodgkin's disease.

The malignant lymphomatous cell is sometimes difficult to differentiate from an oat-cell carcinoma, except for its absence of cellular molding.

The cells also should not be confused with large pools or clusters of mature lymphocytes occasionally seen in some inflammations (tuberculosis) or after a traumatic scraping during bronchoscopy of a lymphoid organ (tonsils or adenoids).

FALSE-POSITIVE PITFALLS

A false-positive diagnosis is more commonly rendered in interpretation of an adenomatous cell than of a squamous one. The main culprits are the cells originating from the nasopharynx and benign alveolar cell hyperplasia or resulting from irritation in certain diseases (asthma, atelectasis).

Chemotherapy or irradiation to the chest produces abnormal, enlarged cells, that, if the history is unknown, are often mistaken for cancer cells.

Figure 134

(A) Cluster of tumor cells in the sputum of a young man with a histologically proven ulcerated bronchial adenoma. Note the absence of cilia, scantiness of cytoplasm, and the regularity of the hyperchromatic nuclei. ($\times$ 450) (B) Bronchogenic adenocarcinoma cells in the sputum and (C) in the bronchial washing of a woman with hemoptysis of two days' duration. Note the cellular molding, the abundance of the vacuolated cytoplasm, and the irregularity of the nuclei and prominent nucleoli. ($\times$ 1300) (D) Tissue section of the resected lung. ($\times$ 450)

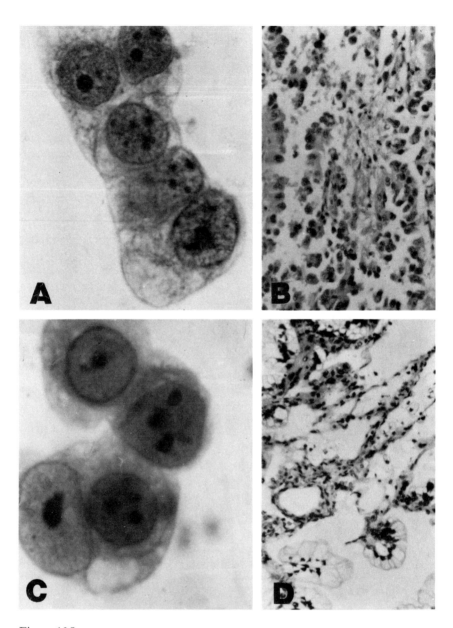

Figure 135
(A) Terminal bronchiolar carcinoma cells in the sputum of a man with a peripheral coin lesion. Note the absence of the cilia, the vesicular cytoplasm, the large round-to-oval nuclei, and the prominent nucleoli, variable in size and number. (× 1300) (B) Histologic section of the tumor. (× 250) (C) Alveolar carcinoma cells in an inflammatory sputum specimen from a woman. The cytologic diagnosis was later confirmed at autopsy. Note the cluster of cells with adequate cytoplasm, irregular nuclear membrane, and prominent nucleoli. (× 450) (D) Histologic section of the tumor. (× 250)

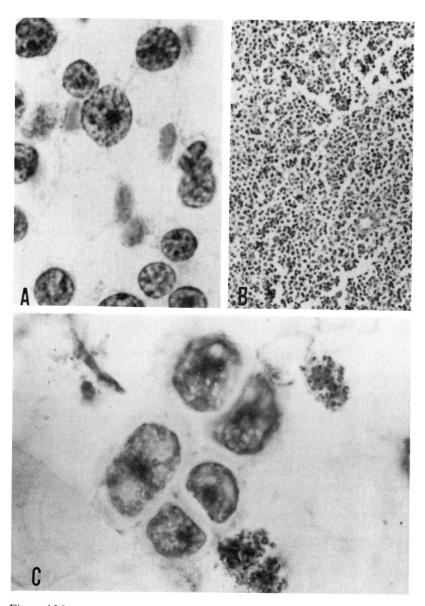

Figure 136
(A) Malignant lymphocytes in the sputum of a patient with generalized lymphosarcoma. ($\times$ 1300) (B) Histologic pulmonary section. ($\times$ 125) (C) Vegetable cells in a sputum specimen. They resemble tumor cells, except for the capsule surrounding them. ($\times$ 850)

Squamous metaplastic cells with hyperkeratinized cytoplasm, such as those seen in tuberculosis or heavy cigarette smokers, may be mistaken for well-differentiated squamous cell carcinoma.

Anucleated degenerated cells originating from the buccal mucosa can be confused with bronchial ghost cells diagnostic of a well-differentiated squamous cell carcinoma.

As the result of a processing delay, poor technique, or incomplete fixation, the artificial eosinophilia thus produced and the degeneration with pyknosis of the nuclei of certain benign cells (squamous or columnar) may be mistaken for well-differentiated squamous carcinoma cells.

A large number of stripped bronchial cell nuclei, showing some degree of degenerative hyperchromatism can be mistaken for an oat-cell carcinoma.

The contamination of a sputum smear by malignant cells belonging to another previously stained specimen and floating in the fixative, staining, or mounting medium is always possible. This is one reason that more than one group of malignant cells should be found in a pulmonary specimen before a positive diagnosis is rendered.

When a sputum contains malignant cells originating from an oral or esophageal carcinoma, cytology cannot pinpoint the true origin of the malignant cells, and such cases are often mistakenly reported as positive for lung cancer.

Certain vegetable contaminating cells may resemble tumor cells (Fig. 136C).

References and Supplementary Reading

Ali, M. Y. The nature of multinucleated cells in the nasopharynx. *J. Clin. Pathol.* 18:424–427, 1965.

Ali, M. Y., and Shanmugaratnam, K. Cytodiagnosis of nasopharyngeal carcinoma. *Acta Cytol.* 11:54, 1967.

Attar, S., and Naib, Z. M. Exfoliative cytology in solitary peripheral lesion of the lungs. *J. Thorac. Cardiovasc. Surg.* 42:168, 1961.

Bangle, R., Jr., Hohl, E. M., Maple, F., and Hart, S. Technique of esophageal-gastric aspiration for collecting swallowed sputum in diagnostic pulmonary cytology. *Acta Cytol.* 9:362, 1965.

Bibbo, M., et al. Bronchial brushing technique for the cytologic diagnosis of peripheral lung lesions. *Acta Cytol.* 17:245, 1973.

Bryan, W. T. K., and Bryan, M. P. Structural changes in the ciliated epithelial cells during the common cold. *Trans. Am. Acad. Ophthalmol. Otolaryngol.* 57:297–303, 1953.

Burke, M. D., and Melamed, M. R. Exfoliative cytology of metastatic cancer in lung. *Acta Cytol.* 12:61, 1968.

Burton, R. D. Early detection of nasopharyngeal cancer. *J. Mich. Med. Soc.* 62:395–397, 1963.

Cahan, W. C., Melamed, M. R., and Frazell, E. L. Tracheobronchial cytology after laryngectomy for carcinoma surgery. *Obstet. Gynecol.* 123:15, 1966.

Cardozo, P. L., et al. The results of cytology in 1000 patients with pulmonary malignancy. *Acta Cytol.* 11:120, 1967.

Chang, S. C., and Russell, W. O. A simplified and rapid filtration technique for concentrating cancer cells in sputum. *Acta Cytol.* 8:348, 1964.

Eisenstein, R., and Battifora, H. A. Malignant giant cell tumor of bone: Exfoliation of tumor cells from pulmonary metastases. *Acta Cytol.* 10:130, 1966.

Fennessy, J. J. A method for obtaining cytologic specimens from the periphery of the lung. *Acta Cytol.* 10:413, 1966.

Frable, W. The relationship of pulmonary cytology to survival in lung cancer. *Acta Cytol.* 12:52, 1968.

Gray, B. Sputum diagnosis in bronchial carcinoma—A comparative study of two methods. *Lancet* 1:549, 1964.

Gupta, P. K., and Verena, K. Calcified (psammoma) bodies in alveolar cell carcinoma of the lung. *Acta Cytol.* 16:59, 1972.

Haam, E. von. A comparative study of the accuracy of cancer cell detection by cytological methods. *Acta Cytol.* 6:508–512, 1962.

Hattori, S., et al. Some limitations of cytologic diagnosis of small peripheral lung cancers. *Acta Cytol.* 9:431, 1965.

Hoch-Ligette, C., and Eller, L. Significance of multinucleated epithelial cells in bronchial washings. *Acta Cytol.* 7:258–261, 1963.

Jarvi, O. H., et al. Cytologic diagnosis of pulmonary carcinoma in two hospitals. *Acta Cytol.* 11:477, 1967.

Johnston, W. W., and Amatrelli, J. The role of cytology in the primary diagnosis of North American blastomycosis. *Acta Cytol.* 14:200, 1970.

Juan, M., and Oritz-Picon, J. M. Contribution à la cytologie des cancers bronchiolo-alveolaires. *Arch. Anat. Pathol.* 15:161, 1967.

Kern, W. H. Cytology of hyperplastic and neoplastic lesions of terminal bronchioles and alveoli. *Acta Cytol.* 9:372, 1965.

Koprowska, I., et al. Cytologic patterns of developing bronchogenic carcinoma. *Acta Cytol.* 9:424, 1965.

Kress, M. B., and Allan, W. B. Bronchiolo-alveolar tumors of the lung. *Bull. Hopkins Hosp.* 112:115, 1963.

Kuper, S. W. A. Exfoliative cytology in the diagnosis of cancer of the bronchus. *J. Clin. Pathol.* 16:399, 1963.

Lange, E., and Hoeg, K. Cytologic typing of lung cancer. *Acta Cytol.* 16:327, 1972.

Leilop, L., Garret, M., and Lyons, H. A. Evaluation of technique and results for obtaining sputum for lung carcinoma screening. *Am. Rev. Resp. Dis.* 83:803, 1961.

Lorange, G. Etude comparative de la cytologie dans les expectorations et les secretions bronchiques. Observation de 618 cas de cancer bronchique. *Union Med. Can.* 96:694–698, 1967.

Makowska, W., and Zawisza, E. Cytologic evaluation of the nasal epithelium in patients with hay fever. *Acta Cytol.* 10:564, 1975.

Naib, Z. M. Primary carcinoma of the lung, cytological detection. In *Transactions, Seventh Annual Meeting of the Intersociety Cytology Council, 1959.* Pp. 294–298.

Naib, Z. M. Giant cell carcinoma of the lung: Cytological study of the exfoliated cells in sputa and bronchial washings. *Dis. Chest* 40:69, 1961.

Naib, Z. M. Exfoliative cytology in fungus diseases of the lung. *Acta Cytol.* 6:413, 1962.

Naib, Z. M. Pitfalls in the cytologic diagnosis of oat cell carcinoma of the lung. *Acta Cytol.* 8:34, 1964.

293

Naib, Z. M., and Attar, S.   Exfoliative cytologic and clinical study of a case of endobronchial hamartoma of the lung. *Dis. Chest* 41:468, 1962.

Naib, Z. M., and Goldstein, H. G.   Exfoliative cytology of a case of bronchial granular cell myoblastoma. *Dis. Chest* 42:645, 1962.

Naib, Z. M., Stewart, J. A., Dowdle, W. R., Casey, H. L., Marine, W. M., and Nahmias, A. J.   Cytological features of viral respiratory tract infection. *Acta Cytol.* 12:162, 1968.

Nasiell, M.   Abnormal columnar cell findings in bronchial epithelium. A cytologic and histologic study of lung cancer and non-cancer cases. *Acta Cytol.* 11:397, 1967.

Nunez, V., Melamed, M. R., and Cahan, W.   Tracheo-bronchial cytology after laryngectomy for carcinoma of larynx. *Acta Cytol.* 10:38, 1966.

Parker, R. E., and Reid, J. D.   Five-year survey of results of cytological examination for lung cancer. *N.Z. Med. J.* 59:68–72, 1960.

Pearson, F. G., and Thompson, D. W.   Occult carcinoma of the bronchus. *Can. Med. Assoc. J.* 94:825, 1966.

Pfitzer, P., and Knoblich, P. G.   Giant carcinoma cells of bronchiogenic origin. *Acta Cytol.* 12:256, 1968.

Pharr, S. L., and Farber, S. M.   Cellular concentration of sputum and bronchial aspirations by tryptic digestion. *Acta Cytol.* 6:447, 1962.

Pierce, C. H., and Hirsch, J. G.   Ciliocytophthoria; relationship to viral respiratory infections of humans. *Proc. Soc. Exp. Biol. Med.* 98:489–492, 1958.

Pierce, C. H., and Knox, A. W.   Ciliocytophthoria in sputum from patients with adeno-virus infections. *Proc. Soc. Exp. Biol. Med.* 104:492–495, 1960.

Plamenac, P., Nikulin, A., and Kahvic, M.   Cytology of the respiratory tract in advanced age. *Acta Cytol.* (Baltimore) 14:526, 1970.

Plamenac, P., Nikulin, A., and Pikula, B.   Cytologic changes of the respiratory epithelium in iron foundry workers. *Acta Cytol.* 18:34, 1974.

Russell, W. O., et al.   Cytodiagnosis of lung cancer. *Acta Cytol.* 7:1, 1963.

Saccomanno, G., et al.   Cancer of the lung: The cytology of sputum prior to the development of carcinoma. *Acta Cytol.* 9:413, 1965.

Skitarelic, K., and Haam, E. von.   Bronchial brushings and washings: A diagnostically rewarding procedure. *Acta Cytol.* 18:321, 1974.

Suprun, H., and Koss, L. G.   The cytological study of sputum and bronchial washings in Hodgkin's disease with pulmonary involvement. *Cancer* 17:674, 1964.

Takeda, M., and Burechaico, R. T.   Smooth muscle cells in sputum. *Acta Cytol.* 13:696, 1969.

Tassoni, E. M.   Pools of lymphocytes: Significance in pulmonary secretions. *Acta Cytol.* 7:168, 1963.

Tsuboi, E., Ikeda, S., Tajima, M., Shimosato, Y., and Ishikawa, S.   Transbronchial biopsy smear for diagnosis of peripheral pulmonary carcinomas. *Cancer* 20:687, 1967.

Umiker, W.   Typing of bronchogenic carcinoma. *Arch. Pathol.* 71:295, 1961.

Effusions

12

The cellular composition of aspirated fluids from the right and left pleural, pericardial, or abdominal cavities is identical. It is impossible to determine by cytology from which body cavity an effusion specimen has been aspirated. Any appreciable accumulation of fluid in these cavities, although often nonspecific, is an indication of a disease. Physiologically the cavities should contain only a few milliliters of serum needed as a friction lubricant.

These body cavities are normally lined by a single layer of flattened epithelium made of mesothelial cells, lying upon a supportive connective tissue. This mesothelial membrane reacts to any type of injury by becoming stratified with the flattened cells changing into cuboidal cells and by the accumulation of a protective effusion. The mesothelial cells possess the ability, similar to that of histiocytes, to change their features. In reaction to different stimuli, they can imitate any type of cell, benign or malignant. Furthermore, the fluids are good culture media, and the exfoliated cells, benign or malignant, can continue to live and divide in these fluids. This explains why cells in mitosis are more likely to be seen in effusions than in any other type of specimen. The fact that the effusions are good culture media may also be the reason that most of the squamous carcinoma cells in effusions will appear more mature than the cells collected directly from the primary lesions.

Except for the primary mesothelioma and a few sarcomas, the neoplasms involving the serosal membrane are secondary in type, and their detection seldom ensures a favorable prognosis. A negative cytologic interpretation is of moderate value (35% false negative), while a positive report (false positive, less than 1%) is important in the differential diagnosis of the causes of these effusions and in determining the treatment of the patient.

Technique for Collecting Specimens

The aspirated fluids should be received either in a vacuum tube prepared for the collection of blood, containing acid, citrate, and dextrose solution (ACD) or in a container to which is added about 5 units of heparin for every cubic centimeter of aspirated fluid. A complete description of the specimen should be included on the back of the requisition, giving the volume, the time elapsed, the color, and other pertinent data. This is important because, as summarized in Table 12, a tentative statement about the sig-

Table 12. Significance of the Gross Aspect of Effusions

Aspect	Significance
Watery and clear	Transudate
Yellowish-white	High white blood cell count
Milky, white or green	Chylous fluid
Watery and brown	Transudate bilirubin
Dark brown to red	Hemorrhagic-neoplastic
Viscid	Mesothelioma

nificance of the effusion can be made after macroscopic inspection of the fluid submitted. If the specimen is too abundant (more than 500 ml), it can be allowed to sit without being disturbed for one-half hour to allow the sedimentation of the heavy tumor cells. The surface fluid is discarded gently until about 200 ml are left and processed in the same manner as when an adequate specimen has been initially received. Any recognizable tissue particles in the fluid should be removed, immediately smeared, and fixed or paraffin-embedded for sectioning. After the effusion has been shaken in its original container, it can either be passed through a membrane filter with a 5 μ pore opening or centrifuged at 4000 rpm for 10 minutes, the supernatant decanted, and 4 smears made from the upper layer of the sediment. If the remaining centrifuged button is large enough, a few drops of old plasma and thrombin solution are added, mixed, and allowed to form a coagulated block (Fig. 137). The cells may be fixed slowly by pipetting a mixture of 95% ethyl alcohol and 5% formalin along the side of the tube, taking care not to disturb the button, which is to be embedded in a paraffin block for sectioning.

The very cellular effusion should be stained separately, and all the staining solutions filtered afterward. Otherwise one can expect to find reactive mesothelial cells (for example) contaminating all or some of the other specimens stained on the same day. This could lead to numerous and disastrous false-positive interpretations.

If the aspiration is done during the night or over the weekend, the fluid can be mixed with the same volume of 70% alcohol and placed in the refrigerator until processing time. If the specimen is very bloody, variable amounts of Carnoy's solution, depending on the amount of red blood cells present, can be added, mixed, and left to stand undisturbed for 10 minutes. The supernatant, containing the hemolyzed red blood cells, is decanted; saline solution is added and mixed with the cells, and the solution is passed through a membrane filter.

Speed in preparation and fixation of effusions is not as critical as for other specimens because fluids are good culture media and the cells remain well preserved for a long time. The presence of degenerate cells and signet ring vacuolated cells probably means that these cells were shed long before and have been present in the fluid for some time before aspiration. In such cases a repeat specimen is recommended.

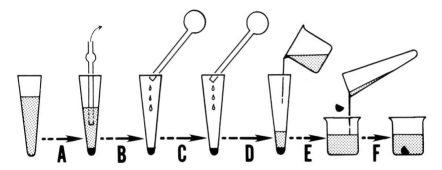

Figure 137
Cell block technique in effusion:
A. Centrifuge at 2000 rpm for 5 min.;
B. Decant the fluid;
C. Add 4 or 5 drops of plasma;
D. Add and mix with 4 or 5 drops of thrombin solution (5000 units of thrombin in 10 ml water). Let clot;
E. Add 10% formalin;
F. Transfer clot in 95% alcohol, fix for 30 min., then process clot in tissue paper.

If a rapid diagnosis is needed, a few droplets of toluidine blue can be added to the centrifuged smeared cells and examined immediately, wet. Most of the positive cases can be easily detected this way. If the specimen has been allowed to clot, the clotted, jellylike material should be placed on a Petri dish and the excess fluid squeezed out until a small, firm mass is obtained that may be fixed in formalin and processed as a cell block.

Transudates

Transudates are the fluids accumulated as a result of an increased venous or lymphatic pressure, usually from a disturbance of cardiac function (congestive cardiac failure). Transudates can also be caused by malnutrition, cirrhosis, or heart or kidney diseases. The aspirated fluid is low in protein, with a low specific gravity of 1.005 to 1.015, and has little tendency to clot. The smears show a clear background with a very low cell content. The scattered, round, single mesothelial cells of serosal origin are small, about the size of a leukocyte. A few histiocytes and inflammatory cells, mainly lymphocytic in type, are also present.

Exudates

Exudates are fluid accumulations resulting from irritation of the serosal mucosa by such agents as cancer, inflammation, a foreign body, or hemorrhage. The exudate has a high protein content and specific gravity (above 1.015). It is rich in fibrin and clots readily. The background of the smear is usually crowded and granular with a large amount of protein deposit. Abundant, single, or in clusters, the mesothelial cells are generally larger

Table 13. Differential Characteristics of Effusions

	Transudate	Exudate
Specific gravity	Less than 1.015	More than 1.015
Rivalta's test	Negative	Positive
Protein	Less than 2 gm/dl	More than 2 gm/dl
Fibrin	Poor	Rich
Glucose	High	Low
Cells	Few	Abundant
Appearance	Watery and clear	Variable

than usual. They are reactive and have large nuclei and prominent nucleoli. In exudates there is also an increase in inflammatory cells (leukocytes), fresh and old red blood cells, and, occasionally, colonies of microorganisms. The differences between transudate and exudate effusions are summarized in Table 13.

Cytology of Benign Effusions

NONREACTIVE MESOTHELIAL CELLS (FIGS. 138, 139)
Nonreactive mesothelial cells are found singly (85%) or in small clusters with occasional clumping that gives the impression that the cells are piling up on each other like a stack of dishes, with slight molding between the cells (rouleau formation). The cell size varies moderately ($15 \pm 5 \mu$), and they are uniformly round or oval in shape.

The cytoplasm is foamy, delicate, pink, and abundant when the cells are young (Fig. 138A) and vacuolated when old (Fig. 138B). Occasionally, the cytoplasm is basophilic, staining dark purple, almost with the same intensity as the nucleus. The cytoplasmic borders are often distinct.

The nuclei are usually single, round, or oval and are centrally located, with fine granular, uniformly distributed chromatin and sharp, well-defined, smooth nuclear borders.

Occasional, small, round, usually single nucleoli may be found.

With periodic acid–Schiff (PAS) stain, a few small red granules may be seen in the periphery of the cytoplasm. This can be used in the differentiation of mesothelial cells from phagocytes that usually have no such granules or from certain adenocarcinoma cells with large PAS-positive intracytoplasmic masses.

HISTIOCYTES
Variable amounts of histiocytes may be present and are sometimes difficult to distinguish from the mesothelial cells. They are similar in size and shape and are also found singly. Their differentiation is based mainly on the occasional, eccentric, bean-shaped or lobular nuclei of the histiocytes, on their more delicate, foamy, and cyanophilic or pale cytoplasm, and on the pres-

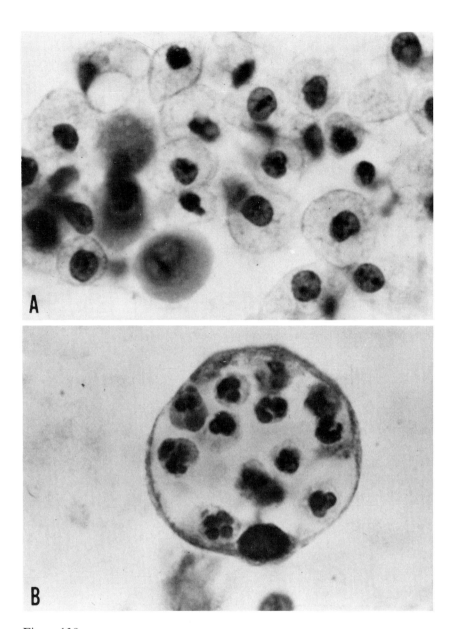

Figure 138
(A) Physiologically exfoliated, benign young mesothelial cells. Note the regularity of their shape and staining variation of their cytoplasm. (× 850) (B) Benign older mesothelial cell as a signet ring cell. Note the vacuole distending its cytoplasm and the phagocytizing neutrophils. (× 850)

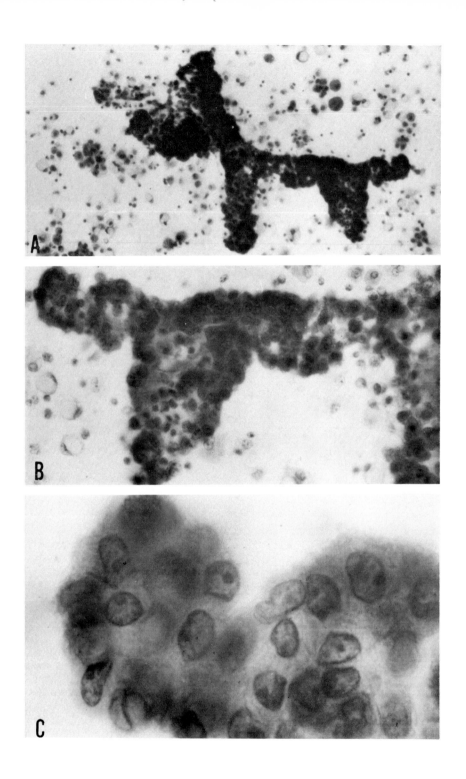

ence of intracytoplasmic phagocytized inclusions. The cytoplasmic vacuoles are phagocytic rather than secretory.

EOSINOPHILS
The greatest numbers of eosinophils are found in pneumothorax, certain types of pneumonia (e.g., Löffler's), systemic parasitic and allergic diseases, and some malignant neoplasms.

Their presence should be an indication for more careful screening of the specimen.

Traumatic pericarditis, infarction, or allergic diseases (rheumatic pericarditis and allergic vasculitis) may also produce an increase of eosinophils in the pericardial fluid. When the presence of these eosinophils cannot be explained, they are often called idiopathic.

NEUTROPHILS
Increased numbers of neutrophils may be found obscuring the slide in cases of pneumonia, pulmonary infarction, abscess, tuberculosis, and emphysema.

REACTIVE ATYPICAL MESOTHELIAL CELLS (FIG. 140)

Causes
1. An infarction of any organ adjacent to the cavity containing the effusion;
2. A liver disease, the most common being liver cirrhosis;
3. Irradiation effect;
4. Degenerative systemic diseases (rheumatic fever, lupus erythematosus);
5. Collagen diseases;
6. Traumatic irritation (foreign bodies, neoplasms);
7. Chronic inflammation of very long duration.

Appearance
Although atypical and enlarged, these mesothelial cells still resemble the normal nonreactive mesothelial cells.

They shed singly or in small or large clusters, often stacked on each other. Their shape varies from round to oval (85%) to irregular (15%). Their size (60 ± 30 μ) varies according to the age and maturation of the cells.

The cytoplasm is variable, scanty to abundant, usually thick, homogeneous, stained deep purple with occasional inclusions (debris, pigment, lipids, blood, or hypersecretory vacuoles). The cytoplasmic membranes are usually sharp, smooth, and heavy.

The nuclei are either central (80%) or eccentric (20%), as seen in the signet ring type cell. The nuclear size is enlarged and varies moderately (25 ± 15 μ). The chromatin is prominent, coarsely clumped, uniformly

---

Figure 139
Traumatically exfoliated (needle-tract) benign mesothelial cells. (A: $\times$ 250; B: $\times$ 850; C: $\times$ 1300)

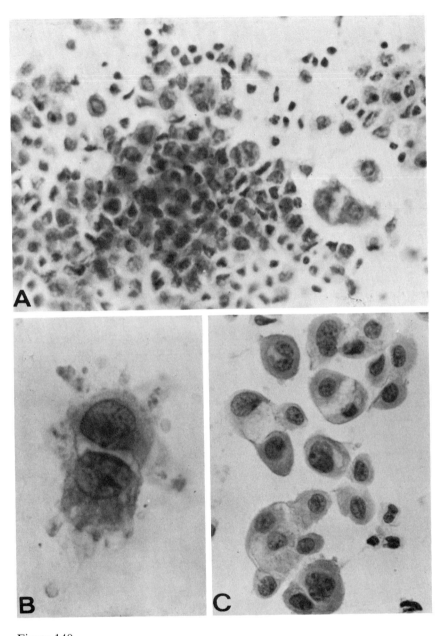

Figure 140
(A) Cluster of reactive mesothelial cells in the ascitic fluid of a patient with liver cirrhosis. (× 450) (B) Benign reactive mesothelial cells. Note the prominent nucleoli, large nuclei, and scanty cytoplasm. (× 1300) (C) Cluster of reactive mesothelial cells in the pericardial fluid of a man with a recent myocardial infarction. (× 450)

distributed (85%), and varies moderately from cell to cell. The nuclear borders are regular and sharp (85% of the cells) or irregular with indentation.

The nucleoli are spherical and prominent and vary in number (1–4) in 90% of the cells.

Multinucleation (2–4 nuclei) occurs (20%) and numerous normal or slightly atypical mitoses may be found.

The background of the smear usually shows large amounts of leukocytes, protein deposits, and cellular debris.

In patients with a history of surgery, talcum-starch crystals may be found half engulfed in large vacuolated histiocytes often seen in clusters. They can be mistaken for mucus-secreting adenocarcinoma. Examination of the smears with a polarized light will show the true nature of these crystals.

Differential Diagnosis

Because of these reactive mesothelial cells, one should be very conservative in applying the general criteria for malignancy described in Chapter 6 of this text to the single cells or clusters of neoplastic cells represented in an effusion. The differentiation of single mesothelial cells, histiocytes, and tumor cells from one another is difficult. It is summarized in Table 14.

In the presence of a definite sheet or acinus, however, the diagnosis of a metastatic neoplasm should always be suspected, even though the individual component cells may appear benign. In such a case, one should be careful in differentiating the true neoplastic sheet or acini from the pseudoacinic formation.

The cytoplasm and the nucleus of the single reactive mesothelial cell usually stain with almost equal intensity, whereas in the single malignant cell the nucleus usually stains far more densely than the cytoplasm.

For further help in the differentiation of the single cells, PAS stain may be employed. The mesothelial cells will contain a coarsely granular ringlike formation of PAS-positive substance near the border of their cytoplasm, whereas certain neoplastic cells will show large droplets of PAS-positive material mainly within their central endoplasm.

Cytology of Malignant Lesions

The mesothelial lining of the body cavities becomes involved by malignant neoplasms, either by direct extension of the tumor or from distant carcinomatous metastasis. Reacting to the presence of these foreign tumor cells, variable amounts of fluid may accumulate in one or several cavities. The site of the primary neoplasm often determines the cavity in which the fluids accumulate. For example, a lung carcinoma usually produces pleural, rather than abdominal fluid.

The tumor cells are usually seen in tight sheets or acinic formation. These structures are usually spherical in shape with smooth common external borders, a result of their being repeatedly rolled between two serosal surfaces with every breath, heartbeat, or peristaltic movement.

Table 14. Differentiation of Single Cells in Effusions

	Mesothelial Cells	Phagocytes	Malignant Cells
Shape of cells	Round to oval	Round to oval	Aberrant
Shape of nuclei	Round to oval	Kidney-shaped	Irregular
Location of nuclei	Central	Peripheral	Variable
Chromatin pattern	Variable	Finely granular	Irregular
Nucleolus	Distinct	Indistinct	Variable
Nuclear membrane	Regular	Regular	Irregular
Cytoplasm	Basophilic, acidophilic, perinuclear halo	Cyanophilic, vesicular	Variable
Periodic acid–Schiff (PAS) stain	Fine granules at periphery (weakly positive)	Negative	Large droplet (strongly positive)
Mucicarmine stain	Negative	Negative	Occasionally positive
Peroxidase reaction	Negative	Occasionally positive	Negative
Neutral red stain	Few scattered granules	Rosette-like formation	Rosette-like formation

Contrary to that which is seen in other cytologic specimens, an increase of normal and abnormal mitoses in benign, as well as in malignant cells, is often found, an indication that most of the mesothelial and neoplastic cells continue to grow, live, and function even after their exfoliation.

The malignant cells are usually well preserved and often imitate the cellular arrangement of the tissue from which they originated.

The background of the smears is bloody in at least one-third of all malignancies.

The possibility of malignancy should be considered when the following secondary criteria are present: hemorrhagic effusion, lymphocytosis, increased histiocytes and lipophagocytes (Sudan III stain), and an increase of intracytoplasmic neutral red–stained structures.

PRIMARY MESOTHELIOMAS (FIG. 141A, B)
Primary mesotheliomas are rare tumors that occur slightly more frequently in asbestos workers. They are difficult to diagnose by effusion cytology, especially if the mesothelioma is fibrous in type.

An increased number of reactive mesothelial cells, sheets, in the form of multiple, large, round, spherical papillary clusters of cells with smooth external borders (60%) are present.

The neoplastic cells resemble the large reactive mesothelial cells, except for the more extensive variation in their size and shape. Distinct malignant features should be present with retention of some benign mesothelial features.

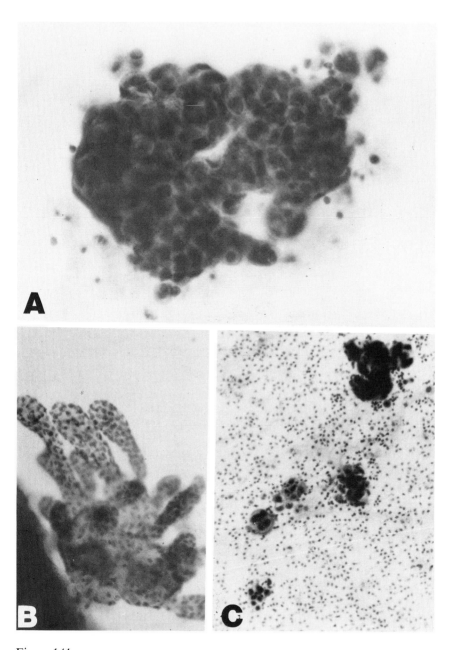

Figure 141
(A) Ascitic fluid in a primary mesothelioma of the peritoneal cavity. Note the morphologic similarity of the sheet of neoplastic cells to the normal meso-thelial ones. (× 450) (B) Papillary structure in a primary mesothelioma of the pleura. (× 450) (C) Pleural fluid with metastatic neoplasm. The clusters of tumor cells that stain dark are easily detected, even by low-power magnifica-tion examination. (× 125)

The cytoplasm is dense, often scanty with no vacuolization. On occasion, toward the periphery of some cells (exoplasm), the cytoplasm has a foamy appearance.

The nuclei are enlarged, hyperchromatic, dense, central, with coarsely irregular chromatin pattern and prominent nucleoli. The nuclei are often elongated and irregular in shape when found in sheets. Multinucleation is frequent (2–4 nuclei), and occasional abnormal mitoses can be seen.

In rare instances, psammoma bodies can be found surrounded by rings of mesothelial cells.

The possibility of a mesothelioma should be considered if the aspirated fluid is viscid. An increased amount of hyaluronic acid also suggests the presence of a diffuse malignant mesothelioma. This increase is suspected if heavy turbidity of the fluid is produced when a 50% glacial acetic acid solution is mixed with the effusion.

POORLY DIFFERENTIATED SECONDARY CARCINOMA

In the presence of poorly differentiated carcinoma, the nature and origin of the primary neoplasm is very difficult to determine.

The malignant cells are uniformly small (15 ± 5 μ) and in tight, often palisading arrangement with indistinct cellular borders (95% of the time). They are easily overlooked because of their size, which is often smaller than that of the benign reactive mesothelial cells.

The malignant cells vary extensively in shape but are fairly uniform in size.

The cytoplasm is scanty, basophilic, and semitransparent with irregular borders.

The nuclei are relatively large, occupying seven-eighths of the surface of the cell. Their size and shape vary extensively. The nuclear membrane is irregular in thickness and shape with multiple indentations (wrinkling).

The chromatin is very dense, almost pyknotic, or irregularly clumped with pointed projections. The distribution of the chromatin is extremely irregular.

The abnormal nucleoli are minute, and frequent mitoses may be present.

SQUAMOUS CELL CARCINOMA

The occurrence of this neoplasm in pleural effusion is often associated with the expansion of a primary bronchogenic carcinoma to the pleural serous membrane. The squamous cancer cells in the pleural fluid are few and often very difficult to interpret correctly. The diagnosis may be missed, especially if the cells exfoliate singly (false-negative rate, about 60%). They may show marked variation in size and shape. Often they are poorly differentiated, and cytoplasmic keratin is seldom seen.

In approximately 30% of the cases, although the primary tumor is a well-differentiated squamous cell carcinoma, the malignant cells seen in the pleural fluid may resemble and may be interpreted as adenocarcinomatous cells with large nuclei, prominent nucleoli, and vacuolated cytoplasm.

Occasionally, especially in direct extension of the tumor into the cavity, sheets of well-differentiated squamous cells forming pearl-like structures can be seen with abundant keratin accumulation in their cytoplasm (deep orange stain). The nuclei of these cells are hyperchromatic, dense, and pyknotic with marked variation in size and shape. The smear usually shows large amounts of blood, inflammatory cells, and protein deposits in the background.

ADENOCARCINOMA OF BREAST ORIGIN (FIGS. 142A, 143B)

The amount of diagnostic, neoplastic cells varies. Usually they are abundant, especially in *infiltrating duct carcinoma.* They shed in tight clusters, in the form of numerous round flat-as-a-pancake or spherical ball-like cell nests. Their size varies. The individual component cancer cells are small and often fairly uniform in size and shape ($20 \pm 5 \mu$). Their cytoplasm is scanty to adequate, and pink-to-purple uniformly stained. Occasional small cytoplasmic vacuoles are present. The nuclei are moderately hyperchromatic and are molded against the adjacent cells. The nuclear membrane is usually smooth. The chromatin is uniform and granular. The usually inconspicuous nucleoli may occasionally be prominent (10%). The diagnosis of most metastatic infiltrating duct carcinomas is relatively easy and specific.

The other breast cancers are more difficult to type. The *medullary carcinoma* cells are shed singly and vary in size and shape. They are often binucleated or multinucleated. The cells show most of the malignant features described in Chapter 6. In *lobular carcinoma* the exfoliated, mainly single tumor cells are easy to miss. These small cells are often found in Indian file or in pseudo-pearl-like formation. Their cytoplasm is scanty and their nuclei hyperchromatic. These cells are often confused with reactive lymphocytic cells.

OVARIAN ADENOCARCINOMA (FIG. 142C, D)

In ovarian adenocarcinoma, especially in serous cystadenocarcinoma, numerous malignant cells may shed singly or in the form of acini, papillae, or rosettes. Their size and shape vary dramatically, depending on the type of carcinoma (pseudomucinous adenocarcinomas give the largest cells).

Their cytoplasm is usually abundant, well defined, and gray to deep purple. In mucinous adenocarcinoma the cytoplasm may be overdistended by large secretory vacuoles (pouched out type), to be differentiated from the smaller, more regular signet ring cell of mesothelial and histiocytic origin.

Their nuclei are enlarged and round or oval, except when molded by neighboring cells or vacuoles. Frequent nuclear overlapping is present. The nuclear borders are often well defined, sharp, and irregular in shape and thickness.

Their chromatin is usually increased with irregular clumping and pointed projections. Mitosis is frequent and often abnormal.

The nucleoli are prominent and vary in size, shape, and number.

Structures like psammoma bodies surrounded by tumor cells are rare but may be seen in certain neoplasms (papillary cystadenocarcinoma).

In rare instances, ciliated cylindrical epithelial cells, single or in palisade

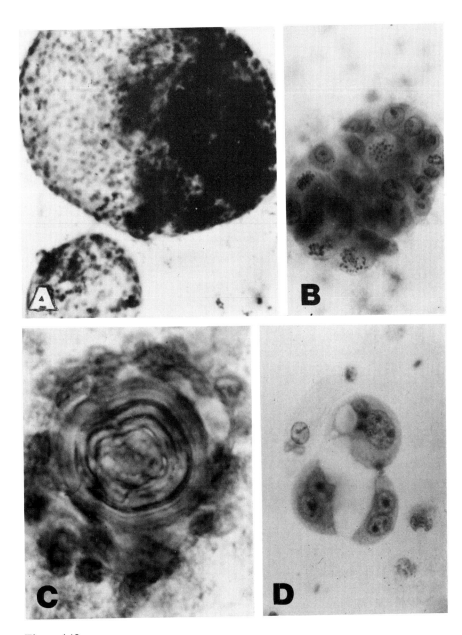

Figure 142
(A) Round sheets of tumor cells in the pleural fluid of a patient with breast carcinoma. (× 250) (B) Metastatic gastric adenocarcinoma cells in the pleural fluid of a patient with recurrent effusion. Note the multiple abnormal mitoses. (× 450) (C) Ovarian adenocarcinoma in abdominal fluid. Note the diagnostic laminated layer of the psammoma body. (× 450) (D) Pulmonary adenocarcinoma in pleural fluid. (× 450)

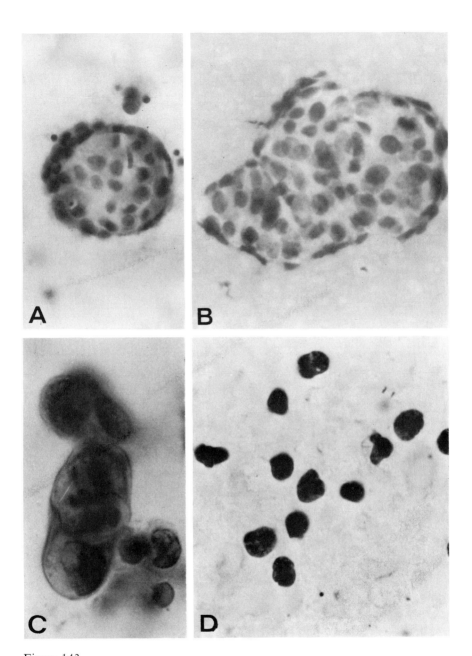

Figure 143
(A) Malignant acinus in the pleural fluid of a woman with a breast cancer.
(× 450) (B) Papillary neoplastic cells in the ascitic fluid of the same woman.
(× 450) (C) Papillary neoplastic cells in the ascitic fluid of a woman with pancreatic adenocarcinoma. (× 750) (D) Lymphosarcoma cells in the pleural
fluid of a patient with generalized disease involving both lung and pleura. Note
the shape and size variation of the cells. (× 1300)

formation, may be found. These can indicate the presence of a benign or a malignant serous cystadenoma.

In *pseudomyxoma peritonei* the bowel becomes embedded in thick mucinous material secreted by benign or malignant mucinous cyst cells. The tumor cells are seldom seen in the aspirated fluid.

GASTROINTESTINAL CARCINOMAS (FIGS. 142B, 143C, 144)

In gastrointestinal carcinomas, the neoplastic cells are not as abundant as in ovarian carcinomas. The cells are shed singly or in a papillary arrangement.

Their cytoplasm is usually scanty and shows no distinctive characteristics, except in mucus-producing adenocarcinomas, in which large mucin-containing vacuoles overdistend the cytoplasm.

Their nuclei are usually hyperchromatic and irregularly shaped with malignant chromatin clumping and occasional prominent nucleoli. The nuclear atypia is more pronounced in the cells exfoliating from a gastric metastasis than from a colonic one.

The presence of an atypical tubular formation made of multilayers of tall, malignant columnar cells is suggestive of colonic adenocarcinoma. An increase in signet ring cells, single or in clusters, is also common but is seen especially in mucus-producing carcinomas. Their shape is usually round.

LYMPHOCYTES AND EFFUSIONS (FIG. 143D)

Variable amounts of mature lymphocytes, irregular in size and shape, can be seen in long-standing effusions and in acute tuberculosis, sarcoidosis, and viral inflammation. The lymphocytes are mature in type and are seen singly or in clusters. They show no prominent nucleoli.

In *lymphosarcoma,* there is marked variation in the size (6–15 μ) of these lymphocytes. Their shape is usually round. Their very sparse cytoplasm stains pale with no vacuoles. Their nuclei are enlarged and vary in size; they are usually round with prominent nucleoli and abnormal, coarse clumping of the chromatin with heavy irregular and indented nuclear membrane. Few mitotic figures can be seen.

In *reticulum cell sarcoma,* numerous large, single cells are often seen, irregular in size (10–30 μ) and shape, and often elongated, with broad, transparent cytoplasm with faint vacuoles and hyperchromatic nuclei with prominent, irregular, multiple nucleoli. The chromatin is more prominent than in other lymphomas.

In *Hodgkin's disease,* the fluids may occasionally contain Reed-Sternberg cells, recognized by their binucleation and prominent nucleoli. A variable amount of immature lymphocytes is usually present in the background, as it is in lymphosarcoma.

In *myeloid leukemia,* azurophilic granules may be seen in the background of the smear.

The changes in *lymphocytic leukemia* are the same as those in lymphosarcoma.

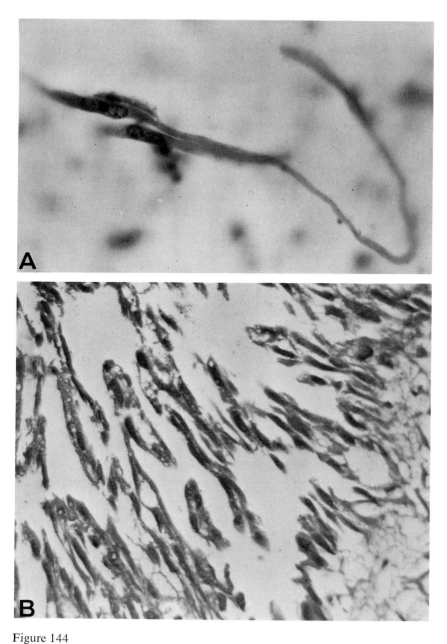

Figure 144
(A) Elongated spindle cells seen in the pleural fluid of a man with a large primary pleural fibrosarcoma. (× 750) (B) Histologic section of the resected tumor. Notice the morphologic similarity of the neoplastic cells to the exfoliated ones. (× 750)

Table 15. Cytology of Synovial Fluid

Diagnosis	Volume	Color	Viscosity	Mucin Clot Test	Inflammatory Cells per Milliliter
Normal	Scanty	Clear	++++	++++	10–20
Septic arthritis	Abundant	Gray, purulent	++	0	+100,000
Traumatic arthritis	Variable	Hemorrhagic	+++	+++	2000–3000
Villonodular synovitis	Variable	Hemorrhagic	±	±	3000
Rheumatoid arthritis	Variable	Yellow	±	±	30,000
Rheumatic fever	Scanty	Cloudy	±	+++	10,000
Osteoarthritis	Scanty	Clear	+++	++	40
Gouty arthritis	Variable	Yellow	±	±	15,000
Pseudogout	Variable	Yellow	±	+++	8000
Systemic lupus erythematosus	Scanty	Yellow	+++	+++	10,000
Neoplasm	Variable	Hemorrhagic	++	++	200

Cytology of Synovial Fluid
(Table 15)

Aspirate as much fluid as possible from the diseased joint using a needle with a gauge larger than 20. Except for one or two droplets that are placed directly on a slide and not smeared, the fluid is mixed with an anticoagulant (25 units of heparin for every milliliter of fluid). At this point the amount, color, transparency, and gross appearance of the aspirated synovial fluid are carefully noted.

An immediate *viscosity test* (Fig. 145) may be done by touching the droplets of fresh synovial fluid on the slide and observing, before they break, the length of strands formed. If the strands break before they reach approximately 2 cm in length, the viscosity of the fluid is considered to be abnormally low.

Another useful and simple procedure is the *mucin clot formation test,* which is done by adding and mixing to the same drops of synovial fluid a few drops of 5% acetic acid solution and allowing it to stand about one minute. The protein hyaluronate of the synovial fluid coagulates and forms

Lymph %	Poly %	Histio-cyte %	Syno-vial Cells	Carti-laginous Cells	Inclu-sions	Crystals	Other
25	10	65	+	0	0	0	
5	90	5	+	+	+	0	Presence of microorganism
20	70	10	++	++	0	0	
10	30	60	++	0	0	0	Foamy histiocytes and foreign-body giant cell
10	80	10	±	0	+++	++	Ragocytes (RA)
40	50	10	++	0	++	0	Immunofluorescence; inclusions
70	10	10	+++	++++	0	0	
10	70	20	+	±	0	++++	At pH 9, crystals are dissolved
20	75	5	+	+++	0	++	Calcium pyrophosphate crystals
80	10	10	+	0	++	0	Presence of LE cells
30	60	10	+	±	0	0	Presence of tumor cells

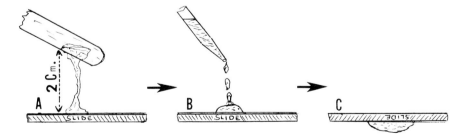

Figure 145
(A) Viscosity test of fresh joint fluid. (B, C) Mucin clotting test. Mixture with 5% acetic acid produces a firm clot that can be inverted without distortion.

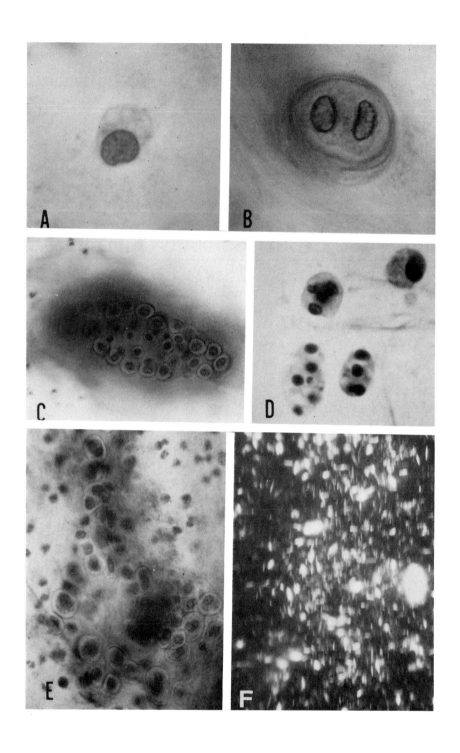

a firm homogeneous clot that normally should remain attached to the slide without distortion when inverted.

The remaining heparinized synovial fluid can then be diluted with a few milliliters of saline and passed through one or several membrane filters. Stained with routine Papanicolaou stain, the fluid is examined for the amount and type of inflammatory, synovial, cartilaginous, and neoplastic cells.

To determine the presence of intracellular inclusions, which may be diagnostic of Reiter's disease and rheumatoid arthritis, a high-power examination ($\times 900$) of some of the cells is necessary. Finally, the smears are examined with polarized light for the recognition and differentiation of the microcrystals found in some joint diseases.

In *normal persons* and in patients with *Baker's cyst and ganglion* or in patients with *nonarticular diseases* resulting from prolonged bed rest, the amount of fluid aspirated is usually scanty (Fig. 146A). Transparent, straw-colored droplets will string out more than 2 cm before breaking in the viscosity test. The mucin clotting test is also strongly positive, and the normal fluid will not coagulate in the absence of heparin. Light microscopy shows few scattered inflammatory cells, no more than 10 to 20 cells in each milliliter. They are mainly mononuclear histiocytes (65%). Few round synovial cells are also present.

In *septic arthritis,* aspirated fluid is usually abundant, gray, and purulent. Viscosity is not decreased, and the mucin clotting test is negative. The fluid will not coagulate, even in the absence of heparin. Large numbers of neutrophils—more than 100,000 per milliliter—are seen masking most of the specimen.

In *traumatic arthritis* (Figs. 146E, 147A), viscosity and mucin clotting tests are usually strongly positive. The amount of inflammatory cells, mainly polymorphonuclear cells (70%), varies from 2000 to 3000 per milliliter. Sheets of cartilage and synovial cells are present.

In *villonodular synovitis* (Fig. 147B), aspiration produces variable amounts of serohemorrhagic brown fluid. The viscosity and mucin clotting test depend on the duration of inflammation. The fluid contains about 3000 inflammatory cells per milliliter with a preponderance of mononuclear cells (over 60%). The presence of multinucleated giant cells and round histiocytes with foamy cytoplasm is diagnostic. No crystals or inclusions are present.

In *rheumatoid arthritis* (Fig. 146D), the amount of yellowish-green, clear-to-cloudy aspirated fluid varies with the duration of the infection and the articulation involved. The fluid will rapidly coagulate if no heparin is used. Its viscosity is decreased and the mucin clotting test produces only

---

Figure 146
Synovial fluid. (A) Normal synovial cell. ($\times 450$) (B) Cartilage cell. ($\times 450$) (C) Cartilage cells in a case of osteoarthritis. ($\times 150$) (D) Ragocytes with cytoplasmic inclusions in a case of rheumatoid arthritis. ($\times 850$) (E) Sheet of cartilage in traumatic arthritis. ($\times 250$) (F) Urate crystals in a case of gouty arthritis seen with a polarized light. ($\times 250$)

an abnormal small, friable, irregular clot. Diagnosis is based on the presence of monolobular or polylobular neutrophils (ragocytes, or R.A. cells) containing 2 to 15 round, dark inclusions in their cytoplasm, which are best seen when a green filter is placed in front of the microscopic light source.

In arthritis associated with *rheumatic fever,* small amounts of yellow, cloudy fluid are usually produced. Final specific diagnosis is best made by immunofluorescent staining. The background of the smear usually contains large amounts of debris and fibrin with occasional histiocytes and giant cells.

The most remarkable diagnostic feature of *osteoarthritic* (Fig. 146C) synovial fluid is the presence of many well-preserved, normal-looking multinucleated cartilaginous cells, which are more often found in sheets or clusters than singly. The background of the smear is clear.

In *gouty arthritis* (Fig. 146F), the fluid is dense, yellow, and cloudy with low viscosity. The mucin clotting test produces a fragile, small, abnormal clot. Examination of the smear by polarized light is diagnostic, revealing many birefringent needle-shaped monosodium urate crystals. Large numbers of inflammatory neutrophils are present.

The diagnosis of *pseudogout,* or *chondrocalcinosis,* is based on recognition of calcium pyrophosphate crystals with polarized light microscopy. These large crystals have rectangular or elongated shapes with sharp parallel borders and square extremities.

Recognition of articular *lupus erythematosus* (LE) (Fig. 147C) is based on the presence of LE cells. In the fully developed stage, the normal chromatin structure of the nucleus of the diagnostic neutrophil is compressed by a pink-to-purplish, homogeneous, amorphous, round intracytoplasmic mass that varies in size.

In *neoplastic effusion* (Fig. 147D), the diagnosis of neoplasm is made by finding sheets or clusters of neoplastic cells with features foreign to the lining membrane of the joint. One should suspect such a lesion when the effusion is of a recurrent type and reforms rapidly after its aspiration.

EFFUSIONS FROM OTHER SITES

The study of aspirated fluid from other cavities of the body (hydroceles, cysts) can be very rewarding. Failure to examine an aspirated fluid for its cellular content is as bad as failure to examine a resected tissue histologically. Normally these fluids show only few phagocytes, lymphocytes, and other inflammatory cells. Occasionally, semitransparent, flattened, benign-looking lining cells are present. The recognition of a neoplasm is quite simple. The presence of papillary, sheet, or acinic formations is diagnostic. The ease of cytologic interpretation is aided by the absence of reactive mesothelial cells.

PITFALLS

False-Positive

Sheets or acini of endometrial cells may migrate transtubally into the ascitic fluid, as seen occasionally in patients with recent dilation and curettage.

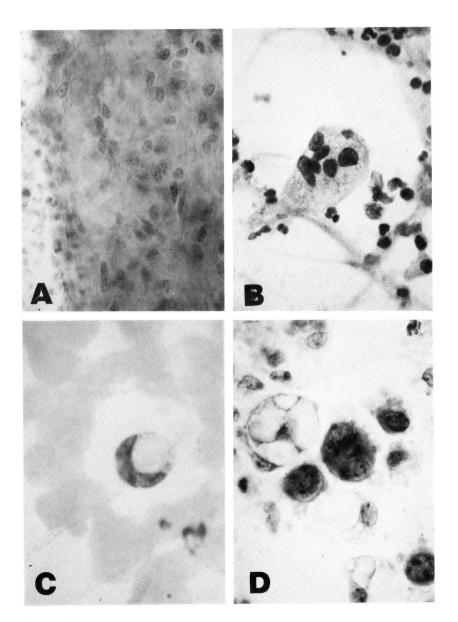

Figure 147
(A) Fibrinous and osteoid matrix in the synovial fluid aspirated from a traumatized knee joint. (× 250) (B) Giant cell in villonodular synovitis. (× 350)
(C) Lupus erythematosus (LE) cell in synovial fluid. (× 750) (D) Metastatic
prostatic adenocarcinoma cell in knee joint fluid. (× 350)

An endometrial cyst may rupture, liberating fresh and old blood, histiocytes, and variable amounts of sheets or acini of benign endometrial cells. The rupture of a benign ovarian cyst may show anucleated squamous cells from a dermoid cyst or papillary projections from a benign papillary tumor.

A history of previous surgical exploration may produce benign squamous cells and esophageal flora in the pleural effusion originating from a surgically produced esophageal fistula, for example.

The presence in the aspirated effusion of sheets of transitional cells, squamous epithelium, striated muscle, ciliated epithelium, and liver cells may result from the traumatic removal of pieces of serous membrane, skin, muscle, lung, and liver by the needle used for the aspiration of the fluid. The cytologic diagnosis of a malignant neoplasm in an effusion should never be based on the presence of one single sheet of cells; several should be found.

Contaminants of all sorts (pollen, talcum) may imitate tumor cells.

False-Negative

Certain neoplasms may produce an accumulation of effusions without invading the serosa and consequently without shedding any cells in the fluid.

The serosal cavities may be segmented into multiple separate pockets by fibrous adhesions, some containing neoplastic cells and others only benign mesothelial cells. The aspiration of the latter will produce false negative cytologic reports.

Some tumors may exfoliate only single cells, which are hard to differentiate from the single reactive mesothelial cell.

References and Supplementary Reading

Bakalos, D., Constantakis, N., and Tsicrieas, T. Distinction of mononuclear macrophages from mesothelial cells in pleural and peritoneal effusions. *Acta Cytol.* 18:20, 1974.

Berge, T., and Grontoff, O. Cytologic diagnosis of malignant pleural mesothelioma. *Acta Cytol.* 9:307–212, 1965.

Berge, T., and Hellsten, S. Cytological diagnosis of cancer cells in pleural and ascitic fluid. *Acta Cytol.* 10:138, 1966.

Billingham, M. E., et al. The cytodiagnosis of malignant lymphomas and Hodgkin's disease in cerebrospinal, pleural and ascitic fluids. *Acta Cytol.* (Baltimore) 19:547, 1975.

Calle, S. Megakaryocytes in an abdominal fluid. *Acta Cytol.* 12:78, 1968.

Cardozo, P. L. A critical evaluation of 3,000 cytologic analyses of pleural fluid, ascitic fluid, and pericardial fluid. *Acta Cytol.* 10:455, 1966.

Ceelen, G. H. PAS reaction as an aid in the interpretation of pleural and abdominal cytology. In *Transactions of the Fifth Annual Meeting Inter-Society Cytology Council,* 1957, pp. 131–138.

Ceelen, G. H. The cytologic diagnosis of ascitic fluid. *Acta Cytol.* 8:175–185, 1964.

Clarkson, R. Relationship between cell type, glucose concentration, and response to treatment in neoplastic effusions. *Cancer* 17:914, 1964.

Danner, D. E., and Gmelich, J. T. A comparative study of tumor cells from metastatic carcinoma of the breast in effusions. *Acta Cytol.* 19:509, 1975.

Foot, N. C. Identification of types and primary sites of metastatic tumors from exfoliated cells in serous fluids. *Am. J. Pathol.* 30:661, 1954.

Foot, N. C. Identification of neoplastic cells in serous effusions. *Am. J. Pathol.* 32:961–977, 1956.

Foot, N. C. The identification of mesothelial cells in sediments of serous effusions. *Cancer* 12:429–437, 1959.

Fullmer, C. D., and Morris, R. P. Primary cytodiagnosis of unsuspected mediastinal Hodgkin's disease. *Acta Cytol.* 16:77, 1972.

Grunze, H. The comparative diagnostic accuracy, efficiency, and specificity of cytologic technics used in the diagnosis of malignant neoplasm in serous effusions of the pleural and pericardial cavities. *Acta Cytol.* 8:150–163, 1964.

Jarvi, O., Kunnas, R., and Tyrkko, J. The accuracy and significance of cytologic cancer diagnosis of pleural effusions. *Acta Cytol.* 16:152, 1972.

Johnson, W. D. The cytologic diagnosis of cancer in serous effusions. *Acta Cytol.* 10:161, 1966.

Juniper, K., and Chester, C. L. A filter membrane technique for cytological study of exfoliated cells in body fluids. *Cancer* 12:278–285, 1959.

Klempman, S. The exfoliated cytology of diffuse pleural mesothelioma. *Cancer* 15:691–704, 1962.

Konikov, N., Bleisch, V., and Piskie, V. Prognostic significance of cytologic diagnosis of effusions. *Acta Cytol.* 10:335, 1966.

Mavrommatis, F. S. Some morphologic features of cells containing PAS positive intracytoplasmic granules in smears of serous effusions. *Acta Cytol.* 8:426, 1964.

McGowan, L., and Bunnag, B. Morphology of mesothelial cells in peritoneal fluid from normal women. *Acta Cytol.* 18:205, 1974.

McGowan, L., et al. Cytologic differential of pelvic cavity aspiration specimens in normal women. *Obstet. Gynecol.* 30:821, 1967.

McGowan, L., Stein, D. B., and Miller, W. Cul-de-sac aspiration for diagnostic cytologic study. *Am. J. Obstet. Gynecol.* 96:413, 1966.

Naib, Z. M. Cytology of synovial fluid. *Acta Cytol.* 17:302, 1973.

Naylor, B. The exfoliative cytology of diffuse malignant mesothelioma. *J. Pathol. Bacteriol.* 86:293, 1963.

Naylor, B., and Schmidt, R. W. The case for exfoliative cytology of serous effusions. *Lancet* 1:711, 1964.

Pfitzer, P., and Huth, F. Alcian-PAS positive granules in mesothelioma and mesothelial cells. *Acta Cytol.* 10:205, 1966.

Spriggs, A. I., and Meek, G. A. Surface specializations of free tumor cells in effusions. *J. Pathol. Bacteriol.* 82:151, 1961.

Suprun, H., and Mansoor, I. An aspiration cytodiagnostic test for gouty arthritis. *Acta Cytol.* 17:198, 1973.

Thompson, M. E., Bromberg, P. A., and Amenta, J. S. Acid mucopolysaccharide determination with effusions. A useful adjunct for the diagnosis of malignant mesothelioma with effusion. *Am. J. Clin. Pathol.* 52:335, 1969.

Ultmann, J. E. Diagnosis and treatment of neoplastic effusions. *Cancer* 12:42–50, 1962.

Woyke, S., Domagaza, W., and Olszewski, W. Ultrastructure of hepatoma cells detected in peritoneal fluid. *Acta Cytol.* 18:130, 1974.

319

The Gastrointestinal Tract   13

Introduction

Cancer of the gastrointestinal tract is second only to lung cancer in men and breast cancer in women as a cause of cancer deaths. Cytology could play an important role in its early detection. The mucosal lining of the gastrointestinal tract consists mainly of:

1. Nonkeratinizing stratified squamous epithelium (oral cavity, esophagus, and anal canal);
2. Single columnar epithelium (stomach, small intestine, colon, and rectum).

Oral Cavity

ANATOMY AND HISTOLOGY (FIG. 148)
The mouth contains the instruments of mastication and the organ of taste. It is bounded in front by the lips, laterally by the internal surface of the cheeks, above by the hard palate and teeth of the upper jaw, below by the tongue and teeth of the lower jaw, and posteriorly it continues with the pharynx, from which it is partially separated by the soft palate. The oral mucosa is made of stratified squamous epithelium, which varies in the degree of surface cornification according to location. The hard palate, the gingiva, the dorsal surface of the tongue, and the floor of the mouth and lateral cheeks are covered with a nonkeratinized stratified squamous epithelium. The rest of the oral mucosa is keratinized.

NORMAL CYTOLOGY (FIGS. 149A, B, 150A, B)
Physiologically the cells exfoliate only from the surface of the keratinized or nonkeratinized epithelium. The deep parabasal cells are seen only when an ulceration is present. Most of the exfoliated cells are lost by autodigestion.

In contrast to the vaginal mucosa, the oral epithelium is only slightly influenced by the different hormones and shows little variation in the degree of its maturation. The thickness and composition of the epithelium varies in different sites, depending on its function and the mechanical irri-

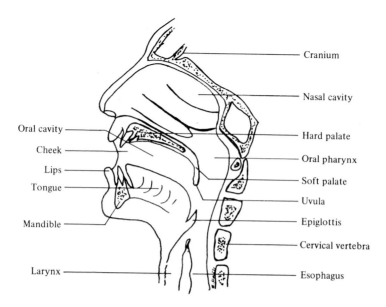

Oral cavity

Cheek

Lips

Tongue

Mandible

Larynx

Cranium

Nasal cavity

Hard palate

Oral pharynx

Soft palate

Uvula

Epiglottis

Cervical vertebra

Esophagus

Figure 148
The oral cavity and its surrounding structures.

tations of mastication to which it is subjected. The oral epithelial cells show no rhythmic alteration in connection with the menstrual cycle and should not be used as a test for the estimation of hormonal ovarian function.

Large numbers of intact anucleated cornified squamous cells originating from the sheltered areas of the buccal cavity (palate, gingiva, tongue, cheeks, floor of the mouth, and tonsillar crypts) are always present. They have an abundant, thin cytoplasm that is polygonal in shape. The cells vary in size and can be found singly or in clusters or sheets. They stain very poorly, yellow to pink, and often show degenerative changes in the form of vacuolization or dark brown intracytoplasmic pigment inclusions.

Also usually present are numerous superficial squamous cells with central nuclei that show a variable degree of pyknosis. They originate from the hard palate and the gingiva and resemble the superficial cells found in the vaginal smear.

Numerous intermediate polygonal squamous cells are shed singly or in large sheets. They have an abundant cytoplasm that is uniformly stained pink to blue, with a small central vesicular nucleus (7–10 μ). Their nuclei-cytoplasm ratio is approximately 1 to 6. They resemble the intermediate squamous cells of the vaginal smears. Occasional binucleation is present.

Occasional parabasal cells with large nuclei and scanty cytoplasm may be found in sheets. They are the result of traumatic exfoliation of an ulcerated area or of an excessively energetic scraping.

Various amounts of different inflammatory cells may also be seen. Small amounts are normal but large numbers indicate an inflammation.

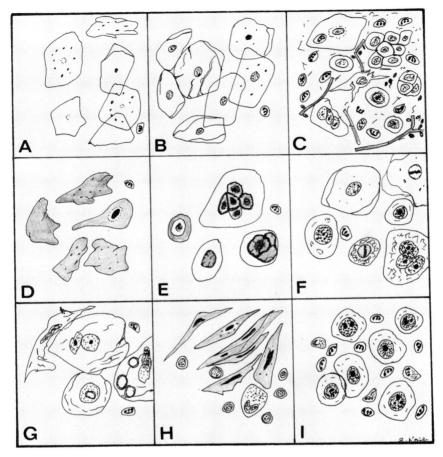

Figure 149
Exfoliated cells from the normal mucosa and benign lesions of the oral cavity.
(A) Normal squamous cell from the scraping of the hard palate. (B) Normal
squamous cells from the cheek, floor of the mouth, and certain portions of the
tongue. (C) Cells from an ulcerated lesion with secondary *Candida* inflammation. (D) Cells of leukoplakia. (E) Cells from herpetic inflammation. (F) Cells
in pernicious anemia. (G) Postirradiation cells. (H) Cells of lichen planus. (I)
Cells from pemphigus.

Fibroblasts with oval to elongated nuclei and coarse but regular chromatin in an elongated spindle-shaped and cyanophilic cytoplasm may only
be found in cases of deep ulceration in which the surface mucosa has been
destroyed.

Histiocytes of various sizes and shapes may be seen in cases of erosion,
chronic inflammation, and foreign-body reaction. Various ingested materials may be seen in their cytoplasm.

Various pathogenic and nonpathogenic fungi, parasitic bacteria, and
other saprophytes may be present in the mouth, often forming large
colonies.

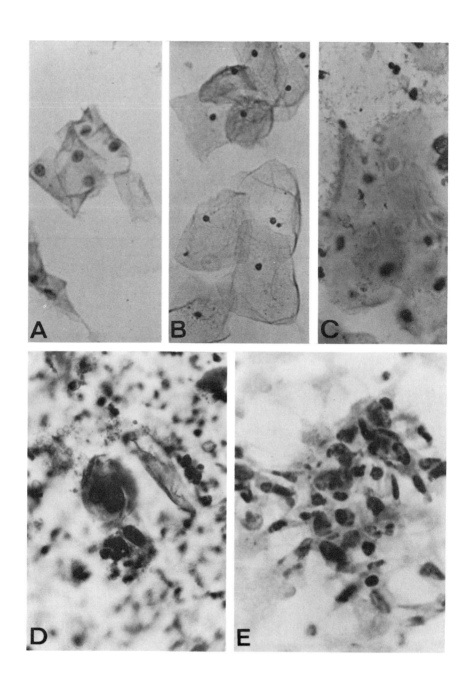

BENIGN LESIONS OF THE ORAL CAVITY

The benign lesions may be hard to differentiate clinically from the early malignant ones. They are often the result of repeated irritation, and they can cover a surface too large to be adequately biopsied.

Hyperkeratosis or Leukoplakia (Figs. 149D, 150C)

Considered a possible precancerous lesion, hyperkeratosis (leukoplakia) may appear grossly as a white geographic zone of thickened epithelium anywhere in the oral cavity. The surface smear will show an increased number of anucleated cells with glassy-appearing, dense, orangeophilic, abundant, hyperkeratotic cytoplasm. The cells are found singly or in large sheets. Occasionally an abnormally enlarged pyknotic or hyperchromatic nucleus is seen. Some inflammatory cells (leukocytes) are almost always present.

Papillomatosis (Papillary Hyperplasia) (Fig. 151A, B)

Found usually under ill-fitting dentures, numerous papillae may cover the mucosa of the hard palate. Variable amounts of sheets of hypertrophic, squamous parabasal cells and intermediate cells may be found. They usually have a scanty to moderate elongated basophilic cytoplasm, rarely keratinized, with heavy, smooth borders. The nucleus is central, large, round or oval, with mildly irregular borders. The chromatin pattern is pale or bland. Occasionally, prominent single or multiple nucleoli may be seen in the reactive nuclei of parabasal cells. These cells could be mistaken for cancer, except for the regularity of the chromatin. Keratotic superficial cells from adjoining irritated areas are also often seen.

Dyskaryosis (Fig. 150D)

Dyskaryosis, a premalignant lesion, sheds cells similar to the ones desquamated from dysplastic lesions of the cervix. They show, besides an abnormally keratinized cytoplasm, marked irregularity of their nuclei. Occasionally they are difficult to differentiate from the cells of a low-grade squamous cell carcinoma. Their presence in a buccal smear is an indication for biopsy.

Anemia—Pernicious, Sprue, Sickle Cell (Fig. 149F)

In these forms of anemia, about 40% of the normal basal and intermediate exfoliated squamous cells may be hypertrophic with an abundant cyto-

---

Figure 150
(A) Intermediate squamous cell and one anucleated cell from scraping of normal buccal mucosa, inner cheek. ($\times$ 450) (B) Superficial squamous cells from normal buccal mucosa, hard palate. ($\times$ 450) (C) Hyperkeratotic anucleated cells from the scraping of an extensive gingival leukoplakia. ($\times$ 450) (D) Dyskaryotic squamous basal cells from the scraping of an ulcerated lip lesion of a heavy pipe smoker. Note the inflammatory cells and enlarged pyknotic nuclei often difficult to differentiate from an early carcinoma. ($\times$ 450) (E) Squamous carcinoma cells obtained from the scraping of a lesion of the inner cheek that was clinically benign in appearance. ($\times$ 450)

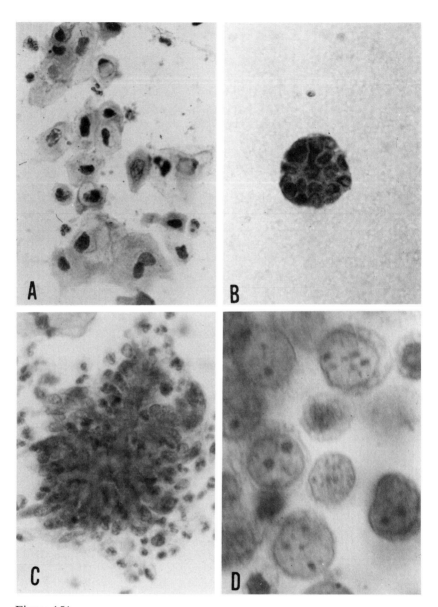

Figure 151
(A) Denture reaction. Note the abnormal nuclei of some of the squamous cells.
(× 450) (B) Scraping of papillomatosis produced by an ill-fitting denture.
(× 450) (C) Cluster of phagocytes in a buccal smear from a patient with a
benign ulcer of the palate. (× 450) (D) Regenerating basal cells in a healing
ulcer. (× 1300)

plasm and an enlarged central nucleus (6–18 μ). The chromatin is often abnormally coarsely granular and uniformly distributed (80%). It may be condensed to the periphery, leaving a blank space in the middle (18%) or may produce an unusual linear clumping in the center of the nucleus (2%). Multinucleation and binucleation may occur. The variation or decrease of this nuclear size may be used to assess the result of antianemic therapy.

Parasitic Infection

Entamoeba gingivalis parasites may be encountered in the buccal smear in over 60% of patients with poor dentition and oral hygiene or following irradiation. Morphologically, the parasites (10 to 25 μ in diameter) are similar to *Entamoeba histolytica,* except for the presence of leukocytes and basophilic nuclear debris in the cytoplasm. They stain poorly (gray to pale blue) and are often confused with epithelial cytoplasmic degenerative debris.

Fungus Infection

Most of the superficial fungi infecting the skin and other mucosa may be found in the oral cavity. One of the most common is *Candida,* often seen in large colonies in sputum or gastric washings. The other fungi that may be seen are those causing actinomycosis, nocardiosis, and blastomycosis and also various nonpathogenic saprophytic fungi.

Viral Infection (Fig. 152A, B, C)

With viral inflammation of the oral cavity, the changes in the exfoliated cells resemble the ones in viral lesions of the upper respiratory tract (see p. 237). The most common inflammation is the herpetic stomatitis mainly due to herpesvirus type I. Large numbers of multinucleated epithelial cells containing prominent intranuclear inclusions are shed. In hand, foot, and mouth disease due to coxsackievirus, the scraping of the buccal mucosa may show hypertrophic squamous cells with round to oval intracytoplasmic eosinophilic inclusions surrounded by prominent halos. The scraping of a suspected Koplik ulcer may help in the diagnosis of early measles. The smear will contain large multinucleated cells packed with intracytoplasmic and intranuclear granular inclusions.

Denture Reaction (Fig. 151A)

The epithelium in contact with a denture becomes atrophic after a while, with variable parakeratotic changes. A scraping of these areas will show a decrease of anucleated squamous cells and a relative increase of hyperkeratinized intermediate and superficial cells, often with bizarre shapes.

Pemphigus (Figs. 149I, 152E)

The cause is unknown. The scraping of the tender pemphigus vesicular lesions shows an abundance of single or large clusters of hypertrophic mononucleated squamous parabasal cells from the stratum spinosum. Their relatively scanty and granular cytoplasm is thick and abundant, basophilic (60%) or acidophilic with sharp, well-defined borders. Occasionally there

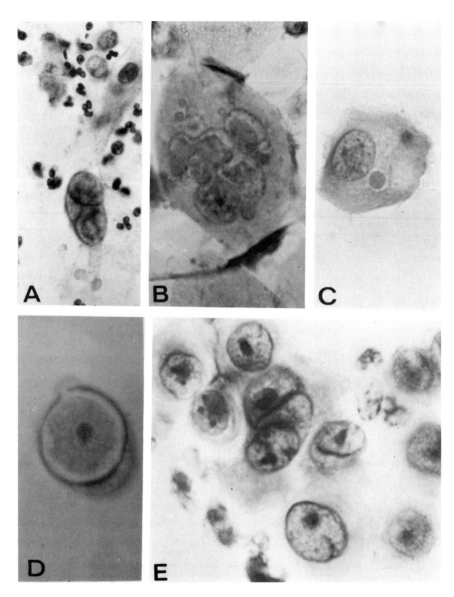

Figure 152
(A) Multinucleated cells obtained from the scraping of a recurrent oral herpetic lesion. Note the molding of the nuclei and blandness of the chromatin. (× 450). (B) Large squamous cell with multiple irregular intracytoplasmic inclusions obtained from the scraping of a gingival ulceration in a recently vaccinated boy. The ulceration had been proved virologically to be caused by vaccinial virus. (× 450) (C) Squamous cell with a single intracytoplasmic inclusion from the scraping of a lesion virologically proven to be hand, foot, and mouth disease. (× 750) (D) Scraping of a Darier-White disease lesion. (× 450) (E) Cluster of basal cells with prominent nucleoli obtained from the scraping of an oral pemphigus. (× 1300)

is a marked increase in the eosinophilia of their cytoplasm, especially when secondary inflammation is present. Perinuclear halos are commonly seen. Their nuclei are central, large (12 μ), round with coarsely granular but uniformly distributed, increased chromatin. Their eosinophilic nucleoli are enlarged and prominent. Multinucleation and mitosis are occasionally seen, as are epithelial pearls, foreign-body giant cells, and various inflammatory cells. They can be mistaken for malignant cells because of their enlarged nuclei and prominent nucleoli. The cells of pemphigus are differentiated by the regular distribution of their chromatin pattern and their uniformly thick nuclear membrane.

In *pemphigus foliaceus,* numerous loose, small, round or spindle-shaped hyperkeratinized cells are present. Their nuclei, oval to elongated, are often smudged; no nucleoli are present. This lesion should be suspected when increased eosinophilia of the small exfoliated cells is noted.

Lichen Planus (Fig. 149H)

Lichen planus is a common primary lesion of the oral mucosa, characterized grossly by a punctated and dendritic appearance of the tongue and lateral mucosa. In some cases ulceration may be present. The smears show a large number of dense, elongated, orangeophilic squamous cells of the intermediate type with irregular cytoplasmic borders and elongated single nuclei with finely granular chromatin and no visible nucleoli. The cells are shed singly or, more commonly, in large sheets. A large number of mature lymphocytes is also usually present with occasional small or large macrophages laden with melanin pigment.

Stomatitis (Fig. 150D)

In stomatitis, the inflammation of the oral mucosa can be specific (smoker's palate) or a result of traumatic or chemical irritation. The smears show variable numbers of enlarged, superficial, and intermediate acidophilic squamous cells. When ulceration is also present, deeply basophilic hypertrophic parabasal cells with irregular projections (repair cells) are seen. Large numbers of mixed inflammatory cells are always present. These reactive squamous cells can be mistaken for the cells desquamated from a leukoplakia. The differentiation is made by the presence of a large number of inflammatory cells in stomatitis. Various degrees of cellular degeneration of the epithelial and nonepithelial cells are usually present. When ulcerated, healing regenerated epithelial cells may show marked variation in the size and shape of their nuclei and staining density of the cytoplasm. Nucleoli are usually prominent, and chromatin is coarsely granular. Besides these nonspecific reactive cells, no other morphologic changes can be seen that may be specifically attributed to the inhalation of tobacco smoke.

Hereditary Benign Intraepithelial Dyskaryosis—Darier Disease (Fig. 152D)

In hereditary benign intraepithelial dyskaryosis, the changes in the exfoliated cells, although nonspecific, help in the diagnosis of these rare, usually asymptomatic, lesions. Large numbers of dyskaryotic cells, some with an abundant, deep orange–stained cytoplasm and central, almost pyknotic, irregular, elongated nucleus suggest this diagnosis. Other cells have scanty cytoplasm that stains a deep orange and a hyperchromatic, almost pyknotic, round-to-oval dark nucleus; they can be mistaken for squamous carcinoma cells. Pseudocannibalism is frequent. Occasionally in the single cells, dark brown cytoplasmic granules form a ring around the nucleus (*corps ronds*).

Erythema Multiforme
The scraping produces large numbers of enlarged squamous parabasal cells in clusters or sheets with densely and irregularly stained adequate cytoplasm containing occasional large degenerative vacuoles (ballooning degeneration). The nuclei are round to oval with uniform chromatin patterns and occasional small nucleoli. Pseudopearl-like formation and pseudocannibalism is frequent. A moderate number of inflammatory cells are present.

Irradiation and Chemotherapy (*Fig. 149G*)
With irradiation and chemotherapy, the cellular changes resemble the ones described at the beginning of Chapter 9. Cytoplasmic and nuclear degeneration are mainly seen in the form of pyknosis and karyorrhexis, whereas regeneration produces marked enlargement of their cytoplasm and nuclei, multinucleation, and thick multiple cytoplasmic vacuolization. Moderate increase of inflammatory cells, especially the phagocytes, is usually present. Moderate amounts of increased keratinization are often seen, with frequent structures like those appearing in pseudocannibalism. The cells may resemble the cells exfoliated from a benign intraepithelial dyskaryosis or a low-grade squamous carcinoma. With a history of irradiation or chemotherapy, the interpretation of the morphological cellular changes should be very conservative.

BENIGN NEOPLASMS OF THE ORAL CAVITY
Most of the benign tumors in the oral cavity arise from nonepithelial structures and are usually covered by normal oral epithelium. Cytology is not diagnostic, except when an ulceration exists.

Pleomorphic adenoma may produce clusters or sheets of small cells with scanty basophilic or transparent cytoplasm surrounding a round or oval nucleus that is uniform in size (10 μ) and shape (round to oval). Nucleoli may be uniformly prominent. The chromatin is finely granular.

The scraping of an ulcerated *ameloblastoma* (Fig. 153A) will produce small, spindle-shaped, single neoplastic cells with poorly stained cytoplasm and elongated nuclei with coarse chromatin clumping and distinct round nucleoli.

In *odontogenic myxoma* (Fig. 153B), the exfoliated cells are large with pale, adequate cytoplasm and large, round, pale nuclei with prominent eosinophilic nucleoli.

MALIGNANT NEOPLASMS OF THE ORAL CAVITY

Well-Differentiated Squamous Cell Carcinoma (*Fig. 150E*)
This is the most common malignant neoplasm of the oral mucosa. The exfoliated neoplastic cells resemble the malignant cells found in vaginal smears of patients with cervical invasive squamous carcinoma. The cells

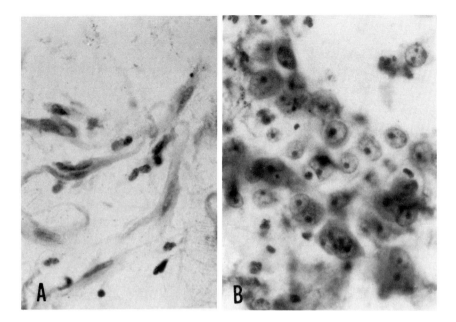

Figure 153
(A) Scraping of an ulcerated ameloblastoma. (× 450) (B) Scraping of an odontogenic myxoma. Note the prominent nucleoli in the round, pale nuclei. (× 450)

are shed singly or in sheets with marked variation in size and shape (polygonal, spindle, tadpole, pearl formation). Their nuclei, always irregular,
can be pyknotic or vesicular. The chromatin, when discernible, is irregularly clumped with pointed projections and occasional prominent nucleoli.
The nuclear membrane is usually thick, irregular in outline, often with multiple indentations. Occasional binucleation or multinucleation is seen. The
cytoplasm of the cells is thick and orangeophilic and shows marked variation in size and occasional keratohyalinic granular precipitate forming
perinuclear rings. Anucleated, heavily keratinized (ghost-like) cells are
common in very well differentiated squamous carcinomas.

Poorly Differentiated Squamous Cell Carcinoma
With this neoplasm, the cells shed singly, in clusters (80%), or in sheets
(20%). They have scanty-to-adequate nonkeratinized cytoplasm and are
evenly stained deep blue-purple, often with indistinct borders. Their nuclei
are hyperchromatic and centrally placed, and the chromatin shows a coarse
but irregular pattern, with large reddish nucleoli in more than 60% of the
cells. Variable amounts of inflammatory cells, degenerate cellular debris,
and protein deposits are usually found in the background of the smears.

Transitional Cell Carcinoma
These are rare tumors, most often found in the tonsillar area. They are
characterized by cells shed singly (80%) or in small tight sheets. Their

331

round, nonkeratinized cytoplasm is adequate and irregularly stained, pale blue to purple. They usually have indistinct borders. Their elongated, vesicular nuclei are often lobular or multiple with an irregular chromatin clumping and prominent multiple nucleoli $(2-6~\mu)$.

Large amounts of debris usually clutter the background of the smear.

In Situ Carcinoma
In the oral cavity no distinctive exfoliated cells diagnostic of an in situ carcinoma are recognized that are similar to those seen in the uterine cervix. The exfoliated cells are similar to the ones diagnostic of an invasive squamous cell carcinoma, except possibly for the tendency of the oral in situ cells to shed as single cells, rather than in sheets, and of their cytoplasm to be slightly scantier and more granular. The amount of inflammatory (polymorphonuclear) cells and cellular debris is less abundant than in invasive squamous carcinoma. Occasionally, large numbers of mature inflammatory lymphocytes may be seen.

Basal Cell Carcinoma
This tumor is rare in the oral cavity and is more commonly found on the upper lip. The scraping produces sheets of small, uniform, basal cells with dense, scanty, eosinophilic cytoplasm and round, hyperchromatic nuclei that are uniform in size $(8~\mu)$ and shape (round to oval). Nucleoli are not distinct. Palisadelike alignment of the cells may be seen at the periphery of these sheets. Occasional mitosis may be seen. Variable amounts of necrotic debris are usually seen in the background.

Adenoid Cystic Carcinoma
This slow-growing tumor occurs mainly in the accessory salivary glands of the palate. Large numbers of small cuboidal cells in single file or glandular structures with poorly preserved cytoplasm are diagnostic. The eccentric nuclei are uniformly large $(12~\mu)$ and round to oval. The chromatin is coarsely granular and irregularly distributed, and the nucleoli are relatively large and prominent. The smooth nuclear membrane shows only a minimum of irregularities. The adequate-to-scanty cytoplasm is often poorly stained and contains no vacuoles or inclusions.

Mucoepidermoid Carcinoma
This mucus-secreting, epidermoid carcinoma occurs in the accessory glands of the palate and in the parotid or submaxillary salivary gland. Clusters of epithelial cells with squamous and columnar characteristics are diagnostic. Their cytoplasm may contain keratin or secretory vacuoles. Monocytic inflammatory cells are often present in the background.

Adenocarcinoma
In the oral cavity these are rare tumors, usually secondary to a neoplasm from another site.

Lymphoma, Leukemia
Large numbers of malignant leukemic or lymphomatous cells may be present in the buccal smear, never in sheets or clusters but often mixed with

sheets of hyperplastic squamous basal cells. The marked variation in the size of the atypical lymphocytes and the occasional prominence of their eosinophilic nucleoli are diagnostic. In reticulum cell sarcoma, the cells are more pleomorphic with occasional spindle or caudate cells, occasional multinucleation, and frequent mitoses. Their nucleoli are consistently prominent. Large pools of mature lymphocytes may also originate from ulcerated hyperactive tonsils or adenoids.

Fibrosarcoma

This rare tumor exfoliates only in its advanced stage when ulcerated. Large numbers of pleomorphic cells, with poorly defined, red-purple stained cytoplasm, are seen in the midst of an abundance of cellular debris and red blood cells. Their often multiple nuclei are large with lobulation. Their chromatin varies from finely granular to irregularly clumped. Nucleoli are large and vary enormously in shape and number.

Parotid Gland

The determination of the nature of parotid gland lesions may be done either by direct needle aspiration of the gland or by the study of smears obtained by aspiration of parotid secretion from the orifice of Stensen's duct with a small polyethylene tube attached to a 10 ml syringe. The secretion may be increased if the patient sucks on a lemon before the procedure.

The normal secretion may contain cells from:

1. The lining of Stensen's duct: tall, columnar epithelium, often seen in sheets;
2. The lining of intercalary ducts: simple, low, cuboidal epithelium, seen singly or in sheets;
3. The lining of secretory alveolar glands: peripheral nuclei and abundant cytoplasm with red granulation and mucus-containing vacuoles in secretory cells.

The secretion may also contain inflammatory cells, but they are generally few in number.

In inflammation, various amounts of inflammatory cells, microorganisms, and cellular debris may be seen. No specific changes are seen in cases of mumps.

In mixed tumors, one can find clusters of enlarged cells with a high nuclei-cytoplasm ratio. The mildly increased chromatin is finely granular and uniformly distributed. Multiple prominent nucleoli are often seen. Large secretions in the cytoplasm displace and mold the nuclei.

In squamous carcinoma, the typical malignant cells with cytoplasmic keratinization may be readily recognized.

Pharynx

ANATOMY AND HISTOLOGY

The pharynx is a membranous sac that has four passages, one opening upward and forward to the nose, the second to the mouth, the third downward to the trachea and lung, the fourth being continuous with the esophagus. The mucous membrane lining the pharynx may show numerous folds. Approximately 50% of its surface is made of stratified squamous epithelium and 50% of pseudostratified ciliated epithelium lining its lateral wall or forming multiple scattered patches.

NORMAL CYTOLOGY
The squamous cells of the pharynx are the anucleated, superficial, intermediate, and basal types and are similar to the cells described in normal oral cytology. The ciliated columnar cells, seen singly or in clusters or sheets, have abundant, vesicular, round-to-elongated cytoplasm with an eccentric round or oval nucleus averaging 7 μ in diameter. Usually, few lymphocytes, macrophages, and leukocytes are present in the smear.

BENIGN LESIONS OF THE PHARYNX
The benign lesions of the pharynx are similar to the ones found in the oral cavity, except for the abundance in the pharynx of noncarbon-bearing macrophages in cases of acute or chronic inflammation. Lymphocytes are rarely seen, except when the tonsils are ulcerated and scraped energetically.

MALIGNANT LESIONS OF THE PHARYNX
The most common neoplasm is the well-differentiated squamous cell carcinoma. The cells are the same as those from the oral squamous neoplasm (see p. 330). Cytology cannot determine the origin of the cells.

The undifferentiated carcinoma (lymphoepithelioma) is characterized by a mixture of reactive lymphocytes and anaplastic squamous cells, both showing marked variation in the size of their nuclei. The malignant epithelial cells are similar to the undifferentiated squamous carcinomas of the oral mucosa (see p. 331), except for the larger variation in the size of their nuclei (10–80 μ in diameter). The nuclei are usually very hyperchromatic and vesicular with prominent nucleoli. A large number of the cells lose their cytoplasm early, and they appear in the smear as stripped nuclei.

The incidence of cytologic false-negative interpretation is relatively high because the surface of the pharyngeal carcinoma, even if fungating, often remains covered with normal mucosa, which prevents the exfoliation of malignant cells. Biopsy is still the method of choice for the detection of these cancers.

Esophagus

ANATOMY AND HISTOLOGY (FIG. 154)
The esophagus is a long membranous tube that begins in the pharynx and extends behind the trachea, heart, and lung and pierces the diaphragm to terminate in the stomach. It is composed of an outer muscular coat and an inner mucosa. The mucosa is made of stratified noncornified squamous epithelium that, at the junction with the cardia of the stomach, changes abruptly into a simple columnar epithelium. This junction appears grossly as an irregular jagged line. This irregularity is due to islands of gastric glandular mucosa found toward the terminal region of the esophagus.

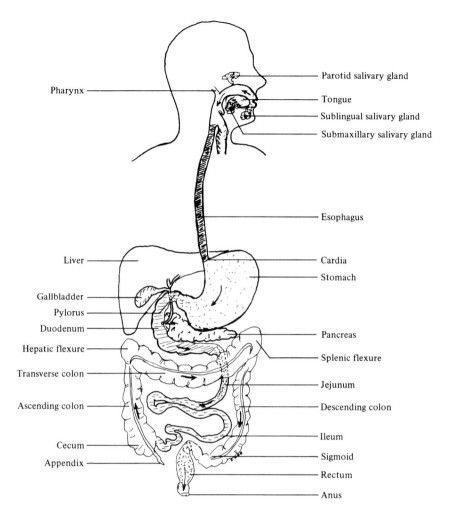

Figure 154
The gastrointestinal tract.

TECHNIQUE FOR COLLECTING ESOPHAGEAL SPECIMENS

Esophageal Washing

To collect a good specimen for cytology, one should first localize the suspected lesion by esophagoscopy or an x-ray barium swallow. A Levin catheter tube to which several extra openings have been added is introduced through the mouth to a distance below the site of the suspected lesion, with the patient sitting upright and having fasted for at least 12 hours (48 hours after the barium x-ray examination). The patient is allowed to drink fluid slowly (e.g., 200 ml of Ringer's solution), which is reaspirated as it enters the stomach. The specimen should be collected in iced containers and immediately processed by passing it through a filter membrane with a 5 μ pore opening.

Esophageal Brushing

Brushing of the lesion under direct vision may also be done with excellent results by using the Gastro-Esophago-Fiberscope and a small brush. This method is gentler and safer than a biopsy and reduces the risk of perforation of the thin-walled esophagus. Barium contamination of the smears may be avoided if the lesion is washed with saline solution before brushing. After smearing, the brush should be rinsed and agitated in 10 ml of saline solution, which, in turn, may be passed through a filter membrane.

CYTOLOGY OF THE NORMAL ESOPHAGUS (FIG. 155A)

The majority of the exfoliated cells of the esophagus consists of benign squamous cells originating from the buccal or esophageal mucosa. It is impossible to differentiate them. They are mainly superficial or intermediate in type, similar to the cells found in vaginal smears. Occasionally, basal squamous cells are seen, which indicate either the presence of an ulcerative lesion or a traumatic esophagoscopy. The structures of these basal squamous cells are similar to the structures of the ones described for the vaginal smear.

Occasionally, gastric cells, singly or in large sheets, are present, indicating either that the tube has passed into the stomach or that the gastric mucosa is abnormally extended into the esophagus (in hiatal hernia, for example).

Large numbers of upper respiratory tract cells may also be present, especially if the tube has been inserted through the nose. These cells, if not degenerate, are recognized by their large, round, purple cytoplasm, oval nuclei with large nucleoli, and the presence of a terminal bar with elongated cilia. If they are degenerate and have lost their cilia, they can become diagnostic problems by being confused with the cells desquamated from low-grade adenocarcinomas.

Occasionally, variable numbers of ciliated or nonciliated bronchial cells originating from swallowed sputum can be seen, although these appear less commonly than in gastric aspirations. The bronchial cells may indicate the presence of an esophagobronchial fistula. Similarly, variable numbers of phagocytes may be present. Some from the respiratory tract contain phagocytized carbon particles; others, smaller, without any carbon particles, come from the gastric content.

Variable amounts of inflammatory cells can be present. They may originate from the oral cavity, esophagus, or stomach.

The presence of foreign material (food particles) may indicate faulty technique or an obstruction of the esophageal or gastric lumen.

BENIGN LESIONS

Chronic Ulcerative Esophagitis (Fig. 155B)

This inflammation can be the result of corrosive agents or hiatal hernia or secondary to a mediastinitis or a specific infection (tuberculous, fungal, or parasitic). It is one of the rare esophageal lesions that shed atypical-looking, degenerate or regenerate cells that may be confused with can-

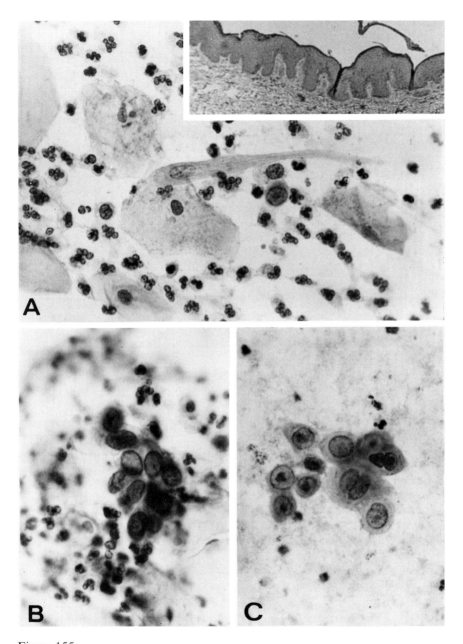

Figure 155
(A) Insert of a histologic section of normal esophageal mucosa. (× 125) (B) Cluster of reactive basal cells found in the esophageal washing in a man with long-standing chronic esophagitis. The absence of marked chromatin irregularities and the uniformity of shape and size differentiate them from carcinoma cells. (× 450) (C) Cluster of cells with occasional intranuclear inclusions and multinucleation diagnostic of a herpetic esophagitis. (× 450)

337

cerous ones. Numerous small distorted parabasal squamous cells are found singly, in clusters, or in sheets, with prominent cytoplasmic and nuclear molding. The cytoplasm is scanty, basophilic (80%), or eosinophilic, with thick, poorly defined borders. The nuclei are central and enlarged two or three times, and show moderate irregularity of shape. The chromatin is coarsely clumped with a tendency to migrate to the periphery. The nucleoli are prominent.

Occasional large, atypical histiocytes may be seen.

In *achalasia,* the sphincter muscle of the cardia fails to stop the gastric fluid from reaching the lower esophagus. The ulcerated chronic esophagitis produced often exfoliates sheets of highly reactive hyperchromatic parabasal cells with large nuclei and prominent but regular nucleoli, which can be mistaken for gastric adenocarcinoma cells. The uniformity of their features helps in their recognition.

Large amounts of inflammatory cells, cellular debris, and protein deposits usually obscure the background of the smear. The presence of specific microorganisms, fungi, or protozoa is helpful in the diagnosis.

Herpetic Infection (Fig. 155C)

This is a relatively common viral inflammation of the esophagus, especially in debilitated patients or those receiving massive cortisone therapy. Besides the large amount of mixed inflammatory cells, numerous typical multinucleated giant cells with molded, bland nuclei or with centrally located inclusions are easily seen in the smears. These cells are similar to the ones exfoliated from lesions of the genital or respiratory tracts. Their presence in the esophageal washings does not necessarily mean that the viral lesion is in the esophagus. A contamination from mucosal cells of the buccal or respiratory tract is always possible.

Irradiation

The four classic cellular changes of irradiation—hypertrophy, multinucleation, cytoplasmic vacuolization, and cellular degeneration—may be seen in the exfoliated esophageal cells of a patient who has received mediastinal irradiation. They can become a pitfall in diagnosis only if the history of irradiation is not given to the cytopathologist.

Ectopy

Occasionally the lowest third of the esophagus is lined by gastric columnar-type cells instead of squamous epithelium. Brushing of such a mucosa will remove sheets of reactive gastric columnar cells and almost no squamous cells.

Benign Tumors

The most common benign tumors of the esophagus are mesenchymal tumors (fibroma, lipoma, neurofibroma), which usually do not exfoliate except if ulcerated. The smears of needle aspiration of such tumors may be helpful in ascertaining their benign nature.

Leukoplakia

In leukoplakia, large numbers of hypermature keratinized superficial squamous cells with abundant, thin orangeophilic cytoplasm, anucleated or with large irregular pyknotic nuclei are found. Exfoliative cytology cannot differentiate them from similar cells originating from an oral lesion.

Pernicious Anemia

As in the case of the cells from the oral cavity, in pernicious anemia the esophageal squamous intermediate or parabasal cells may show general hypertrophy of their cytoplasm and nuclei. The frequently binucleated round or oval vesicular nucleus may show its chromatin clumped toward the periphery, leaving a bland center or occasionally forming a central linear structure. Their uniform cytoplasm is thin, pink or blue, with normal keratinization. These changes are not specific and can also be found in other types of anemia.

Scleroderma

Histologically the lesion is seen as a fibrosis of the submucosa. On occasion, the overlying epithelium becomes atrophic, and the esophageal washings, usually scanty in cells, show an increased number of single, small, parabasal cells with marked cytoplasmic eosinophilia.

MALIGNANT LESIONS
Esophageal cancers are:

1. More common in older males (over the age of 50) of any race. Four percent of all cancer deaths are due to esophageal cancer;
2. Located mainly in the lowest third of the esophagus (50% of the time) and in the middle third (30% of the time);
3. Asymptomatic when early, then dysphagia develops in more than 90% of the cases, followed by weight loss and regional pain;
4. Detectable by cytology in more than 95% of the cases;
5. Of poor prognosis (less than 3% of the patients survive 5 years).

Squamous Cell Carcinoma (Fig. 156A)

The squamous cell carcinoma is the most common malignant tumor (90%) encountered in the esophagus. The tumors vary in type from well-differentiated to anaplastic to small-cell poorly-differentiated. Usually the tumors are well differentiated and shed numerous large diagnostic cells with abundant orangeophilic cytoplasm and irregular pyknotic nuclei, similar to the ones exfoliated from the squamous cell carcinoma of other sites. The majority of the tumors are ulcerated, which explains the large amount of inflammatory cells, cellular debris, and protein deposits usually seen in the background of the smears. Occasionally the squamous cell carcinoma is poorly differentiated in type, and it sheds single or large sheets of cells with scanty basophilic cytoplasm and large hyperchromatic irregular nuclei with abnormal chromatin clumping and wrinkled membrane. It is usually very

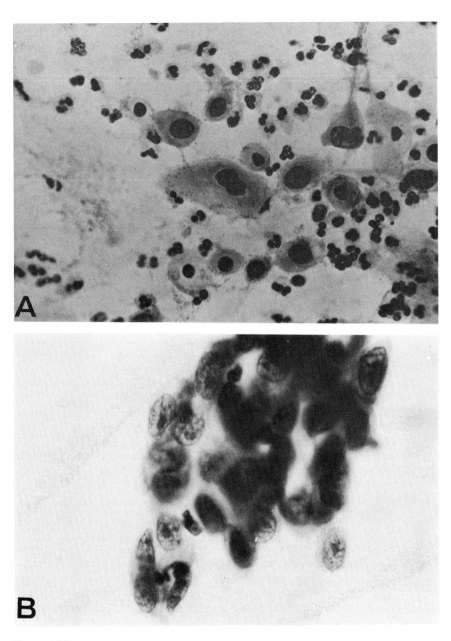

Figure 156
(A) Squamous carcinoma cells in the esophageal washing from a man with a lesion in the lower third of his esophagus. (× 450) (B) Poorly differentiated carcinoma cells in the esophageal washing. Note the abnormal nuclei with occasional prominent nucleoli and the scantiness of the cytoplasm. (× 750)

difficult to differentiate the cells of anaplastic carcinoma of the esophagus from the poorly differentiated adenocarcinomatous cells of gastric origin.

An in situ carcinoma will exfoliate the same cells as an invasive one, and it is impossible to tell by cytology alone the stage of the lesion.

Adenocarcinoma (Fig. 156B)

Adenocarcinoma of the esophagus can be a primary neoplasm or, occasionally, the result of an upward expansion of a gastric adenocarcinoma. They account for about 7% of the esophageal cancers and are often located toward the cardia. The exfoliated cells are seen singly (30%), in clusters (40%), or in sheets (30%). Their cytoplasm is either scanty basophilic (80%) or hypervacuolated (signet ring type, 20%). The nuclei of these cells are enlarged (three to four times) and hyperchromatic. The chromatin is coarsely clumped with the irregular nuclear membrane showing multiple indentations. The nuclei usually have one or several prominent nucleoli. Occasionally, the malignant glandular cells mixed with well-differentiated, orange squamous dysplastic or neoplastic cells may indicate the presence of a mixed tumor (mucoepidermoid carcinoma or adenoacanthoma).

Secondary Tumor of the Esophagus

The exfoliated cells of a secondary tumor of the esophagus resemble the primary neoplasm. On occasion, a squamous bronchogenic carcinoma desquamates into the esophagus through a fistula, and cytology is unable to pinpoint the true origin of these cells. Some of the secondary neoplasms are the result of the infiltration of the wall of the esophagus by a metastatic neoplasm contained in the cervical or mediastinal lymph nodes. The exfoliated cells are usually in large sheets or clusters, and their structure depends on the nature of the tumor.

Sarcoma

Sarcomas of the esophagus are very rare tumors, mainly leiomyosarcomatous or fibrosarcomatous in type. They may, when the tumor is necrotic or ulcerated, shed large sheets of atypical elongated cells. These spindle-shaped cells have transparent, vesicular cytoplasm with large, elongated, single or multiple, irregular nuclei, often bland in appearance, with occasional prominent nucleoli. Most of the time these neoplasms do not shed any diagnostic cells and are rarely detected by cytology.

Primary malignant melanoma of the esophagus has been reported. The finding of melanin-containing, round melanoblasts with typical, often multiple, hyperchromatic nuclei and prominent nucleoli is diagnostic.

Lymphoma

A lymphoma can be a primary neoplasm of the esophagus or an extension of a gastric lesion. Large pools of lymphocytes and lymphoblasts, variable in size with prominent nucleoli, can be seen in the smears. They should be differentiated from the similar-looking but benign traumatically exfoliated cells from the lymphoid tissue (tonsils).

PITFALLS

1. Chronic esophagitis: hypertrophy of the cytoplasm and nucleus, with occasional nucleoli. Nuclear membrane remains smooth and regular, and the amount of the chromatin is not increased;
2. Herpetic esophagitis: nuclear hypertrophy and giant cell formation but with characteristic bland nuclei or intranuclear inclusions, or both;
3. Achalasia: reactive degenerate squamous basal cells with large nuclei and prominent nucleoli that could be mistaken for poorly differentiated carcinoma cells, except for the regularity of their abnormal features;
4. Contamination of the specimen by cells from other sites (nasal and buccal cells, sputum, food, and ingested vegetable cells, reactive lymphocytes from ulcerated tonsils, and so on).

Stomach

ANATOMY AND HISTOLOGY (FIG. 154)

The stomach is situated on the left side of the abdomen immediately below, and in contact with, the diaphragm. It has two openings, one with the esophagus (cardia), the other with the duodenum (pylorus). The cardia is the opening of the esophagus. The fundus is the bulbous upper region. The body is the middle region, and the antrum is situated just above the pylorus. The cardiac region is lined by mucus-producing glandular epithelium. The fundus and body are mostly lined by enzyme-secreting (papain, renin, lipase) chief cells and acid-secreting (HCL, intrinsic factor needed for the absorption of vitamin B_{12}) parietal cells. The pyloric region is lined by mucus-producing pyloric glands.

TECHNIQUE FOR OBTAINING SPECIMENS FOR CYTOLOGY

It is not difficult to diagnose gastric carcinoma by exfoliative cytology. The main problem lies in obtaining good specimens.

Contrary to the function of other hollow organs, the stomach does not act as a reservoir for the exfoliated cells, and, furthermore, these cells are rapidly cytolyzed by the action of the different gastric enzymes (autodigestion). This necessitates a rapid removal and fixation of the cells.

Also, the wall of the gastric mucosa is normally lined with thick mucus, which often prevents the cells from exfoliating easily. This mucus must be liquefied or removed before the collection of the cells.

Numerous techniques have been used. None of them are perfect, and the results obtained are in direct relation to the dedication, experience, and desire of the technician to collect a good specimen.

Preparation

The patient must fast overnight but be kept well hydrated by being allowed to drink water or tea during the night and, especially, one hour before the procedure. In cases of obstructive gastric retention, the gastric content should be aspirated, a continuous suction should be left overnight, and the patient hydrated with intravenous fluids. This initial gastric aspirate should not be used for cytology. If barium has been given to the patient, a delay of at least one week is needed before a satisfactory washing is possible. The barium, besides contaminating the smear with large opaque droplets, may also prevent the diagnostic cells from exfoliating, especially if an ulceration, in which the barium may be retained for weeks, is present.

Abrasion Method

This technique produces a large number of well-preserved gastric cells with very little contamination, but has several disadvantages. The patient may have difficulty in swallowing the balloon or brush used, and the possibility is always present that it may cause the perforation of an already thin, ulcerated mucosa. Furthermore, the abrasion method will not reach certain areas of the stomach (bottom part of the rugae or an ulcer). The following technique, which has given us good results for a number of years, is recommended.

Selected surface scraping under direct vision of a flexible gastrofibroscope will produce well-preserved diagnostic cells with a minimal amount of contaminants from exfoliated normal cells or debris. The cells collected by the brush can either be smeared directly on a slide or can be rinsed in saline solution and passed through a filter membrane. This is the method of choice in experienced hands.

Gastric Lavage

The Levin tube, which should be large enough to permit the passage of the gastric content, is introduced through the mouth or the nose to about the 75 cm mark. No lubrication should be used, but the passage is facilitated if the tube is moistened and placed in ice. Repeated portions of Ringer's solution are injected under pressure, aspirated, and discarded until the return is clear. Then 500 ml of fresh Ringer's solution is injected under maximum pressure and reaspirated while the abdomen is vigorously massaged. The aspirated fluid should be placed in a container in ice and sent immediately to the laboratory to be processed. If the fluid is clear and contains no gross particles, it can be immediately passed through one or several membrane filters with pore openings of 5 μ or spun down at 6000 rpm for 10 minutes, several smears made from the button, and immediately fixed in 95% alcohol. A sediment block may be made if enough sediment is obtained.

CYTOLOGY OF THE NORMAL STOMACH

Gastric Cells (Fig. 157A, B)

The gastric cells, tall and columnar in type, may shed singly or in large sheets, often with a honeycomb appearance. Their size and shape are fairly uniform. Their cytoplasm is often transparent and stained pink to blue, and the amount of the cytoplasm depends on the preservation of the cells. Degenerative cytoplasmic lysis, producing various numbers of bare nuclei, is very common, especially if direct scraping is the technique used. The nuclei are usually single, eccentric, round or oval, and show very little variation in size (7 μ), shape, or chromatin structure. The nuclear borders are well defined, sharp, and smooth. The chromatin is usually finely granular and

343

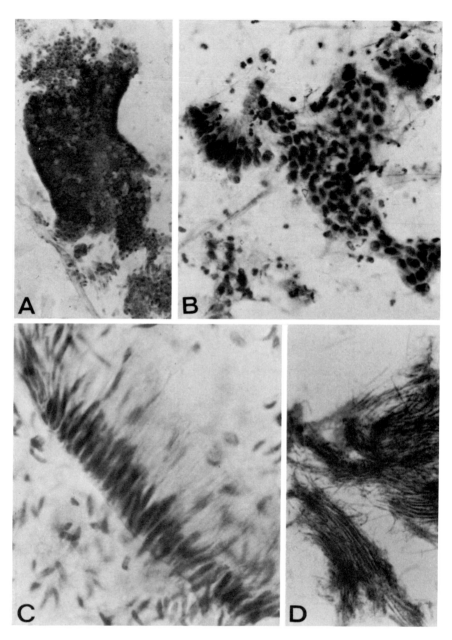

Figure 157
(A) Low-power magnification view of an exfoliated benign gastric gland acinus. (× 125) (B) Cluster of benign gastric surface epithelial cells. Note the regularity of the size and shape of the cells. (× 450) (C) Well-preserved surface gastric cells in palisade formation. (× 450) (D) Contaminant in a gastric washing. (× 450)

homogeneous: they can also be clumped into large fuzzy masses as a result of nuclear degeneration. The nucleoli, when present, are minute.

Parietal Cells (Fig. 157C)

The parietal cells exfoliate rarely and are found singly or in sheets. Their size varies from 15 to 20 μ in diameter. They have an adequate-to-abundant finely vacuolated intensely acidophilic cytoplasm with well-defined borders. Their nuclei (8 μ), occasionally multiple, are central and large with sharp, well-defined nuclear membranes and coarsely clumped chromatin. The nucleoli are single and prominent.

Chief Cells

These cells (15 μ), rarely seen, are cylindrical in shape and usually shed singly. Their cytoplasm is basophilic, scanty, and poorly stained, with multiple, small, uniform vacuoles. The nuclei are eccentric and spherical with finely granular uniformly distributed chromatin and occasional small nucleoli.

Squamous Cells

Anucleated, superficial, intermediate or parabasal squamous cells may originate from the mouth, pharynx, or esophagus. They can be the preponderant cells in the smear if a poor technique has been used. This is usually caused by the failure to discard the saliva-rich first lavage aspirate.

Respiratory Cells

In smears from the gastric washing, it is common to see carbon-bearing histiocytes and ciliated or nonciliated columnar cells originating from ingested sputum from the upper or lower respiratory tract. Occasionally, one can make the diagnosis of an unsuspected pulmonary carcinoma by the recognition of swallowed squamous carcinomatous cells.

Inflammatory Cells

Large numbers of mixed inflammatory cells may be present in smears from the gastric washing, especially in cases of inflammatory lesions of the gastric, esophageal, oral, or respiratory tract mucosa.

Pools of lymphocytes may be found originating from the traumatic exfoliation of the tonsils or adenoid glands. The possibility of benign reactive lymphoreticular hyperplasia of the gastric mucosa or malignant lymphoma should be considered. The specimen obtained by direct-vision brushing usually contains far fewer inflammatory cells, making screening easier (a definite advantage).

Red Blood Cells

When fresh, red blood cells may indicate a trauma resulting from the passage of the tube to obtain the specimen and thus have no significance. When old and in the presence of hemosiderin-laden histiocytes, they often indicate a bleeding gastric ulcer or neoplasm.

Bacteria and Fungi

Bacteria or fungi may be found as a result of contamination from an infection in the oral cavity, esophagus, or respiratory tract. They usually form large colonies, which occasionally may obscure the smears and become a handicap in the diagnosis. They are increased in gastric retention, chronic ulceration, carcinoma, and in debilitated patients. Free-living ameba normally found in soil and water, such as *Acanthamoeba*, may be found as contaminants, often originating from the tubing used to obtain the specimen. Autoclaving of the tubing after each use is strongly advised.

Foreign Cells—Food Particles (*Fig. 157D*)

Various food particles (vegetable or animal) may be seen, indicating a poor technique or gastric obstruction. They can become a pitfall in the diagnosis either by imitating malignant cells or obscuring them.

BENIGN LESIONS OF THE STOMACH

Acute Gastritis

Large amounts of inflammatory cells (leukocytes) are the main cellular components of the gastric washing. The epithelial cells are few, mainly single. The occasional intact gastric cells show marked edema and hypertrophy, with degenerative vacuoles filling their cytoplasm. Their nuclei are usually pale, round, and enlarged, with prominent nucleoli. A larger number of nuclei of stripped epithelial cells is seen in the background of the smear, mixed with fresh and old red blood cells. The benign nature of these cells is recognized by the uniformity of their size and shape. On occasion, sheets of traumatically sloughed gastric mucosa may be seen, with small dark cuboidal cells containing very little mucus. The proportion of exfoliated parietal cells is slightly higher than normal.

Atrophic Gastritis

Atrophic gastritis is a chronic inflammation found in elderly patients or patients with pernicious anemia. This condition is often associated with a two to four times higher than normal incidence of gastric carcinoma. The gastric washing is usually rich in epithelial cells, with a predominance of large goblet cells resulting from the frequent intestinal metaplasia that may occur in these often long-standing diseases. The cytoplasm of these cuboidal cells is hypervacuolated, with single or multiple vacuoles overdistending it. Some of these goblet cells may contain migrating polymorphs in their cytoplasm and are not to be confused with mucus-producing adenocarcinoma cells. Their nuclei remain benign. They are usually oval and eccentric, with a uniform, finely granular or bland, washed out, chromatin pattern. The nuclear membrane is finely wrinkled. The enlarged cells often have a straight end resembling a terminal bar, but no cilia are ever seen. No parietal or chief cells can be found in the smear. There is usually an increased number of lymphocytes, plasmocytes, and inflammatory eosinophils. Few leukocytes can be found, except in the presence of secondary inflammation. Variable numbers of plasma cells can be found, however. Various amounts of fresh or old red blood cells, some ingested in phagocytes, are usually present.

Peptic Ulcer (Fig. 158A)

In the patient with suspected acute or chronic peptic ulcer, because of the usual increased acid output, the gastric washing specimens should be processed even faster than usual to prevent degeneration of the cells. This rapid degeneration of the cells explains in part the relatively high rate of cytologic false-negative results in cases of peptic ulcers with malignant change. Large numbers of stripped epithelial cell nuclei, uniform in size, round or oval in shape, are usually present. There is a decrease in the number of parietal cells. Small amounts of debris, hemosiderin-laden histiocytes, and foreign-body giant cells may be seen. Goblet cells resulting from the frequent intestinalization of the mucosa may be found in chronic peptic ulcer.

Gastric cells with various degrees of atypia and originating from the mucosal margins of the chronic peptic ulcer may present some difficulty in interpretation. The cells with mild atypia, almost always present with peptic ulcer, show moderate nuclear hypertrophy with coarse chromatin clumping and one or more small, round nucleoli. The regularity of their size and shape and their smooth nuclear membranes easily differentiate them from malignant cells.

Gastric Polyp

A small fraction of gastric carcinomas may arise from adenomatous polyps. The surface cells, which exfoliate readily, have often been subjected to trauma and irritation and show reactive enlarged nuclei with prominent nucleoli and coarse chromatin condensed toward the nuclear rim. This cytologic picture may lead to a false-positive diagnosis. The absence of extreme chromatin and nuclear membrane irregularity helps prevent such a diagnosis. Inflammatory cells and degenerate cellular debris often obstruct the background.

Granulomatous Gastritis

The presence of multinucleated giant cells (Langhans' cells) associated with cells previously described in atrophic gastritis may indicate a tuberculous disease (Crohn's) or luetic (syphilitic) gastritis. Elongated spindle fibroblastic cells with oval, eccentric nuclei and scanty, poorly defined cytoplasm may also be present.

Intestinal Metaplasia (Fig. 158B)

As a result of chronic nonspecific irritation (pernicious or Addison's anemia or chronic gastritis), the normal gastric cells may be replaced by goblet columnar cells, often having at one end a straight line resembling a terminal bar, but with no demonstrable cilia. Their more or less well-preserved cytoplasm is often overdistended with secretory vacuoles (signet ring cells), containing often well-preserved polymorphonuclear leukocytes. Their dark nuclei are uniform in size and shape and have a prominent nuclear membrane. The nucleoli, when present, are not prominent.

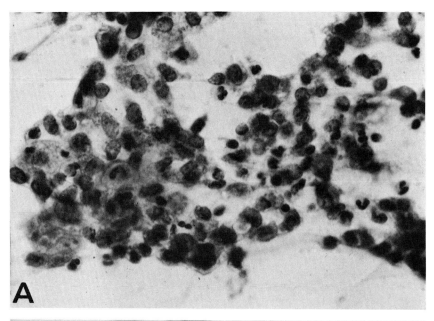

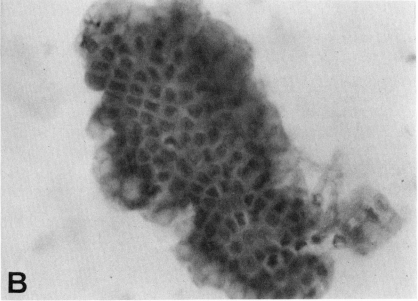

Figure 158
(A) Gastric washing in a patient with a peptic ulcer. Note the loose epithelial cells with uniform round or oval nuclei, the histiocytes, and the inflammatory cells. (× 450) (B) Intestinal metaplasia. Note the secretory vacuoles. (× 450)

Whipple's Disease

This is a rare metabolic disease that results from a faulty carbohydrate-protein metabolism and involves the gastric and intestinal mucosa. The gastric washing shows large numbers of single mononucleated or binucleated phagocytes with multiple, large intracytoplasmic vacuoles containing lipid and glucoprotein deposits.

MALIGNANT TUMORS OF THE STOMACH

Poorly Differentiated Adenocarcinoma (Fig. 159)

This carcinoma usually sheds numerous cells singly (85%), in loose clusters (10%), or in sheets (5%). The shape of the cells varies: oval or round (60%), elongated (10%), and irregular (30%); but they retain a columnar pattern. The size varies only moderately ($30 \pm 10 \mu$).

The scanty cytoplasm is often poorly preserved, transparent, and basophilic with ill-defined borders.

The nuclei are enlarged ($12 \pm 5 \mu$) and show marked irregularity of shape (85%). The nuclear membrane is heavy with frequent indentations or lobulations. The chromatin is coarsely granular, often unevenly distributed.

The nucleoli are prominent ($3–4 \mu$ in diameter) in 60% of the cells.

Abnormal mitosis may be seen in the sheets of cells obtained by the direct brushing method. Some of these cells may have either no vacuoles or multiple small ones.

Large amounts of bacteria, inflammatory cells, and fibrinous cellular debris often obscure the background of the smears.

Well-Differentiated Adenocarcinoma of the Stomach (Fig. 160B)

The diagnostic cells are scanty, except when the lesion is fungating and ulcerated. The cells are usually shed in tight clusters or large sheets (80%). The size of the cells varies ($60 \pm 20 \mu$). The shape is irregular, with marked intracellular molding.

Their cytoplasm is abundant, often hypervacuolated, and contains well-preserved neutrophils. The mucin-containing vacuoles often mold the nucleus into a crescent shape (signet ring cells).

The cytoplasmic membrane is delicate, often well defined, and irregular.

The eccentric nuclei are single or multiple (2–4 in 40% of the cells) and are hyperchromatic with sharp, irregular nuclear borders. The increased chromatin is coarsely clumped, with irregular, pointed projections. The nucleoli are prominent and large ($2–5 \mu$).

The amount of bacteria, inflammatory cells, and cellular debris in the background is variable.

Linitis Plastica (Fig. 160C, D)

In linitis plastica, the diagnostic epithelial cells are very scanty, even in a good washing. No more than 10 to 12 good cells may be found scattered in a smear. They are always single with very scanty, transparent-to-light-blue cytoplasm. The cells are irregular in shape with poorly defined borders.

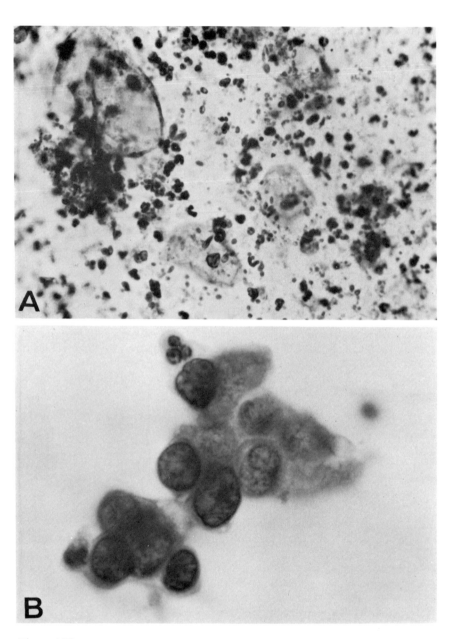

Figure 159
(A) Low-power magnification view of a positive gastric washing. Filter preparation showing the abundance of the dark, single and clustered, irregular adenocarcinoma cells. ($\times$ 250) (B) Cluster of poorly differentiated adenocarcinoma cells in a gastric washing. Note the irregularity of the chromatin, nuclear membrane, and shape and size of the cells. ($\times$ 1300)

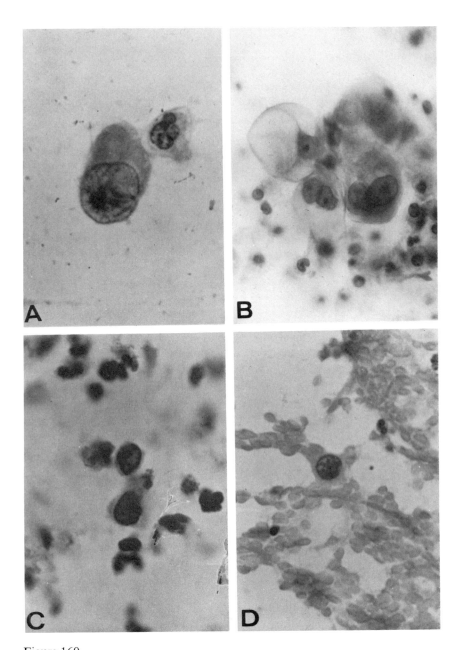

Figure 160
(A) Single malignant cell in a gastric washing originating from an anaplastic adenocarcinoma. Note the prominent nucleoli and the adequate cytoplasm. ($\times$ 1300) (B) Group of malignant cells diagnostic of a well-differentiated mucus-secreting gastric adenocarcinoma. Note the prominent mucus-containing vacuoles distending the cytoplasm, the irregular nuclei, and the bland chromatin and prominent nucleoli. ($\times$ 450) (C, D) Scattered single small malignant cells in the gastric washing in 2 patients with linitis plastica. Note the scantiness of the cytoplasm and the hyperchromatism of the nuclei. ($\times$ 750)

The nuclei are central, round or oval, and hyperchromatic, with little variation in size ($10 \pm 5 \mu$). The chromatin is coarsely clumped, with occasional prominent, pointed projections. The nucleoli are prominent in 60% of the cells and show some irregularity in shape and number.

These cells, especially if degenerate, are easy to confuse with the numerous single histiocytes and plasmocytes frequently seen in the background of most gastric washings.

Moderate amounts of cellular debris, protein deposits, and mixed inflammatory cells obscure the background of the smear.

Lymphoma

The presence of large numbers of scattered single (80%) cells or of clusters of lymphoblasts evenly distributed in the smear, with usually one predominant cell type, is diagnostic. The cytoplasm is scanty and basophilic, with poorly defined borders. The nuclei are central and single, with coarsely clumped chromatin and irregular, thick, distinct, nuclear membranes. The usually single nucleolus is prominent in 60% of the cells. Mitotic figures are rare in lymphosarcoma, but are increased in the reticulum cell type. These malignant lymphomatous cells may be difficult to differentiate from the small cell poorly differentiated adenocarcinoma or from the clusters of mature lymphocytes seen in certain chronic inflammation (atrophic gastritis) (Table 16). The giant follicular lymphoma, which often involves the stomach, exfoliates only small mature lymphocytes that are difficult to diagnose by cytology.

In reticulosarcoma, the pleomorphic nuclei have multiple indentations (convolutions) or nipple-like protrusions. Occasional multinucleated giant cells are present. Nucleoli are prominent.

In Hodgkin's lymphoma, large binucleated mirror-image cells with abundant cytoplasm are present in the midst of reactive lymphocytes. The chromatin is irregular and a single nucleolus is prominent (its size is about one-third that of the nucleus).

PITFALLS
1. Pernicious anemia: the enlarged cytoplasm and nuclei of the gastric cells are similar to those in low-grade cancer cells but they have a bland appearance and there is little contrast between the cytoplasmic and nuclear staining;
2. Intestinal metaplasia can produce confusing dark but uniform nuclei;
3. Peptic ulcer: the regenerating gastric cells often have reactive nuclear features, but these abnormalities are not extreme;
4. Gastritis: the increased inflammatory cells and debris may hide the cancer cells;
5. Polyp: sheets of reactive gastric cells with increased inflammation may be misinterpreted;
6. Contaminant: some cancer cells from other sites may be found in the gastric washing. Certain ingested food cells may resemble sheets of malignant cells;
7. Technique: the brushing of the surface of a large fungating carcinoma

Table 16. Differential Characteristics of Gastric Carcinoma

Criteria	Poorly Differentiated Carcinoma	Well Differentiated Carcinoma	Linitis Plastica	Lympho-sarcoma
Size	$30 \pm 10 \mu$	$60 \pm 20 \mu$	$15 \pm 5 \mu$	$10 \pm 6 \mu$
Shape	Oval 70%	Irregular 90%	Round to oval	Round
Occurrence				
Single	85%	5%	95%	80%
Clusters	10% (loose)	15% (tight)	2%	20% (loose)
Sheets	5%	80%	3%	0%
Cytoplasmic stain	Basophilic	Basophilic	Pale	Pale
Amount of cytoplasm	Scanty	Abundant	Scanty	Very scanty
Cytoplasmic vacuoles	Minute	Large	None	None
Nuclear size	$12 \pm 5 \mu$	$20 \pm 10 \mu$	$10 \pm 5 \mu$	$9 \pm 6 \mu$
Nuclear shape	Irregular 85%	Irregular 95%	Round 95%	Round 100%
Chromatin irregularity	Moderate	Marked	Moderate	Moderate
Multinucleation	5%	40%	1%	0%
Nucleoli	Prominent 50%	Prominent 95%	Prominent 60%	Prominent 60%
Number of diagnostic cells	Abundant	Less abundant	Scanty	Abundant

may produce only nondiagnostic necrotic debris and inflammatory lesions.

Duodenum

ANATOMY AND HISTOLOGY (FIG. 161)
The second portion of the duodenum is the part that mainly interests the cytopathologist. The hepatic and cystic ducts, after merging to form the common duct, open in the ampulla of Vater, situated about 6 inches from the pylorus, where it joins the pancreatic duct. The epithelium of these ducts is made of a single layer of tall nonciliated columnar cells. The duodenal mucosa contains numerous villi lined by other tall columnar cells.

TECHNIQUE OF COLLECTING DUODENAL SPECIMENS
The cytologic interpretation of duodenal specimens is not difficult, but the collection is. The exfoliated cells that result from secretin stimulation may originate from the hepatic, cystic, common biliary and pancreatic ducts and from the duodenal mucosa.

MATERIAL NEEDED. A Levin tube or, preferably, a weighted radiopaque Diamond double lumen tube, a vacuum pump (120 mm Hg pressure), and an ampule of secretin.

PROCEDURE. The gastric content of a well-hydrated patient who has fasted overnight is aspirated and discarded. A weighted tube (3–5 gm) is then passed through the mouth of the sitting patient to about the 45 cm mark. The patient then lies on his left side with his head elevated and an additional 15 cm of tubing is slowly passed. The patient is allowed to sit up and bend forward several times. Then he lies on his right side, and another 10 cm of tubing is slowly fed. The patient then turns on his back, and 15 cm of additional tubing is passed. The position of the tube may be checked by fluoroscopy to see if the tip is in the duodenum. An ampule of secretin may then be injected intravenously, a pump connected to the tube, and secretions continuously aspirated for 20 to 30 minutes and collected in an ice-cooled vessel (the duodenal contents often degenerate very rapidly). Except in cases of biliary obstruction fresh golden bile should be seen in the aspirated fluid. This specimen may either be passed through a membrane filter with a 5 μ pore opening and fixed or spun at 6000 rpm for 10 minutes and the sediment smeared on several slides and fixed.

CYTOLOGY OF THE BENIGN DUODENAL SECRETION (FIG. 162A)
Numerous small, round or oval, single duodenal cells ($20 \pm 3\ \mu$), in loose clusters or sheets and showing marked degenerative changes, are found. The cytoplasm is granular or vacuolated and transparent, and the nuclei are central, deeply stained, round, and vesicular with smooth borders, fine chromatin, and prominent nucleoli. The sheets of cells are often palisade or honeycomb in appearance.

In better preserved specimens, variable amounts of elongated tall colum-

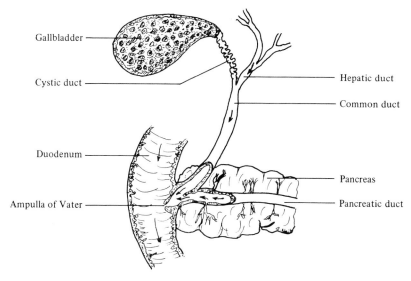

Figure 161
Anatomy of the duodenum and the biliary and pancreatic ducts.

nar cells are also seen (biliary or pancreatic duct cells). These columnar cells have thin, abundant, cyanophilic cytoplasm and are usually single, but often in clusters or palisade formation. The nuclei of these elongated cells are often small (5 μ), oval, vesicular, or pyknotic. The cytoplasm is basophilic, with well-preserved and well-defined borders.

Goblet cells may also occur and may indicate the presence of an inflammatory process. Variable amounts of mixed inflammatory cells, phagocytes containing biliary pigments, and occasional foreign-body giant cells are usually seen.

Plasma cells often are found in abundance in chronic diseases.

Biliary cholesterol crystals, resembling fragments of broken, transparent glass with parallel edges and notched corners, are easy to recognize, especially if polarized. In small quantity they have no particular significance, but in large amounts and in clumps they suggest cholecystitis, cholesterosis, or cholelithiasis. Similarly, golden yellow-brown, finely granular calcium bilirubinate crystals may indicate the same diseases. The presence of both crystals has been related to gallstones.

Contaminants from food particles, cellular debris, and protein or bile precipitate commonly obscure the background of the smears.

Giardia lamblia, a protozoan, may give the appearance of poorly fixed, degenerate cells containing two nuclei. Larva of *Strongyloides stercoralis* may also be seen in rare instances.

MALIGNANT NEOPLASMS OF THE DUODENUM (FIG. 162B)

Well-differentiated, typical squamous carcinoma cells occasionally are found exfoliating from rare cases of primary squamous carcinoma of the

355

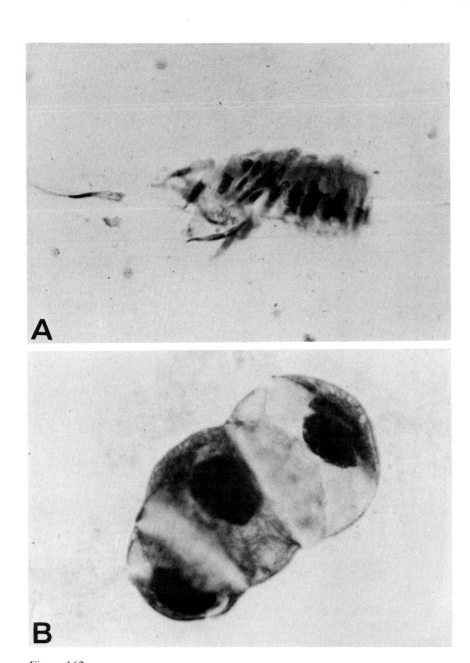

Figure 162
(A) Normal cells originating from the biliary or pancreatic duct. ($\times$ 450) (B) Group of malignant cells seen in the duodenal drainage of a patient with a pancreatic adenocarcinoma. ($\times$ 1300)

ampulla of Vater or gallbladder. Cancer of the duodenum proper is extremely rare.

It is usually easy to recognize the cells that are shed by the well-differentiated or poorly differentiated adenocarcinoma of the pancreas or biliary duct. They are found in clusters, and their size is two or three times larger than the size of the benign cells. Their hyperchromatic, large, irregular nuclei have prominent nucleoli. Their cytoplasm varies in size and occasionally contains large secretory vacuoles in the form of signet ring cells. The recognition of these malignant cells becomes difficult only when excessive, rapidly occurring, degeneration is present.

Papillary adenocarcinoma of the pancreas or biliary duct may exfoliate numerous papillary structures made of neoplastic cells with a common smooth external border.

Colon

ANATOMY AND HISTOLOGY (FIG. 163B)
The colon is divided into six parts: the cecum with the attached appendix, the ascending, the transverse, the descending, and the sigmoid colons, and the rectum. The mucosa usually does not fold and is made of straight tubular glands (Lieberkuhn's) lined by low columnar epithelium rich in goblet cells. Squamous stratified epithelium covers the anal region.

TECHNIQUE FOR COLLECTING COLONIC SPECIMENS
On the night preceding the examination, a normal saline enema should be given to the patient, who has been on a low-residue diet for the preceding two days. In the afternoon of the day preceding the lavage, a strong cathartic is administered. Another cleansing enema is given on the morning of the examination and repeated until the return is clear. No soap or oil solution should be used. A French urinary catheter is introduced beyond the rectosigmoid junction with the patient lying on his left side with his knees drawn up. Using a Dakin syringe, approximately 1000 ml of normal saline at 36° C is administered. The tube is pinched and the patient rotated on his back. The abdomen of the patient is massaged to work the saline around the descending and transverse colon. The patient is then rotated onto his right side with his knees up for about 3 minutes while the ascending colon is massaged. The patient is then turned to the left side with his knees up, and the water is siphoned out with the Dakin syringe. If the enema fluid contains gross particles, the procedure should be repeated. After it has been passed through a large-mesh strainer, the enema fluid is spun down (10 minutes at 5000 rpm) or divided and passed through several membrane filters and the cells immediately fixed.

If the lesion can be visualized by the proctosigmoidoscope or flexible fiberscope, direct brushing of its surface will produce a specimen richer in cells and hence is the method of choice.

CYTOLOGY OF THE NORMAL COLON (FIG. 163A)
Numerous small colonic tall columnar cells shed singly, in clusters, sheets, or a syncytium. They are uniform in shape (triangular) and size (15 ± 2

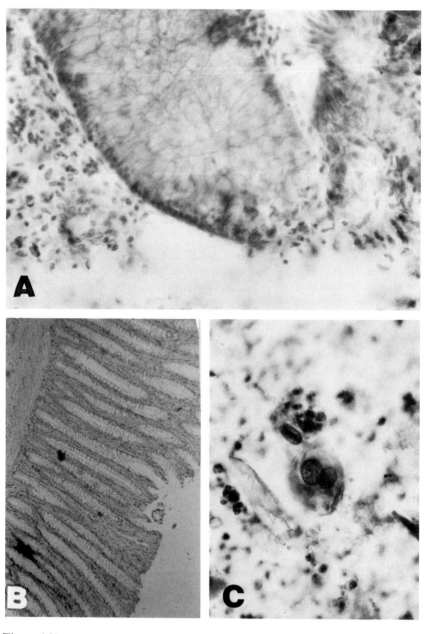

Figure 163

(A) Sheet of normal colonic cells forming an acinus. Note the secretory vacuoles in the goblet cells. (× 250) (B) Histologic section of a normal colon. (× 125) (C) Colonic washing in a patient with ulcerative colitis. Note the abundance of leukocytes infiltrating the cytoplasm of one epithelial cell and the binucleation of the other. (× 450)

μ); nuclei are central and round or oval. The chromatin is finely granular and evenly distributed. A minute single nucleolus is often present.

The goblet cells usually shed singly or in small sheets. Their semitransparent cytoplasm, abundant, basophilic, and hypervacuolated, has an eccentric, benign-looking single round or oval nucleus, often compressed toward the base. The cells are uniform in size (5–7 μ in diameter) and in chromatin pattern. The nuclear borders are even and delicate. Occasional small nucleoli may be seen. No mitotic figures are present.

Variable amounts of small round or oval histiocytes with vesicular semitransparent blue cytoplasm with poorly defined borders, and a central or eccentric, small, round nucleus may be seen. The nuclei are often bean shaped, with regular, finely granular chromatin. Scattered polymorphonuclear cells and lymphocytes are usually present.

Occasionally, squamous cells, anucleated, superficial or intermediate in type, are found originating from the anal region or as contaminants from the perineum.

Variable amounts of contaminants, food particles, debris, and inflammatory cells are present.

BENIG N LESIONS OF THE COLON

Ulcerative Colitis (Fig. 163C)

Large numbers of inflammatory cells (leukocytes and plasmocytes), often seen in clusters, are always present in ulcerative colitis. Markedly degenerate epithelial cells, with their cytoplasm heavily infiltrated with leukocytes, are seen. These epithelial cells are hyperactive and columnar in type. They are enlarged (12 ± 5 μ) and have an elongated dense nucleus with coarse chromatin pattern. The nuclear membrane is very irregular. Occasional prominent nucleoli can be seen. Cells with large, bland, pale nuclei and adequate vesicular cytoplasm are another type often seen. Usually present is a large number of stripped oval nuclei with a bland homogeneous appearance and round nucleoli resulting from the degeneration of the colonic cells. Multinucleation (2–3 nuclei) is common.

Various amounts of cellular debris, protein deposits, and exudate mixed with fresh and old blood generally obscure the background. Numerous histiocytes are also present, with occasional foreign-body giant cells containing hemosiderin or cellular debris.

Amebic Colitis (Fig. 164B)

Variable numbers of *Amoeba histolytica* parasites may be seen in the midst of large amounts of inflammatory cells (leukocytes) and cellular debris. These parasites resemble *Trichomonas,* except for their larger size and multiple pale nuclei (2–4). The small, round nuclei often show a margination of the chromatinic material. Their cytoplasm is poorly stained, gray and finely vacuolated, containing ingested debris or intact round red blood cells. There is usually a large number of protein deposits, red blood cells, and degenerate polymorphonuclear cells in the background. An increased number of mononucleated phagocytes is often seen. An iron stain may be help-

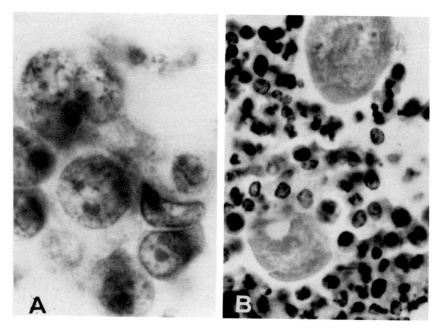

Figure 164
(A) Scraping of a benign polyp of the colon with surface atypia. (× 750) (B)
Amoeba histolytica in the scraping of an amebic colitis lesion. (× 750)

ful in detecting these parasites by staining blue the phagocytized red blood
cells.

Adenomatous Polyp (Fig. 164A)

Adenomatous polyps are usually difficult to diagnose by cytology. Occasionally, large sheets of elongated, intact, benign columnar cells can be
seen, especially after a direct scraping of an adenomatous polyp. These
columnar cells are often spindle and elongated in shape with dark, adequate orangeophilic cytoplasm and elongated central hyperchromatic nuclei
with regular, sharp, nuclear borders. The presence of secretory vacuoles
often indicates reactive surface atypia.

Occasionally, variable amounts of goblet cells are seen, with abundant
vacuolated cytoplasm and nuclei that are round and central or eccentric
and oval, with uniform chromatin and prominent nucleoli. Their presence
indicates a benign lesion rather than a malignant one.

Variable amounts of inflammatory cells, blood, and cellular debris and
phagocytes can also be seen.

MALIGNANT LESIONS OF THE COLON

Colonic adenocarcinomas are common neoplasms (10% of all cancers).
Their occurrence is slightly more frequent in men over the age of 50: about
70% of the adenocarcinomas are found in the rectosigmoidal areas.

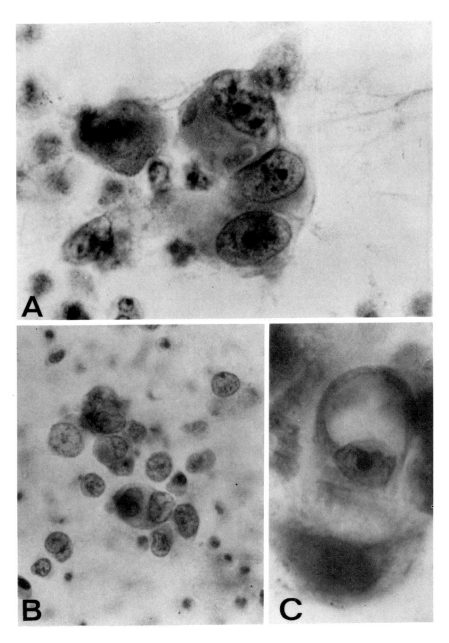

Figure 165
(A) Tumor cells seen in the colonic washing of a man with an adenocarcinoma of the descending colon. Note the enlarged size of the nuclei and the prominence of the nucleoli. (× 1300) (B) Cluster of tumor cells seen in the colonic washing in a man with a poorly differentiated adenocarcinoma of the colon. These cells may be confused with reactive histiocytes, except for their size and the irregularity of the shape of their nuclei and chromatin. (× 450) (C) Degenerate tumor cell in the colonic washing of a woman with a well-differentiated adenocarcinoma. Note the abundant vacuolated cytoplasm. (× 450)

361

Exfoliative cytology is most useful when the lesion is beyond the range of sigmoidoscopy and biopsy or when several benign tumors, too numerous for biopsy, are present (familial polyposis). The correct interpretation of the exfoliated malignant cells is facilitated by the absence of confusing pre-cancerous changes.

The diagnostic cells exfoliate singly or in clusters or large sheets and are often degenerate. They are easy to recognize, if well preserved, because of their large size, which varies ($30 \pm 15\ \mu$). Their cytoplasm can be scanty (80%) or abundant (20%), often with orangeophilic, thick, vacuolated cytoplasm with poorly defined borders (75%). Their nuclei are enlarged and hyperchromatic and vary in size (10–$16\ \mu$) and shape, with irregular chromatin clumping and pointed projections. The nuclear membrane is well defined and irregular with indentations. Their prominent, red nucleoli vary in shape, size, and number. Large amounts of inflammatory cells (leukocytes), often forming big clusters, are usually mixed with an abundance of fresh and old red blood cells, degenerate cellular debris, and food contaminants. There is an increased number of goblet cells.

The rare squamous cell carcinoma mainly arises from the anus and shows the typical well-differentiated squamous carcinoma cells as previously described for other sites.

References and Supplementary Reading

Bennington, J. L., Porus, R., Ferguson, B., and Hannon, G. Cytology of gastric sarcoid–Report of a case. *Acta Cytol.* 12:30, 1968.

Bloziz, G. G. Oral cancer detection. *J. Am. Dent. Assoc.* 70:1472, 1965.

Boon, T. H., and Schade, R. O. K. Exfoliative gastric cytology. *Lancet* 1: 1163–1164, 1961.

Broderick, P. A., Allegra, S., and Corvese, N. Primary malignant melanoma of the esophagus. *Acta Cytol.* 16:159, 1972.

Bryan, M., and Bryan, W. The use of aural cytology in the diagnosis of various inflammations and malignant tumors. *Acta Cytol.* 14:411, 1970.

Cark, A. H., McKee, E. E., and Dixon, D. Identification of trophozoite form of *Entamoeba histolytica* by cytologic techniques. *Acta Cytol.* 16:429, 1972.

Dee, A. L., and Harris, B. An experience in the cytopathologic diagnosis of carcinoma of the esophagus. *Acta Cytol.* 7:236, 1963.

DeKemore, S., and Spasor, S. A comparison of oral and vaginal smears in women with normal menstrual cycles. *Acta Cytol.* 14:31, 1970.

Galambos, J. T., et al. Exfoliative cytology in chronic ulcerative colitis. *Cancer* 9:152, 1956.

Gardner, A. F. Table clinic questions–Oral cancer. *Dent. Surv.* Feb.–Mar. 1962.

Gardner, A. F. The challenge of oral exfoliative cytology. *N.Y. Dent.* 34:381, 1964.

Gardner, A. F. The cytologic diagnosis of oral carcinoma. *J. Calif. Dent. Ass.* 40:1, 1964.

Gardner, A. F., and Axelrod, J. H. A study of twenty-one instances of amelo-blastoma, a tumor of odontogenic origin. *J. Oral Surg.* 21:230, 1963.

Gardner, Frank H. Observations on the cytology of gastric epithelium in tropical sprue. *J. Lab. Clin. Med.* 47:529–539, 1956.

Haam, E. von.   The historical background of oral cytology. *Acta Cytol.* 9:270, 1965.

Johnston, W., Amatulli, J., and Thompson, L. K.   Parotid duct cytology as a diagnostic tool in the evaluation of parotid gland enlargement. *Acta Cytol.* 14:254, 1970.

Kalins, Z. A., et al.   Analysis of cytologic findings in patients with gastric carcinoma. *Acta Cytol.* 11:312, 1967.

Katz, S., Sherlock, P., and Winawer, S.   Rectocolonic exfoliative cytology. *Am. J. Dig. Dis.* 17:1109, 1972.

King, O. H., Jr., and Coleman, S. A.   Analysis of oral exfoliative cytologic accuracy by control biopsy technique. *Acta Cytol.* 9:351, 1965.

Klavins, J. V., and Flemma, R. J.   A method of studying the material of the bile ducts. *Acta Cytol.* 8:332, 1964.

Knoernschild, H. E., and Cameron, A. B.   Mucosal smear cytology in the detection of colonic carcinoma. *Acta Cytol.* 7:233, 1963.

Kobayashi, S., Prolla, J. C., and Kirsner, J. B.   Brushing cytology of the esophagus and stomach under direct vision by fiberscopes. *Acta Cytol.* 14:219, 1970.

Kondo, T., and Momoi, Y.   The clinical value of malignant cells from enema returns in carcinoma of the colon. *Nagoya J. Med. Sci.* 29:271, 1967.

Lance, K. P., and Groisser, V. W.   Utility of cells exfoliated onto the esophagoscope as an aid in the diagnosis of esophageal carcinoma. *Am. J. Dig. Dis.* 10:1, 1965.

MacDonald, W. C., et al.   Esophageal exfoliative cytology. *Ann. Intern. Med.* 59:332, 1963.

MacDonald, W. C., et al.   Gastric exfoliative cytology: An accurate and practical diagnostic procedure. *Lancet* 2:83–86, 1963.

Medak, H., et al.   The cytology of vesicular conditions affecting the oral mucosa: Pemphigus vulgaris. *Acta Cytol.* 14:11, 1970.

Medak, H., et al.   Correlation of cell populations in smears and biopsies from the oral cavity. *Acta Cytol.* 11:279, 1967.

Medak, H., et al.   *Oral cytology.* Washington, D.C.: U.S. Government Printing Office, 1972.

Monto, R. W., Fine, G., and Rizek, R. A.   Exfoliative cells of the oral mucous membranes in patients receiving chemotherapy for malignant disease. *J. Oral Surg.* 21:95, 1963.

Nieburgs, H. E., et al.   The morphology of cells in duodenal-drainage smears: Histologic origin and pathologic significance. *Am. J. Dig. Dis.* 7:489, 1962.

Orell, S. R., and Ohlsen, P.   Normal and postpancreatic cytologic patterns of the duodenal juice. *Acta Cytol.* 16:165, 1972.

Peters, H.   Cytologic smears from the mouth. Cellular changes in disease and after radiation. *Am. J. Clin. Pathol.* 29:219, 1958.

Prolla, J. C., and Kirsner, J. B.   *Handbook and Atlas of Gastrointestinal Exfoliative Cytology.* Chicago: University of Chicago Press, 1972.

Prolla, J. C., Kobayashi, S., and Kirsner, J. B.   Cytology of malignant lymphomas of the stomach. *Acta Cytol.* 14:291, 1970.

Prolla, J. C., Xavier, R. G., and Kirsner, J. B.   Morphology of exfoliated cells in benign gastric ulcer. *Acta Cytol.* 15:128, 1971.

Raskin, H. F., et al.   Gastro-intestinal cancer: Definitive diagnosis by exfoliative cytology. *Arch. Surg.* 76:507–516, 1958.

Raskin, H. F., Palmer, W., and Kirsner, J. B.   Benign and malignant gastrointestinal mucosal cells. *Arch. Intern. Med.* 107:138–150, 1961.

Raskin, H. F., and Pleticka, S. Exfoliative cytology of the stomach. *Cancer* 10:82–89, 1960.

Raskin, H. F., and Pleticka, S. The cytologic diagnosis of cancer of colon. *Acta Cytol.* 8:131, 1964.

Raskin, H. F., and Pleticka, S. Exfoliative cytology of the colon—Fifteen years of lost opportunity. *Cancer* 28:127, 1971.

Roseu, R., Garret, M., and Aka, E. Cytologic diagnosis of pancreatic cancer by ductal aspiration. *Ann. Surg.* 167:425, 1968.

Sasson, L. Unfavorable experiences with the Ayre rotating stomach brush. *Am. J. Dig. Dis.* 9:398, 1964.

Serck-Hanssen, A. Exfoliative cytology in the detection of malignant gastric disease. *J. Oslo City Hosp.* 17:221, 1967.

Silverman, S. The cytology of benign oral lesions. *Acta Cytol.* 9:287, 1965.

Staats, O. J., Goldsby, J. W., and Butterworth, C. E., Jr. The oral exfoliative cytology of tropical sprue. *Acta Cytol.* 9:228, 1965.

Stahel, J., and Wiman, L. G. Exfoliative cytology in otolaryngology. *Acta Otolaryngol.* 55:377, 1962.

Taebel, D. W., Prolla, J. C., and Kirsner, J. B. Exfoliative cytology in the diagnosis of stomach cancer. *Ann. Intern. Med.* 63:1018, 1965.

Thabet, R. J., and MacFarlane, E. W. Cytological field patterns and nuclear morphology in the diagnosis of colon pathology. *Acta Cytol.* 6:325, 1962.

Tiecke, R. W., and Bloziz, G. G. Oral cytology. *J. Amer. Dent. Assoc.* 72:855, 1966.

Wiendi, H., et al. Cytophotometric studies of gastric mucosal smears in malignant and benign diseases of the stomach. *Acta Cytol.* 18:222, 1974.

Witte, S. Gastroscopic cytology. *Endoscopy* 2:88, 1970.

Yamada, T., et al. Clinical evaluation of proteolytic enzyme lavage method in the gastric cytodiagnosis, especially in the detection of early cancer of the stomach. *Acta Cytol.* 8:27–33, 1964.

Zimmermann, E. R., and Zimmermann, A. L. Effects of race, age, smoking habits, oral and systemic disease on oral exfoliative cytology. *J. Dent. Res.* 44:627–631, 1965.

The Urinary Tract and the Prostate **14**

The Urinary Tract

ANATOMY AND HISTOLOGY

The detailed description of the anatomy, physiology, and histology of the urinary tract is very complex and beyond the scope of this manual. However, a minimal knowledge of the different components is needed for the comprehension of the special problems encountered in a urine specimen (Fig. 166).

The kidneys are a pair of organs situated retroperitoneally on each side of the vertebral column. The cross section shows a thin, fibrous external capsule over a reddish-brown cortex, 1.5 cm thick, composed mainly of nephrons containing the tubules and a pale medulla that is made up of multiple pyramids with pointed tips, referred to as papillae. The microscopic study of a section will reveal only a few of the 1 to 2 million nephrons through which the blood is filtered.

Each nephron is made of a glomerulus and a renal tubule. Part of the glomerulus is a mass of tangled capillaries lined with endothelial cells. The other part (Bowman's capsule) is composed of a monolayered epithelium upon a basal membrane. The renal tubule is divided into (1) a proximal tubule, lined by monolayered cylindrical cells with brush borders and round nuclei, (2) the loop of Henle, lined by flattened monolayered cells with thin cytoplasm and round bulging nuclei, and (3) distal tubules, lined by monolayered cylindrical cells.

The collecting tubules, which open first on the papillae in the calices and then in the pelvis, are lined by a single layer of small cuboidal cells with central, round-to-oval nuclei and transparent cytoplasm.

The ureters are retroperitoneal membranous ducts (28 cm in length, 4–6 mm in diameter), which conduct the urine from the pelvis to the bladder. They are lined by a transitional epithelium similar to that found in the renal pelvis and the bladder.

The urinary bladder, situated deep within the pelvis, is a membranous sac, flat when empty and spherical when full. It may contain as much as 600 ml of urine. It is lined by a transitional epithelium made up of several layers of large polyhedral cells. The basal layer is made up of small cuboidal cells with scanty cytoplasm, while the superficial layer is composed of large, often multinucleated, cells with abundant cytoplasm.

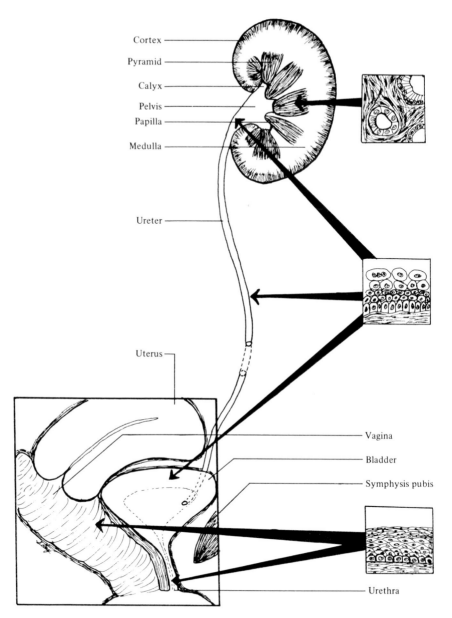

Cortex

Pyramid

Calyx

Pelvis

Papilla

Medulla

Ureter

Uterus

Vagina

Bladder

Symphysis pubis

Urethra

Figure 166
The urinary tract in women.

Around the urethral openings, mucus-secreting glandular epithelium may be found lining the walls of small mucosal crypts.

The length of the *urethra* is 18 cm in men and 3 cm in women. The proximal portion of the mucosa is made of transitional epithelium, except for the lining of the few mucus-producing glands (Cowper's and Littre's glands), and the terminal portion is lined by stratified squamous cells similar to the vaginal cells.

TECHNIQUE FOR COLLECTING SPECIMENS

The number of cells exfoliated from the upper urinary tract is normally scanty. The bladder and urethral transitional cells exfoliate more readily. Controversy exists concerning whether the patient should be hydrated or dehydrated before the collection of the specimen. When the patient is dehydrated, the urine is more concentrated in cells, and a small volume is satisfactory. With hydration, although diluted, the amount of urine and the number of cells is increased. With the use of a membrane filter, dilution is no problem, and preliminary hydration of the patient is strongly recommended.

Exercise prior to collection seems to increase the cellular concentration of the urine. The repeated irrigation of the urinary bladder, ureter, and pelvis with 50 ml of isotonic saline solution after voiding usually yields an excellent specimen, rich in well-preserved cells.

Urine is harmful to the exfoliated cells because of its variable pH and specific gravity. This necessitates rapid processing of the specimen. Urine collected by catheterization is always preferred. Voided urine can be accepted from the male in certain cases, but catheterization is always preferred in the female in order to avoid contamination from the vaginal secretions. The third portion of the urine collected in the morning is richest in cells.

When processing cannot be done rapidly, the cells can be preserved by mixing the urine with an equal volume of 50% alcohol. This process, however, has the inconvenience of precipitating the protein, which, in turn, makes the use of the membrane filter difficult.

Preparation of the Patient

Hydration of the patient is accomplished by administration of one glass of water every 30 minutes for a 3 hour period. Diuretic drugs or intravenous fluid can also be given to increase the urinary output.

Collection of the Specimen

The bladder is catheterized, and the properly labeled urine specimen is sent to the laboratory in ice to prevent cellular lysis. The urine from the pelvis and ureters is collected into tubes labeled *right* and *left*. During the slow withdrawal of the catheters, continuous irrigation of the pelvis and ureters with normal saline solution should be carried out, and this fluid should be added to the urine collected. The specimens are immediately brought to the laboratory or prefixed with an equal volume of 50% alcohol solution.

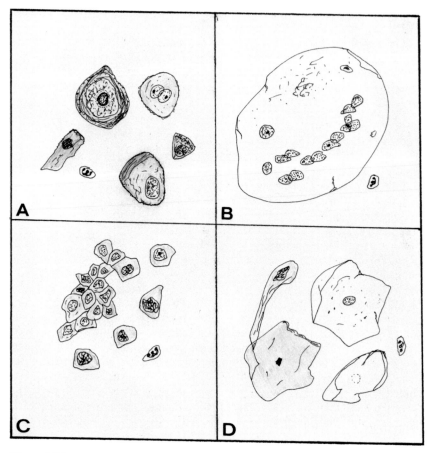

Figure 167
Cells of normal urine. (A) Physiologically exfoliated transitional cells. (B) Multinucleated transitional cells. (C) Traumatically exfoliated transitional cells. (D) Squamous urethral cells.

Processing

CENTRIFUGATION. The urine specimen is spun at 2000 rpm for 10 minutes and smeared on as many slides as possible. The slides may be coated with albumin to aid in the retention of the cells.

FILTRATION. All urine is passed through one or two membrane filters with 5 μ pore openings.

CYTOLOGY OF THE NORMAL URINARY TRACT
Physiologically Exfoliated Superficial Transitional Cells
(Figs. 167A, B, 168B)
These physiologically exfoliated transitional cells from the surface of the bladder and urethral mucosa are usually few in number and shed singly or

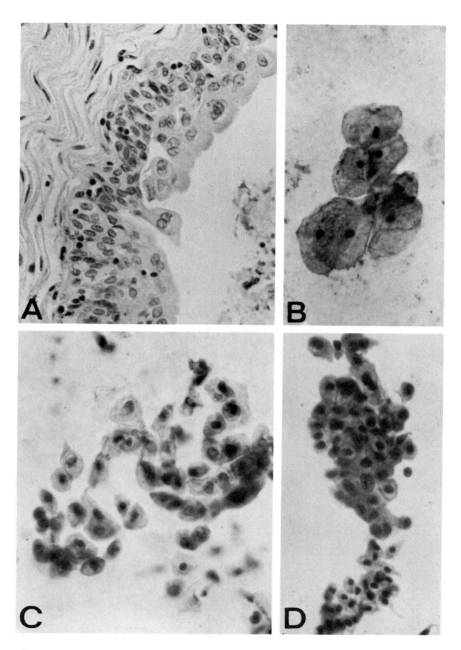

Figure 168
(A) Normal transitional epithelium of the ureter. Note the occasional binucleation. (× 450) (B) Physiologically exfoliated transitional cells in bladder urine.
(× 450) (C) Traumatically exfoliated loose group of transitional cells from
pelvic urine. (× 450) (D) Traumatically exfoliated sheet of normal transitional
cells in a catheterized urine. (× 450)

in clusters (85%), rarely in sheets. There is a marked variation in their size (10–150 μ) and shape (often triangular).

Their abundant cytoplasm is pale and basophilic and is well defined with occasional multiple vacuolization.

Their single or multiple round-to-oval nuclei are uniform in size and shape with finely granular chromatin and sharp borders. Frequent multinucleation (up to 80 nuclei) and giant cell formations (50–300 μ in diameter) may occur in about 10% of the patients with or without neoplasms. Their abundant, frosted-glass-like cytoplasm may contain 2 to 80 central or peripheral, often overlapping, round-to-oval nuclei, variable in size with homogeneous chromatin and occasional nucleoli. These cells are more commonly seen in urethral and pelvic specimens obtained by catheterization (about 15% of the cells lining the convoluted tubules are normally multinucleated). The presence of these multinucleated cells has been related to uremia, chronic acidosis, and administration of laxatives, as well as to various other factors. Other multinucleated giant cells with different features may be caused by radiotherapy, various viral infections, and chemotherapy.

Traumatically Exfoliated Deep Transitional Cells
(Figs. 167C, 168C, D)
The exfoliation of the traumatically exfoliated deep transitional cells is usually abundant as a result of traumatic catheterization, irrigation, or irritation from space-occupying lesions such as calculi or tumors. They are shed singly (20%), in clusters (30%), or in large sheets (50%). The cell size (15 ± 5 μ) and shape are uniform. They are elongated or triangular when single, and polyhedral when in sheets.

The adequate-to-scanty cyanophilic cytoplasm is semitransparent (40%) or deeply stained and dense (60%), with well-defined borders.

Their nuclei are usually single. They vary in size (10 ± 3 μ) but are uniform in shape (oval) and chromatin content. The nuclear membrane is smooth and well defined; the chromatin is coarsely granular but uniformly distributed. Small, prominent nucleoli are often present.

Mild alterations in their shape and size may occur in response to various hormonal stimulations, but in far less degree than with vaginal cells, and they cannot be used for hormonal determination.

It is important to differentiate these sheets of transitional cells from the true pathologic papillae often diagnostic of a papillary neoplasm (Fig. 169). Their folded edges occasionally give the appearance of a peripherally smooth border, but the true papillary projections, in addition to their smooth external border, are characterized by solid molding between the component cells and uniformity of the cellular layers.

Squamous Cells
Variable amounts of intermediate, superficial, or anucleated squamous cells may normally be present. They can originate from a squamous metaplastic area of the bladder mucosa or from the normal epithelium of the distal urethra. They can be difficult to differentiate from contaminant squamous cells desquamating from the vaginal mucosa.

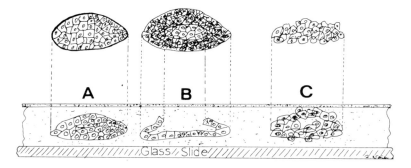

Figure 169
The difference between a true papillary formation and pseudopapillary structures. (A) True papillae. (B) Folded sheet of cells. (C) Cluster of cells.

The epithelial lining of the bladder trigone and of the urethra responds to
the hormonal changes as does the vaginal mucosa. This allows the cells exfoliated from these areas to be used for hormonal evaluation, especially in
cases in which the vaginal smear is too inflammatory for this purpose. The
cells are found mainly in the first few cubic centimeters of a voided urine.

Columnar Cells

Often elongated, some may have terminal plate with attached, short, tufted
cilia. The oval nuclei may have a finely vesicular pale chromatin with a
nucleolus or dark hyperchromatism without nucleoli. The cytoplasm is
translucent. Some mucus-producing cells may also be seen originating from
Littre's glands and periurethral crypts. The significance of the presence of
the cells is controversial.

Inflammatory Cells, Polymorphonuclear Cells,
Monocytes, and Histiocytes

These cells are often present in a normal urine but in small amounts. In
larger amounts they may indicate an inflammation of the urethra (urethritis), bladder (cystitis), pelvis (pyelonephritis), or kidney (nephritis),
or tubular degeneration (nephrosis). Bacteria are scanty in fresh specimens
and abundant in urine in which processing has been delayed. Various fungi
and contaminant yeasts may be seen flourishing in such urine and may become a handicap by obscuring the diagnostic cells.

Glitter Cells

Glitter cells are leukocytes with ingested multiple fine intracytoplasmic
granules showing Brownian movement. The continuous agitation of these
granules is more conspicuous when the urine is hypotonic. Glitter cells are
found more frequently in the urine of patients with kidney inflammation
(pyelonephritis, glomerulonephritis) than with lower urinary tract infection
(cystitis, urethritis).

Spermatozoa

Spermatozoa may occur in small or large numbers, especially after a prostatic massage, prolonged continence, nocturnal emission, or spermatorrhea. They can occasionally be found in the urine of either sex after coitus. Their presence does not need to be reported.

Nonpathologic Urinary Crystals

To see these crystals best, one polarizing filter should be placed in front of the light source and another placed on top of the slide, then rotated until the background becomes dark. The presence of a membrane filter is no handicap since it usually does not polarize.

CALCIUM SULFATE (FIG. 170A). These crystals have no clinical significance. Small transparent needles, they can be found in clusters or scattered singly in an acid urine.

CALCIUM CARBONATE (FIGS. 170B, 172C). These crystals, which occur in alkaline urine, are very small, round or dumbbell-shaped. They have no known significance.

AMORPHOUS URATE AND PHOSPHATE CRYSTALS (FIGS. 170C, 172B). The variable amount of tiny granular crystalline precipitate in the background, pink to purple in color, should be differentiated from cellular debris or protein deposit. Morphologically, urate crystals cannot be distinguished from the phosphate ones. The urate crystals are found in acid urines and dissolve by heating, whereas the phosphate crystals are found in alkaline urines and are dissolved by acidification.

URIC ACID CRYSTALS (FIGS. 170D, E, 173B). These semitransparent, yellow-to-brown, six-sided, starlike crystals may also appear as flat, diamond-shaped or butterfly-like structures. They suggest an elevated serum uric acid, as in gout, or the presence of a uric acid calculus if they are seen in abundance in fresh urine. With a polarized light they may be seen as beautiful rainbow-colored crystals.

AMMONIUM BIURATE CRYSTALS (FIG. 170F). These crystals are spheroid in shape with multiple tail-like structures. They are often seen in urine in which delayed processing has allowed heavy bacterial growth.

BICALCIUM PHOSPHATE (FIG. 170G). These transparent or yellow crystals, often forming starlike structures or single, elongated pyramidal shapes, are soluble in acid. They have no known clinical significance.

TRIPLE PHOSPHATE CRYSTALS (FIG. 170H). Elongated, transparent, often rectangular or diamond-shaped, they suggest an alkaline urine.

CALCIUM OXALATE CRYSTALS (FIG. 170I). They may appear as transparent, diamond-shaped, or small concave, round discs. They may be confused with talcum powder crystals with ordinary light examination, but

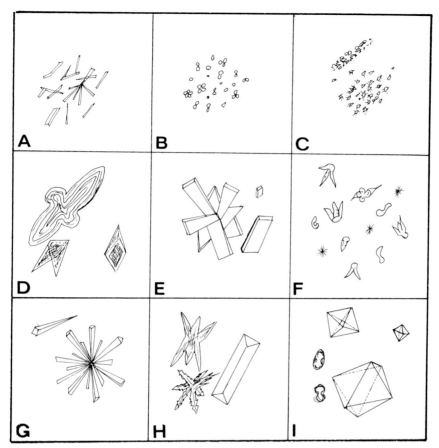

Figure 170

Nonpathologic urinary crystals. (A) Calcium sulfate. (B) Calcium carbonate. (C) Amorphous urate. (D) Uric acid. (E) Uric acid. (F) Ammonium biurate. (G) Dicalcium phosphate. (H) Triple phosphate. (I) Calcium oxalate.

once polarized, they do not show the brilliant crosslike appearance of talcum crystals. Their presence suggests an acid urine. They are seen also after the patient has eaten foodstuffs containing large amounts of oxalate, such as tomatoes, spinach, or apples.

INFLAMMATORY LESIONS OF THE URINARY TRACT

In acute or chronic inflammation of the bladder (cystitis), ureter (ureteritis), or pelvis and kidney (pyelonephritis), and after intravenous or retrograde pyelogram, the urine may show the following cells.

Hyperactive Atypical Transitional Cells

They are shed usually in abundance, singly, in clusters, or in sheets, and show marked variation in size (15–40 μ) and shape.

Their granular or vacuolated cytoplasm is adequate to scanty, staining

373

blue to purple-orange, with marked distortion of shape. The cytoplasmic borders are poorly defined as the result of partial degeneration. Occasionally, leukocytes containing vacuoles may be present.

Their nuclei are hypertrophic and irregular in shape (wrinkled nuclear membrane) with moderate hyperchromatism. The chromatin is coarsely granular with clumping. The nuclei-cytoplasm ratio is elevated. Binucleation and multinucleation are frequent, with occasional extreme increase in the number of nuclei (50+).

Some of the cells show pronounced cellular degeneration with fuzzy nuclear pyknosis and cytoplasmic eosinophilia, not to be confused with the sharp, irregular pyknotic nuclei and keratinized cytoplasm of the squamous cell carcinoma.

Multiple intracytoplasmic pink-to-red inclusion bodies may be seen; they are most probably related to cellular degeneration. Increased amounts of debris, crystals, casts, and leukocytes are also usually seen.

Degenerative Transitional Cells

After inflammation or trauma (e.g., retrograde pyelogram) various numbers of degenerate transitional cells may be seen. Some are without nuclei (karyolysis), with nuclear shrinkage and opacification (karyopyknosis) or with broken nuclei (karyorrhexis). The degeneration of some of the young parabasal transitional cells may produce a dark condensation of their chromatin, mainly toward the nuclear membrane (decoy cells), and mimic hyperchromatic malignant cells.

Inflammatory Cells

Variable numbers of often degenerate neutrophils mixed with monocytes are usually present. These inflammatory cells may obscure the diagnostic transitional cells and necessitate a repeat urine specimen after antiinflammatory therapy.

Fresh and Old Red Blood Cells

The presence of intact red blood cells (round or crenate) or of debris or red blood cells in urine (hematuria) is of clinical significance, especially in a voided urine specimen, and should be reported. Some of them are found ingested in the cytoplasm of mononucleated or multinucleated phagocytes, indicating a prolonged hematuria. Intact red blood cells are sometimes confused with pink-stained, round yeast cells. The contaminant yeast cells will show some budding, variation in size and no crenate forms.

Microorganisms, Yeast, Fungi (Fig. 177C)

Large numbers of microorganisms, yeasts, or fungi may be present, especially in urinary bladder retention or when the specimen is left standing for too long a time without proper preservation. If no inflammatory reaction is present, these colonies of living organisms have no pathologic significance.

Pathologic Urinary Crystals

CHOLESTEROL (FIGS. 171A, 173C). Cholesterol crystals are seen as flat, transparent structures with irregular edges, resembling fragments of broken

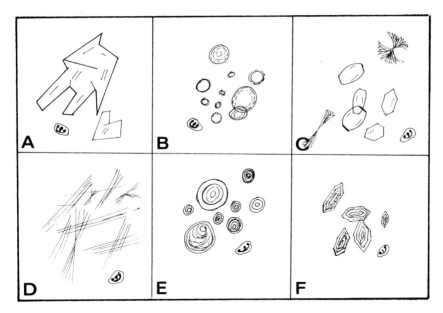

Figure 171
Pathologic urinary crystals. (A) Cholesterol. (B) Fat. (C) Sulfonamide. (D) Tyrosine. (E) Leucine. (F) Cystine.

glass with notches. Their presence in the urine may indicate a nephrosis, and they should be reported.

FAT (FIG. 171B). The presence of multiple and scattered spherical, highly refractile globules stained blue to purple-red, with marked size variation, may indicate a nephrotic syndrome. The cytoplasm of tubular epithelial cells may be seen filled with fat droplets. Some are doubly refractile and shine with a polarized light.

TYROSINE (FIG. 171D) is demonstrated by the presence of fine, needle-like transparent structures in bundles that are often diagnostic of advanced liver disease.

LEUCINE (FIGS. 171E, 172D) is occasionally present as yellowish-brown glob-ules with an apparent concentric ring and radial striations. These also may indicate a liver damage, such as that in acute yellow atrophy. They can be dif-ferentiated from similar-looking crystals of acid sodium urate and of acetyl derivative or sulfonamides by heating the urine to the boiling point. Only the leucine crystals will dissolve.

CYSTINE AND DYCYSTINE (FIGS. 171F, 172A, 173D) may appear as trans-parent, refractile crystals, rhombic in shape and variable in size. They are often seen in clusters and may indicate the presence of a cystine calculus or some metabolic disorder associated with aminoaciduria. They should be differentiated from sometimes similar-looking uric acid crystals.

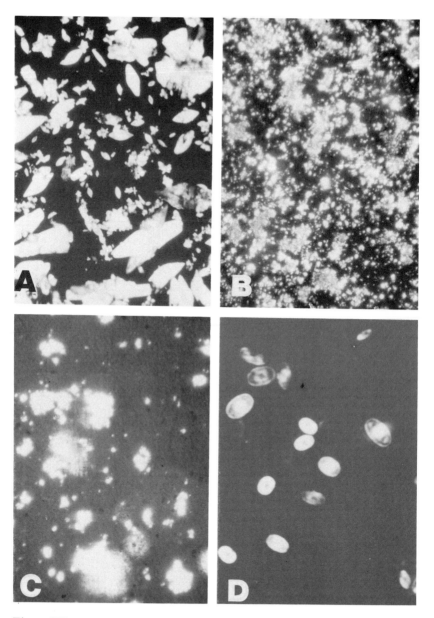

Figure 172
Urinary crystals seen with a polarized light. (A) Cystine. (B) Amorphous urate. (C) Calcium carbonate. (D) Leucine.

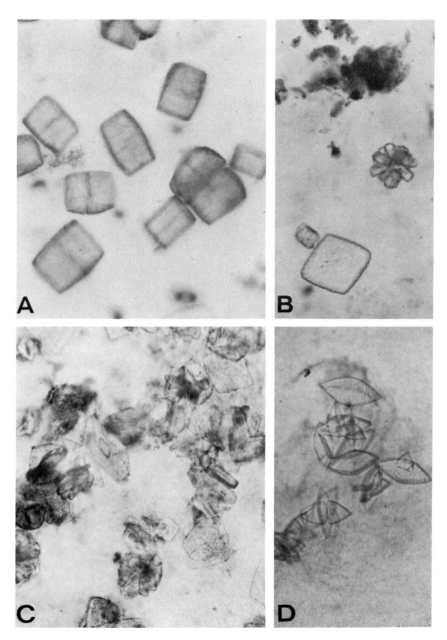

Figure 173
(A) Sulfonamide crystals. (× 450) (B) Uric acid crystals. (× 450) (C) Cholesterol crystals. (× 450) (D) Cystine crystals. (× 450)

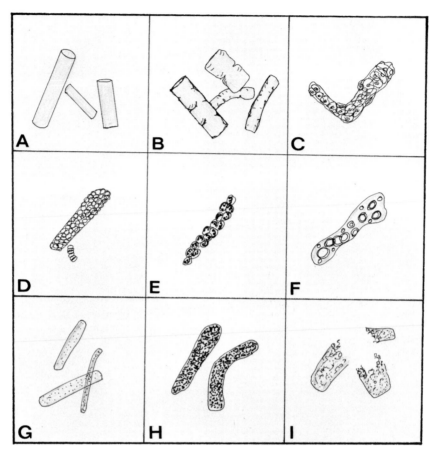

Figure 174
Urinary casts. (A) Hyalin. (B) Waxy. (C) Epithelial. (D) Blood. (E) Pus.
(F) Lipid. (G) Finely granular. (H) Coarsely granular. (I) Diabetic coma
cast.

Urinary Casts

HYALINE CASTS (FIGS. 174A, 175A) appear as homogeneous, semitransparent, cylindrical structures with parallel sides and blunt ends and a high
refractive index. Generally straight, they vary in size. They may indicate
renal damage with a low degree of morbidity. They are associated with
many conditions producing proteinuria.

WAXY CASTS (FIG. 174B). These highly refractile, yellow-to-transparent,
often short casts have sharp, square edges and notches on their sides. They
are homogeneous and in appearance resemble dull ground glass. They may
indicate an advanced nephritis or amyloidosis and are usually found in
uremic patients. They differ from hyaline casts by their higher density and
greater refractive index.

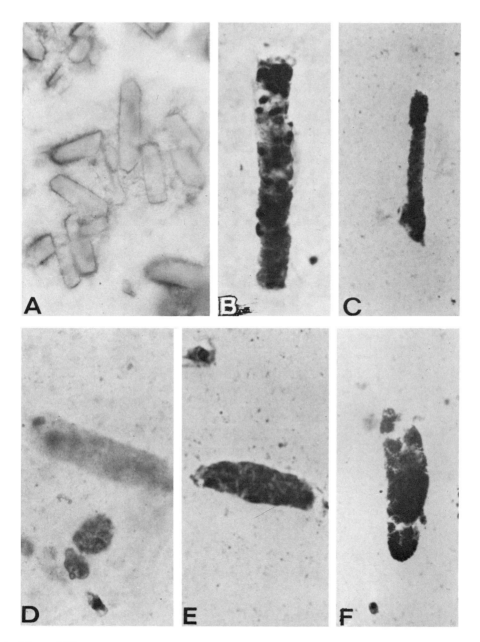

Figure 175
Urinary casts. (A) Hyalin. ($\times$ 450) (B) Epithelial. ($\times$ 450) (C) Coarsely granular. ($\times$ 450) (D) Finely granular. ($\times$ 450) (E) Inflammatory. ($\times$ 450) (F) Diabetic coma cast. ($\times$ 450)

EPITHELIAL CASTS (FIGS. 174c, 175b) are made up of clusters of small, desquamated, tubular cells with scanty cytoplasm and pyknotic nuclei. They originate from the collecting tubule epithelium. They are present in severe nephritis or nephrosis.

BLOOD CASTS (FIG. 174d) appear as red-to-brown cylinders containing fresh and old debris or intact red blood cells. They may indicate severe renal inflammatory disease (acute glomerulonephritis, lupus nephritis) or intrarenal bleeding (trauma, infarction).

INFLAMMATORY CELL (PUS) CASTS (FIGS. 174e, 175e) are made up of clusters of relatively well preserved inflammatory cells, polymorphonuclear cells or monocytes in a hyaline matrix. They may indicate the existence of an acute or chronic pyelonephritis. Occasionally, they may be seen in a noninfectious inflammatory process, as in certain collagen diseases.

LIPID (FATTY) CASTS (FIG. 174f) are brown to red and contain multiple clusters of highly refractile fat globules of variable sizes. They are present in advanced nephrosis. A sudanophilic stain of the smears will confirm the diagnosis by staining the lipid droplets red.

FINELY GRANULAR CASTS (FIGS. 174g, 175d) are often seen at the same time as the hyaline casts and indicate mild nonspecific kidney damage.

COARSELY GRANULAR (BROAD) CASTS (FIGS. 174h, 175c) are probably formed of degenerate and necrotic pus cells and may indicate the presence of a chronic pyelonephritis. They are often found in uremia.

DIABETIC COMA CASTS (FIGS. 174i, 175f).   On occasion, numerous short, mainly granular cylinders with round, irregularly broken granules on one side are found in the urine of patients in diabetic coma.

PSEUDOCASTS.   Elongated strands of mucus, often present in the background of the urine sediment smear, should not be mistaken for casts. The irregularity of thickness and their curled, tapered ends are distinctive. Similarly, hair or cotton fibers may contaminate the specimen and should not be mistaken for casts.

Parasites (Figs. 176, 177)

TRICHOMONAS is occasionally a normal inhabitant of the urethra in both male and female, or the organisms can be a contaminant from the vagina or from prostatic fluids. They may produce in the urethral cells the same cellular changes as those found in the vaginal cells of a patient with an acute genital *Trichomonas* infection (see Chap. 4).

AMOEBA (FIG. 176b).   These protozoans are round to oval in shape, measure 20 to 30 μ in diameter and show two cytoplasmic zones, one

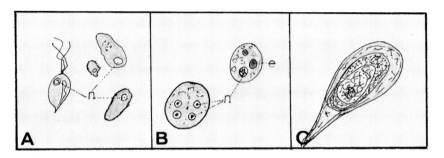

Figure 176
Different parasites that may be encountered in the urine. (A) *Trichomonas.*
(B) *Amoeba histolytica.* (C) *Schistosoma haematobium.* (n = nucleus, e =
erythrocyte)

amorphous and transparent at the periphery (exoplasm) and the other
granular and denser toward the center, with occasional ingested debris and
intact red blood cells (endoplasm). One to 4 small, round, gray central
nuclei must be clearly seen before the diagnosis is made. In the urine,
amebic parasites are usually contaminants from the colonic content.

SCHISTOSOMA HEMATOBIUM (BILHARZIA HEMATOBIUM) (FIG. 176C).   The
ova are characterized by their elongated oval shape, large nuclei, and transparent highly refractile shell with a pointed tail-like terminal spur. Large
granular eosinophilic miracidia occupy the largest parts of the center of
the viable eggs. This parasite is commonly found in the urine of the in-
habitants of certain countries of the Middle East and South America, and
it is associated with a greater incidence of bladder squamous cell carcinoma. The urine specimen contains much blood and occasional ova, especially if taken between 10 A.M. and noon, during which period the ova
excretion reaches its peak.

SCHISTOSOMA MANSONI (FIG. 177A, B).   This parasite has a lateral sharp
spur that makes it easy to recognize. They are found in urine as well as in
rectal and vaginal smears. Degenerated ova and empty egg shells are often
found scattered throughout the smear. Large numbers of neutrophils,
eosinophils, and leukocytes are usually present. Histiocytes and multinucleated giant cells may be present.

Cells with Inclusions

CYTOMEGALIC VIRAL INFECTION (FIGS. 178A, 179A, B).   This viral infection is characterized by the presence of large, round basophilic inclusions
surrounded by a prominent halo in the enlarged nucleus of the purple-
stained, usually single, enlarged, transitional, mainly tubular cell (5–50 μ in
diameter). The coarse chromatin is compressed next to the inner nuclear
membrane surface. No more than one or two such infected cells are usually
found in one urine specimen. Careful screening under high-power magnifi-

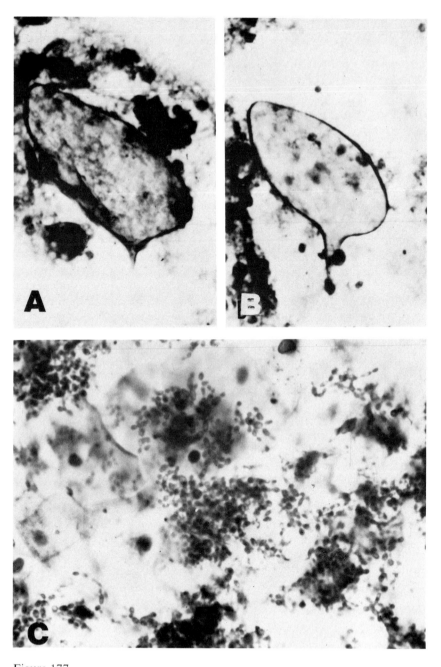

Figure 177
(A) *Schistosoma mansoni* in a urine specimen. Note the lateral spur. ($\times$ 250)
(B) Shell of a *Schistosoma mansoni*. ($\times$ 250) (C) *Rhodotorula* spore, a contaminant fungus, growing in a urine specimen whose processing was delayed.
($\times$ 250)

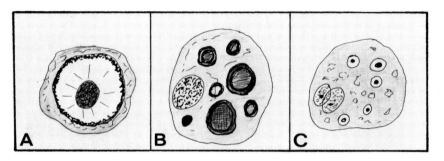

Figure 178
Different cellular inclusions encountered in the urine. (A) Cytomegalic viral inclusion. (B) Malacoplakia inclusion. (C) Phagocytic or degenerative inclusion.

cation and repeated urine specimens are strongly recommended before ruling out the disease.

The urine of patients receiving immunosuppressive drug therapy for renal allograft should be carefully screened for this disease. The changes must be differentiated from those of other viral diseases (e.g., herpes, chickenpox, and adenovirus).

HERPESVIRUS INFECTION.   Infection of the urethra and bladder mucosa with genital herpesvirus is more common than is generally thought. The presence in the urine of single or large multinucleated giant cells with inclusion-bearing nuclei is diagnostic. The main cytologic difference between infected transitional cells and the similarly infected skin or vaginal cells seems to be in the looseness and minimal molding of these nuclei. The granularity of these inclusions and their abundance aid in their differentiation from cytomegalovirus infection.

MALACOPLAKIA (FIG. 178B).   The cause of malacoplakia, a chronic inflammatory disease of women over the age of 30, is unknown. It has been associated with bacterial infections. Multiple basophilic or round PAS-positive cytoplasmic inclusions may be found in large, mononucleated transitional cells or macrophages. These inclusions may be laminated. They vary in size (1–7 μ). Although rare, they are indicative of the disease, which can produce urethral obstruction if undetected and untreated. Large numbers of macrophages are usually present.

PHAGOCYTIC OR DEGENERATIVE INCLUSIONS (FIGS. 178C, 179D).   Multiple small (7–15 μ), round eosinophilic bodies surrounded by small halos may be found in the cytoplasm of normal, atypical, or malignant transitional cells. Their significance is doubtful as they are most probably the retained by-product of cellular metabolism. The same structures can be present in degenerating cells found in effusions and gastric, duodenal, and colonic washings.

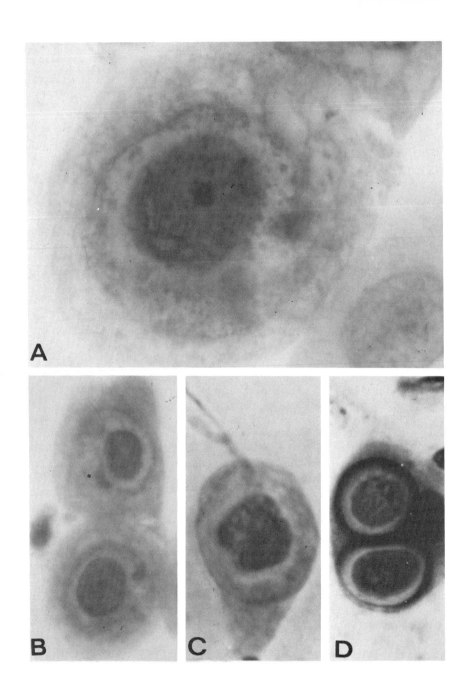

Large, single or multiple intracytoplasmic gray-to-blue inclusions may be found in multinucleated giant cells. They are thought to be made of acid polysaccharide and to represent abnormal changes and aggregation of mucin. They are found more commonly in renal pelvis and urethral urine. They have also been related to the purgative medication that most patients receive before a retrograde pyelogram. No apparent association with viral or neoplastic diseases has been found.

Renal Transplantation

In the first two weeks following renal transplantation, small to moderate numbers of large tubular columnar cells with vesicular oval nuclei, prominent nucleoli, and delicate vacuolated cytoplasm are normally seen. If present in large amounts, these cells may indicate the beginning of the rejection of the graft.

During the recovery phase, these cells are seldom seen if the graft is successful. Because of the various potent drugs administered to these patients to prevent the rejection of the graft, their resistance to various infections is decreased. Their chances of acquiring cytomegalovirus and other viral infections are increased for several months after the renal graft, and the urine should be carefully screened for the detection of such infection.

Rejection of the graft should be suspected if there is an increase of fresh and old red blood cells, lymphocytes, phagocytes, tubular cells, protein deposits, or clustering of the exfoliated cells.

Postpyelography

For about three to four days following an intravenous retrograde pyelogram, an increased number of small sheets of slightly atypical cells may be seen exfoliated into the urine. Some dye may be found engulfed in the cytoplasm of large transitional cells in the form of large gray, poorly stained, round intracytoplasmic inclusions. Most of the free dye may be seen in spherical globules, slightly yellow or colorless, lying on top of the filter membrane. A slight increase in histiocytes and foreign-body giant cells may be seen in the urine in some cases.

Irradiation Changes

After roentgen or radium therapy, for urologic as well as nonurologic pelvic diseases, an irradiation cystitis may develop. The exfoliated cells show various degrees of typical irradiation changes, similar to those encountered in

---

Figure 179
(A) High-power magnification view of an exfoliated transitional cell infected with cytomegalic inclusion virus. Note the central granular inclusion surrounded by an irregular halo, the chromatin and nuclear membrane, and the distinct vacuolated cytoplasm. ($\times$ 2100) (B) Lower-power magnification view of the same cell. ($\times$ 1300) (C) Degenerate exfoliated transitional cell with pyknotic nucleus and perinuclear halo, not to be confused with the inclusion-bearing cells. ($\times$ 1300) (D) Histiocytes in urine with ingested, intact red blood cells, not to be confused with viral inclusions. ($\times$ 1300)

gynecologic material. The transitional cells enlarge to three to four times their normal size. The cytoplasm becomes vacuolated. The hypertrophic nucleus appears hyperchromatic. These changes cause the exfoliated cells to resemble the anaplastic carcinomatous cell, with which they could be confused except for the chromatin that remains finely granular and uniformly distributed. Large amounts of mixed inflammatory cells, debris, and protein deposit are usually found in the background.

Similar changes may be seen after the administration of Cytoxan (cyclophosphamide) or other chemotherapeutic drugs. Myleran therapy may also produce changes similar to those in dysplasia or in situ carcinoma. Most of these irradiation changes begin to regress six weeks after radiotherapy and are seldom seen longer than two months afterward.

Benign Dysplastic Changes
Squamous metaplastic cells and dyskaryotic transitional cells may be found in cases of calculi, prostatic hypertrophy with urinary retention or in any acute or chronic lower urinary tract inflammation. Large reactive mucus-secreting and squamous dysplastic cells are often found in interstitial and glandular cystitis and may be the cause of false-positive cytologic interpretation.

NEOPLASTIC LESIONS (TABLE 17)
A majority of the neoplasms are transitional cell type (90% of bladder carcinomas) and may arise from the urethra, urinary bladder, the ureter, or the pelvic mucosa. Exfoliative cytology cannot, by itself, pinpoint the exact location of a tumor. The number of exfoliated neoplastic cells and the degree of malignant features increase with the histologic grading of the cancer.

Simple Papilloma (Figs. 180B, 181A)
These are transitional tumors with a wide range of differentiation; they are composed of cells with little or no atypia. They are found mainly toward the trigon area of the bladder. They can be sessile or pedunculated. Although structurally considered benign by the pathologist, they are treated as malignant by the clinician because of the ease of their seeding in the urinary mucosa and the frequent recurrence after their removal. Diagnosis by cytology, which relies on cellular and not architectural features, is difficult because of the benign appearance of their exfoliated cells (50% false-negatives).

With simple papilloma there is increased exfoliation of small transitional cells, which are often degenerate. They are found singly (20%), in small clumps (40%), or in large papillary sheets (40%). The cells are fairly uniform in size ($15 \pm 3 \mu$), but they may vary in shape (triangular 20%, spindle 30%, tadpole 10%, and irregular 40%).

The cytoplasm is scanty to adequate and often stretched. The presence of long, tail-like cytoplasmic processes is often indicative of the lesion. Degenerative vacuoles are often seen. There is no keratinization, nor are inclusions present.

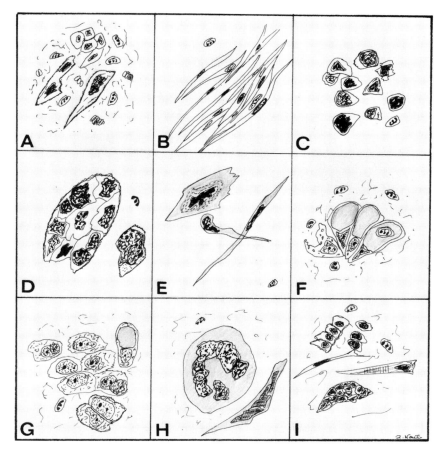

Figure 180
Exfoliated cells in the urine from various malignant neoplasms of the urinary
tract. (A) Intraepithelial papillary carcinoma. (B) Simple papilloma. (C) Nonpapillary in situ carcinoma. (D) Transitional cell carcinoma. (E) Squamous
cell carcinoma. (F) Mucinous adenocarcinoma. (G) Renal adenocarcinoma.
(H) Fibrosarcoma. (I) Wilms' tumor.

The nucleus is usually single, small, uniformly hyperchromatic, regular
in size ($8 \pm 3 \mu$), and often elongated.

The background of the smear may show a moderate increase of inflammatory cells mixed with blood, cellular debris, and protein deposits, especially if the neoplasm is ulcerated (40%).

Papillary in Situ Carcinoma (Figs. 180A, 181B)

By cytology, this tumor is difficult to differentiate from the simple papilloma. The single cell of the intraepithelial papillary carcinoma may be very
difficult to differentiate from a normal but degenerate transitional cell. The
cells may shed in sheets or in papillary or palisade arrangement (85%),

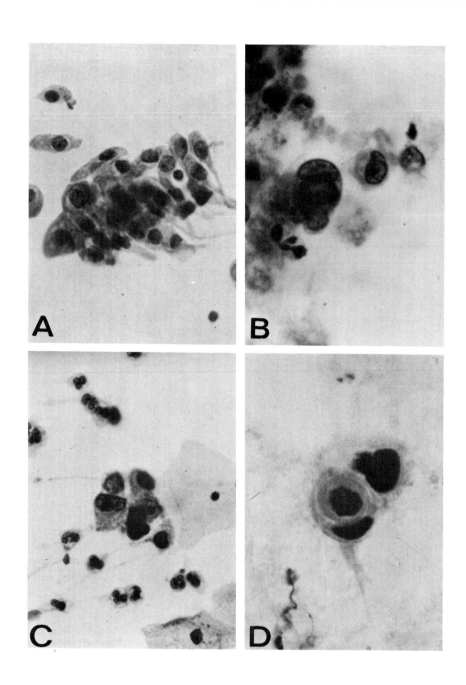

with smooth, common external borders. The individual cell shows only moderate variation in shape (round to elongated). The size usually is uniform, being slightly larger than the normal transitional cell ($14 \pm 4 \mu$).

The cytoplasm is adequate to scanty, transparent or slightly basophilic, often well preserved, and has distinct, sharp cellular borders. Inclusions are usually not found, nor is there vacuolization. The nuclei-cytoplasm ratio is almost normal.

The single, mostly round nucleus is uniform in size and slightly enlarged ($9 \pm 3 \mu$). The nuclei are hyperchromatic with coarsely granular or abnormally clumped chromatin pattern, and with frequent pyknosis. The nuclear membrane is heavy and occasionally irregular (30%) and touches the cytoplasmic membrane at several points. No prominent nucleoli can be seen.

The background of the urine smear is moderately clean with little protein or debris deposit, except in the case of a secondary infection or ulceration.

Nonpapillary in Situ Carcinoma (Figs. 180C, 181C)

This tumor is less common than the papillary one; often it is a recurrence of the treated primary bladder carcinoma. Large numbers of malignant cells are shed singly (40%) or in loose clusters (60%). They resemble the dysplastic basal cells described in the vaginal smear. The cells are relatively uniform in size ($20 \pm 5 \mu$) and shape—round (60%), pear-shaped (20%), or irregular (20%).

The cytoplasm is often adequate, opaque, and deeply basophilic with variable numbers of degenerative vacuoles. The cytoplasmic border is usually poorly defined. The nuclei-cytoplasm ratio remains almost normal.

The nucleus is hyperchromatic with coarse chromatin that is irregularly clumped and occasionally pyknotic. Although the nuclei are usually single (85%), the cells can be multinucleated or binucleated. Their nuclear membrane is well defined and heavy with occasional indentations and marked variation of thickness. No prominent nucleoli can be seen.

The background of the smear is usually clean with few inflammatory cells, red blood cells or cellular debris unless the tumor is ulcerated.

Cells of nonpapillary in situ carcinoma should be differentiated from the ones shedding from atypical reactive transitional mucosa, resulting from repeated catheterizations, treatment, or chronic cystitis (cystitis glandularis).

Figure 181

(A) Cluster of elongated transitional cells in the urine of a patient with bladder papilloma. Note the benign appearance of the cells. ($\times$ 450) (B) Atypical transitional cells in the urine of a patient with an intraepithelial papillary carcinoma. ($\times$ 450) (C) Tumor cells seen in the urine of a man with hematuria but with no visible lesion. Random biopsy of the bladder mucosa showed an extensive in situ carcinoma. ($\times$ 450) (D) Cluster of tumor cells in the urine of a woman with multiple papillary carcinoma of the kidney pelvis and urethra. ($\times$ 450)

Table 17. Differential Cytologic Features of Urinary Tract Carcinomas

Criteria	Papilloma	Papillary in Situ	Nonpapillary in Situ	Transitional Carcinoma	Adenocarcinoma	Squamous Carcinoma
Size	15 ± 3 μ	14 ± 4 μ	20 ± 5 μ	30 ± 15 μ	65 ± 15 μ	40 ± 20 μ
Shape	Spindle 60%	Irregular	Round 60%	Irregular	Round to oval	Irregular
Occurrence	Sheets 40% Clump 40% Single 20%	Sheets 85%	Single 40% Cluster 60%	Single 40% Papillary projections 60%	Single 30% Cluster 70%	Single 40% Sheets 60%
Amount of cytoplasm	Adequate	Adequate	Scanty	Scanty	Adequate to abundant	Adequate to abundant
Cytoplasmic stain	Pale	Acidophilic	Basophilic	Basophilic	Pale	Orange
Cytoplasmic vacuolization	Degenerative	None	None	None	Large	None
Nuclei-cytoplasm ratio	2:10	2:10	2:10	7:10	4:10	Variable
Nuclear size	2 ± 3 μ	9 ± 3 μ	10 ± 5 μ	15 ± 10 μ	15 ± 2 μ	4–100 μ
Nuclear shape	Elongated	Irregular	Irregular	Irregular	Oval	Irregular
Chromatin pattern	Regular hyperchromatic	Hyperchromatic pyknotic	Irregular	Markedly irregular	Finely granular	Pyknotic
Multinucleation	2%	1%	5%	15%	60%	20%
Nucleoli	2%	0%	0%	40%	90%	15%
Number of diagnostic cells	Variable	Variable	Abundant	Variable	Variable	Variable
Tumor diathesis	Mild	Mild	Mild	Pronounced	Pronounced	Pronounced

Invasive Transitional Cell Carcinoma (Figs. 180D, 181D)

With different invasive transitional cell carcinomas, the number of exfoliated malignant cells may vary. They can be scanty or very profuse. They are shed singly (40%), in sheets (20%), or in papillary projections (40%). Their size and shape vary. Often they are elongated. They show variation in the degree of their differentiation.

The dense and opaque cytoplasm is scanty (70%) or adequate. Usually the cytoplasm is well preserved and has well-defined borders. The cytoplasmic and nuclear membranes often touch each other at more than one point.

The nuclei vary in size ($15 \pm 7 \mu$). Often round, they can be very irregular in shape. Usually hyperchromatic, the chromatin is coarsely clumped, often with markedly irregular pointed projections. The nuclear membrane is well defined and irregular. The nucleoli vary in size, shape, and number and may be irregularly prominent in about 40% of the cells.

The background of the smear usually shows large amounts of cellular debris, inflammatory cells, protein deposit, and fresh and old red blood cells.

The presence of abnormal papillary projections with a smooth external border and the more striking malignant changes of the nuclei and cytoplasm of the individual cells indicates a papillary transitional carcinoma.

The undifferentiated tumor cells exfoliate mainly singly or in loose clusters, and they have a very scanty ill-defined cytoplasm. Their nuclei are very hyperchromatic and show marked variation in size. The nucleoli are not as prominent as in well-differentiated carcinoma.

Squamous Cell Carcinoma (Figs. 180E, 182C)

Squamous cell carcinoma can arise from the bladder, ureter, or pelvis. Some may originate as transitional cell carcinomas that later differentiate into a squamous type. In certain countries of the Middle East and South America, the squamous carcinoma of the bladder is associated with *Schistosoma* infection. Others are metastatic or secondary extensions from adjacent organs (cervix). Grossly they are usually nonpapillary tumors arising from the anterior portion of the bladder. The criteria for malignancy of the exfoliated cells are the same as for the squamous cell carcinomas seen in the genital or respiratory tracts.

The exfoliated cells are generally edematous with an apparent decrease in density. They may shed in sheets, in pearl formations or singly, with tadpole, fiber, or snake shapes.

The abundant, distorted cytoplasm stains irregularly orange with keratohyaline rings. The nucleus is hyperchromatic and often pyknotic and irregular in shape and size.

The background of the smear contains a large amount of cellular debris and inflammatory cells. The diagnosis of these tumors is relatively easy to make from the obvious malignant features of their exfoliated cells.

Mucinous Adenocarcinoma (Figs. 180F, 182A)

Mucinous adenocarcinomas of the transitional epithelium are very rare tumors that arise in the bladder, and occasionally in the pelvis, from glandu-

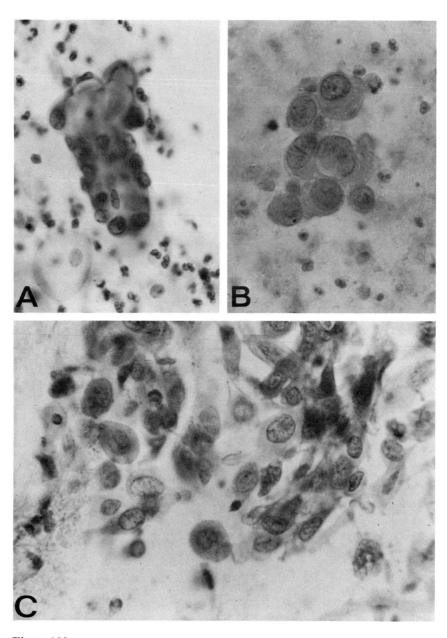

Figure 182
(A) Cluster of tumor cells in the urine of a patient with a bladder adenocarcinoma. Note the papillary structure and secretory vacuoles. ($\times$ 450) (B)
Cluster of tumor cells in the catheterized pelvic urine of a woman with a renal
adenocarcinoma. Note the bland nuclei and occasional nucleoli ($\times$ 450) (C)
Numerous cancer cells in the voided urine of a man with epidermoid carcinoma. Note the polymorphism of the cells. ($\times$ 450)

lar metaplastic or ectopic cells or from intraepithelial mucous cysts (Littre's glands). They shed diagnostic cells singly (30%) or in small clusters (70%). Their size varies (50–80 μ). They often appear as overlapping, irregular, tall columnar or round cells with hypervacuolated cytoplasm containing mucin that is being ingested by a variable number of well-preserved leukocytes.

The nuclei are large (12–18 μ), mildly hyperchromatic and round or oval in shape, with coarsely granular chromatin and prominent irregular nuclei, varying in size, shape, and number. A periodic acid–Schiff stain helps in their recognition by staining the intracytoplasmic mucus red.

Large amounts of inflammatory lymphocytes and neutrophils with protein deposits, mucus, and debris are present in the background.

Renal Adenocarcinoma—Hypernephroma or Clear Cell Carcinoma (Figs. 180G, 182B)

Renal adenocarcinomas exfoliate cells into the urine only in an advanced stage when the pelvis has been invaded. The detection rate by exfoliative cytology is only about 15%.

They may shed few or numerous cells, singly or in clusters. They are large (40 ± 5 μ), and fairly uniform in size and shape.

The delicate and abundant cytoplasm of the clear carcinoma cells may contain several vacuoles, varying in size but usually smaller than the vacuoles of the mucinous adenocarcinoma.

Their enlarged, pale nuclei (15 ± 2 μ) may vary in number (1 to 3). They are often oval or polygonal in shape with minimal irregularities. Their chromatin is finely granular with well-defined borders. Occasionally (2%), a marked polymorphism of the cells is seen with giant multinucleated cells or with large multilobular nuclei. The nucleolus is single or multiple, spherical, and often enlarged.

These cells are sometimes difficult to differentiate from the large, degenerate, reactive transitional cells originating from the bladder or ureter, which may also show a similar cytoplasmic vacuolization and enlarged nuclei. The moderate irregularity of the chromatin pattern, the prominence of the nucleoli, and the moderate degenerative changes of the malignant cells help in their differentiation.

Their recognition is improved with the aid of a fat stain (oil red–O). Bright red granules in the cytoplasm of suspected cells are diagnostic.

Sarcomas—Fibrosarcomas (Fig. 180H)

Sarcomas can originate from the kidney, pelvis, or bladder. They may appear clinically as a large mass protruding into the pelvic or bladder lumen. Often ulcerated, they shed numerous multinucleated giant cells, varying enormously in size (30–100 μ) and each containing 2 to 5 large nuclei. The cells are found singly or in loose clumps.

They have a scanty, thick, uniform, darkish-brown cytoplasm. No cytoplasmic inclusions are usually seen.

The nuclei vary enormously in size, shape, and chromatin pattern. The nuclear membranes are irregular in thickness and occasionally have large irregularly shaped nucleoli. These irregularities differentiate the cells from the multinucleated benign transitional cells normally seen in a urine specimen.

The background of the smear is cluttered by granules of protein, cellular debris, fresh and old red blood cells, and well preserved inflammatory and transitional epithelial cells.

Wilms' Tumor (Nephroblastoma) (Fig. 180I)

This poorly differentiated embryonal tumor of the kidney occurs in children and does not shed cells until its late stage. Single or clusters of poorly differentiated anaplastic, small tumor cells with scanty cytoplasm and hyperchromatic, single, oval, or elongated nuclei are diagnostic in the few cases reported in the literature.

Secondary Tumors

Secondary tumors are mainly local extensions of uterine, prostatic, vaginal, large bowel, or adjacent lymph node neoplasms. The structure of the cells is the same as in the primary tumor. It is often difficult to decide if the presence of tumor cells in a urine specimen indicates extension or contamination from the cancer of an adjacent organ. For example, the presence of squamous cancer cells in the urine of a woman with known cervical cancer may indicate an extension of the tumor to the bladder or contamination from the vagina because of a faulty technique (noncatheterized urine, for example). Large amounts of inflammatory cells and necrotic debris are usually present in bladder tumor extension. The majority of metastases to the upper urinary tract originate from pancreatic or pulmonary tumors and seldom exfoliate.

Lymphomas

Lymphomas can be primary or secondary (rare). Unusually large numbers of pleomorphic immature lymphoblasts with prominent nucleoli are diagnostic.

PITFALLS

False-Positive

Cells with large dark nuclei, originating from the epithelial lining of the seminal vesicles, may be mistaken for poorly differentiated squamous cell cancer. The presence of lipochrome intracytoplasmic pigment, dense but homogeneous nuclei, and large numbers of spermatozoa should help in the differential diagnosis.

Irradiation and chemotherapy changes simulate carcinomatous changes.

Degeneration with pyknosis of nuclei resulting from a delay of fixation of the cells is a common pitfall.

Presence of the naked nuclei of irritated transitional cells with nuclear hyperchromatism may be confusing. (No diagnosis should be based on naked nuclei alone.)

Secondary metaplasia and leukoplakia in urinary lithiasis exfoliates abnormal keratinized squamous cells.

Prostatic hyperplasia or vesicle neck contracture may produce cells with large, atypical nuclei.

Some of the normal but hyperchromatic cuboidal ureteral lining cells and the multinucleated epithelial giant cells may be misinterpreted.

Changes in prostatic and urinary bladder cells of patients receiving various hormones may mimic metaplastic and dysplastic cells.

Acute or chronic inflammation (pyelonephritis, tuberculosis, cystitis) may produce atypical reactive features in the benign exfoliated transitional cells.

False-Negative

Poor technique may result in the loss of the cells during filtration, centrifugation, or staining.

Cellular degeneration due to a delay of fixation of the urine specimen or to surface necrosis of the tumor, especially after irradiation makes their recognition more difficult.

Low-grade transitional papillary carcinoma cells may show very few malignant criteria and may be confused with normal or hyperplastic transitional cells.

Large amounts of inflammatory cells, fresh or old blood, cellular debris, or colonies of microorganisms may obscure the diagnostic cells.

Lack of communication between the tumor and the pelvic or bladder cavities, as in hypernephroma, Wilm's tumor, or some metastatic tumors, prevents the exfoliation of diagnostic cells into the urine.

Cytology of the Prostate and the Seminal Vesicles

PROSTATE

The size and shape of this glandular organ, which surrounds the urethra near the neck of the bladder, vary in relation to the action of different hormones and the age of the patient. The individual folded glands are encased in abundant smooth-muscle and connective-tissue fibers. The ductal mucosa is made up of tall secretory columnar cells and small, deep, basal cells, except shortly before the ducts enter the urethra, where the epithelium changes to a transitional type. The prostate produces a thin whitish secretion thought to be the vehicle used by the spermatozoa after its ejaculation. This secretion contains an acid phosphatase enzyme.

SEMINAL VESICLES

The seminal vesicles contain coiled and convoluted tubes that open into the prostatic urethra. Their size and shape vary in relation to the action of different hormones and the age of the patient. Their mucosa is made of cylindrical columnar cells and small deep basal cells containing intracytoplasmic yellow-brown pigments. These vesicles produce a thick yellow secretion used as a vehicle and nutritive material for the spermatozoa after their ejaculation.

TECHNIQUES FOR COLLECTING SPECIMENS (FIG. 183)

The prostatic secretions are collected by a firm, prolonged massage (more than 3 minutes) of the entire gland, starting from the periphery and repeatedly converging toward the medial portion. Special effort should be

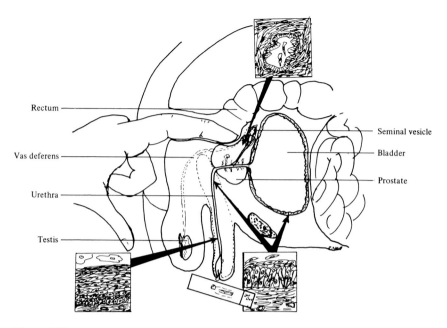

Rectum

Seminal vesicle

Vas deferens

Bladder

Urethra

Prostate

Testis

Figure 183
The lower genital tract in man and technique for obtaining prostatic secretion.

made to avoid the seminal vesicles in order to prevent contamination of the smear by large numbers of spermatozoa.

The first few drops of the secretion appearing at the tip of the urethral meatus should be wiped away because they will mainly contain transitional and squamous cells from the urethra and meatus. The succeeding drops are smeared and immediately fixed. The patient is then allowed to void, and the first few milliliters of urine are processed by the membrane filter technique and examined for prostatic cells. This examination is especially important if no secretion was produced by the massage. If the patient has an indwelling catheter, the inflammation usually present should be controlled before taking the specimen.

In some cases, the cytologic material is obtained by needle aspiration through the rectum with a long needle directed by the finger. The cellular core obtained is placed between two alcohol-moistened, completely frosted slides and smeared by pulling the slides against each other. This traumatic procedure is not recommended for routine use. Besides the possible danger of inoculating the needle track with cancer cells, this method obtains only a sample of cells from a very small area of the prostate.

CYTOLOGY OF THE NORMAL PROSTATE AND SEMINAL VESICLES

Prostatic Cells (Figs. 184A, B, 185)

Prostatic cells are scanty in a healthy prostate. They shed singly (25%), in clusters (65%), or in sheets (honeycomb pattern) (10%). Their shape

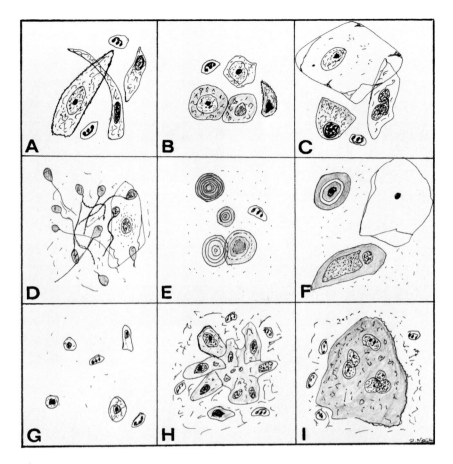

Figure 184
Different structures that may be seen in the benign prostatic secretion. (A)
Cylindrical prostatic cells. (B) Round basal prostatic cells, increased after
castration. (C) Transitional urethral cells. (D) Sperm, seminal vesicle smears.
(E) Corpora amylacea. (F) Metaplastic prostatic cells after estrogen therapy.
(G) Postcastration smear. (H) Degenerate prostatic cells in inflammation. (I)
Large multinucleated syncytial prostatic cells in inflammation.

is elongated, columnar fusiform (85%), small round or oval (10%), or
thin polyhedral (5%). Their size is regular (15–20 μ).

Their cytoplasm is adequate, thin, clear, well preserved, and acidophilic
with multiple small secretory vacuoles containing lipid, mucopolysaccha-
rides, and various acids. The cytoplasmic membrane is thin but well
defined.

Their nuclei are centrally located and round or oval, usually single. The
size varies from 8 to 12 μ. They have a uniform, finely granular chromatin.
The nuclear membrane is smooth, well defined, and regular. Occasionally
(20%), prominent, small, round nucleoli are present.

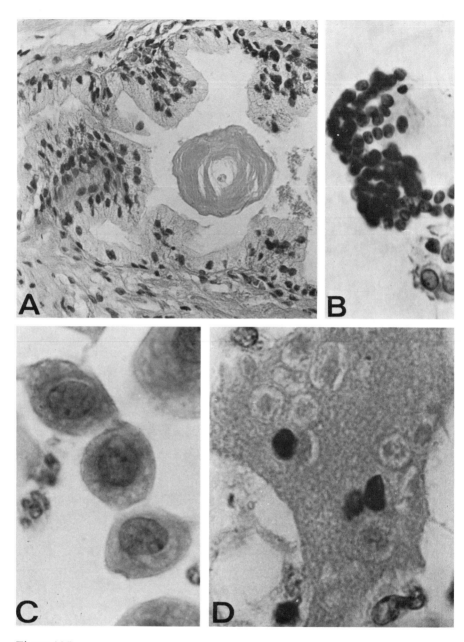

Figure 185
(A) Section of normal prostate gland. Note the accumulated secretion in the lumen. (× 450) (B) Cluster of benign prostatic cells in a prostatic massage specimen. Note the regularity of the nuclei. (× 450) (C) Round basal prostatic cells. (× 1300) (D) Large syncytial prostatic cell seen in the prostatic massage smear in a patient with acute prostatitis. (× 1300)

Transitional Urethral Cells (Fig. 184C)

Transitional urethral cells are found mainly in the first few drops of the secretion. They resemble the benign polyhedral intermediate squamous cells of the vaginal mucosa. These cells can be used for hormonal evaluation. Their shape and maturation vary in relation to hormonal changes. On occasion, small reactive basal cells are also found, with thick basophilic cytoplasm and large round or oval nuclei with coarse chromatin and prominent nucleoli.

Seminal Vesicle Cells

Seminal vesicle cells can be found in about 10% of the prostatic specimens. They are shed singly (50%) or in sheets (50%) and are irregular in size (12–16 μ). Their cytoplasm is either dense and basophilic or abundant and vesicular, containing yellow-brown granular pigment (75%) and multiple small vacuoles. The cytoplasmic membrane is thin, smooth, and well defined. The usually dark, single nuclei vary in size and shape. Occasional multinucleated giant cells with variation in size and shape of their nuclei may be seen. Their abundant chromatin is finely granular and uniformly distributed. Nucleoli are usually prominent. The background is often obscured by large amounts of spermatozoa, stromal cells, mucus, protein deposits, and cellular debris. The giant nuclei of some of these cells should not be confused with malignant prostatic cell nuclei.

Spermatozoa

Spermatozoa are shed singly or in clusters, overlapping each other but with no molding. Some may have retained narrow pink cytoplasmic sleeves at the beginning of their tails (young spermatozoa) and others are seen without tails. They are recognized by their small, oval, dark (often bicolor-stained) heads, and are uniform in size and shape (8 μ).

Spermatozoa are very abundant after a seminal vesicle massage, and they can, by masking the prostatic cells, become a diagnostic handicap.

Abnormality of size, shape, and number of tails, if present in excessive numbers (more than 2%) should be reported if a fertility problem exists.

Corpora Amylacea (Fig. 184E)

These prostatic concretions are round or oval in shape, with well-defined margins. They stain yellow-red to purple and can vary in size (7–20 μ). They possess a central core around which variable amounts of circular, sedimented protein are present in ringlike formations.

Inflammatory Elements

Some leukocytes are always seen, even when no inflammation is present. Various amounts of small phagocytes with typical vesicular, foamy cytoplasm and bean-shaped nuclei are often present with occasional (2%) multinucleated foreign-body giant cells. Ingested heads of spermatozoa may be seen in the cytoplasm of some of these cells. The background of the smear is always covered with an abundance of pink granules resulting from the deposit of mucus.

BENIGN LESIONS OF THE PROSTATE AND SEMINAL VESICLES

Estrogen Effect on the Prostatic Secretion (Fig. 184F)

Estrogen treatment often produces a diffuse squamous metaplasia of the prostatic gland. The previously described normal elongated prostatic columnar cells change into large polyhedral cells with acidophilic cytoplasm and often pyknotic, irregular, centrally positioned nuclei or into small signet ring types with hypervacuolization and nuclear molding. These vacuoles often contain a large amount of glycogen, staining yellow to brown.

Changes due to the estrogen treatment will not only decrease the amount of prostatic secretion but may also prevent the exfoliation of tumor cells, leading to an increase in the number of false-negative reports. Therefore, cytology should not be relied on to rule out the persistence of a prostatic cancer in a patient receiving estrogen therapy.

The malignant cells either shrivel and decrease in size or form, by fusion, large multinucleated syncytial cells with hyalinization of their cytoplasm, making their recognition more difficult.

Effect of Castration on the Prostatic Secretion (Fig. 184G)

Castration produces a general atrophy and decrease in the size of the prostatic glands. Their secretion becomes scantier and difficult to obtain. The normally seen clusters of tall columnar cells disappear almost completely and are replaced by single round, small cells with thick basophilic cytoplasm and small, round, central hyperchromatic nuclei.

Prostatitis (Figs. 184H, I, 185D)

An acute or subacute prostatitis is often secondary to a cystitis, urethritis, or epididymitis. Chronic prostatitis is more common in old age.

In the acute stage, large numbers of inflammatory cells (neutrophils) are present, and there is increased secretion. In chronic prostatitis, aggregates of lymphocytes, macrophages, and foreign body giant cells are more frequently found.

Large numbers of fresh and old red blood cells, necrotic cellular debris, microorganisms, and protein are deposited in the background.

The reactive prostatic cells are usually single, and they are characterized by their irregularly shaped cytoplasm, often elongated and containing multiple inclusions, and by their large single or multiple hyperchromatic or pyknotic nuclei with coarse chromatin, prominent spherical nucleoli, smooth, regular nuclear membrane, and the more or less prominent perinuclear halo (50%) that indicates degenerative cellular hypertrophy. Large multinucleated syncytial prostatic cells are often seen.

Occasional cases have been reported with prostatitis resulting from *Trichomonas* infestation, where changes in the epithelial cells were similar to the effect of *Trichomonas* on vaginal cells.

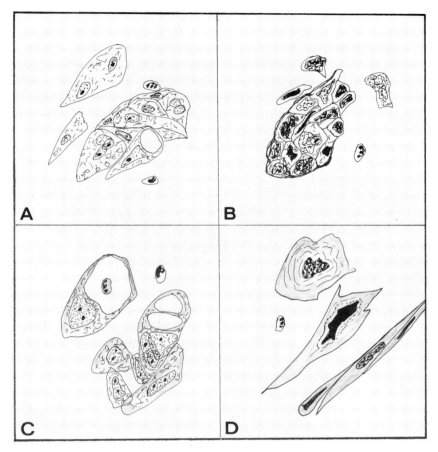

Figure 186
Cytology of prostatic neoplasm. (A) Adenomatous hyperplasia. (B) Poorly differentiated adenocarcinoma. (C) Well-differentiated adenocarcinoma. (D) Squamous cell carcinoma.

The specific granulomatous prostatitis (tuberculosis) is difficult to diagnose by cytology.

Nodular Hyperplasia—Adenomatous or
Benign Prostatic Hypertrophy (Fig. 186A)
This is one of the most common prostatic disorders of the elderly male. There is usually an increased secretion containing large numbers of prostatic cells often shed in sheets with marked cellular molding.

There is a general cellular hypertrophy (50–75 μ in diameter) and marked variation in cellular shape with occasional pseudopodlike prolongations.

The cytoplasm, adequate to abundant, is pink to dark blue and regularly stained, with multiple small vacuoles. The cytoplasm is often flattened and deformed.

The nuclei are hyperchromatic, single, round or oval in shape (12–15 μ) and have few mitoses. The nuclear membrane is well defined, smooth, and regular. A prominent nucleolus, round, red-pink, and usually single, is present in 90% of the cells. Variable amounts of fresh and old red blood cells, few inflammatory cells (polymorphonuclear cells), and occasional corpora amylacea are seen. There is often an unusual abundance of mucus and of cellular debris in the background of the smear.

MALIGNANT LESIONS

Poorly Differentiated Adenocarcinoma (Figs. 186B, 187A, B)

In poorly differentiated adenocarcinoma, the exfoliated cells are usually abundant, in small sheets (60%) with a honeycomb appearance, single (20%), or in loose clusters (20%). These sheets are usually smaller than the ones found in the benign adenoma and are often seen in crowded, cast-like structures. The cells are regular in shape and size with mild hypertrophy (14 ± 2 μ).

The cytoplasm is scanty, vesicular, stains pale blue, and is delicate with a poorly defined border. An increased nuclei-cytoplasm ratio is usually present.

The nuclei are round or oval with occasional irregularity and indentation. They are often hyperchromatic with irregular coarse chromatin clumping and well-defined, thick, regular, nuclear membranes. No prominent nucleoli are usually seen.

Large amounts of inflammatory cells (polymorphonuclear cells and phagocytes), with fresh and old blood and cellular debris, may be present. There is often an increased amount of mucus and of cellular debris in the background of the smears.

Well-Differentiated Adenocarcinoma (Figs. 186C, 187C)

The low-grade, well-differentiated adenocarcinoma is often difficult to diagnose by cytology. The exfoliated cells may resemble the benign adenomatous ones. Usually less than 25% of these neoplasms are detected by cytology. The number of diagnostic cells is scanty. They often shed in sheets or tight clusters. Their moderately hyperchromatic nuclei are round (20%), oval (20%), elongated (40%), or irregular (20%). The nuclei vary in size (12–16 μ) and are usually single. The chromatin is irregular and clumped, with pointed projections or is almost pyknotic. The cells may have 1 to 3 large, irregular nucleoli (3–5 μ). The background of the smear shows the usual deposit of granular secretions.

Squamous Cell Carcinoma (Figs. 186D, 187D)

These rare tumors may shed typical malignant squamous cells with abundant orangeophilic cytoplasm and irregular hyperchromatic or pyknotic nuclei (as described for the squamous carcinoma of the cervix). They may originate from rare primary prostatic squamous carcinoma or, more often, from a secondary tumor. Cytology cannot pinpoint the site of the lesion.

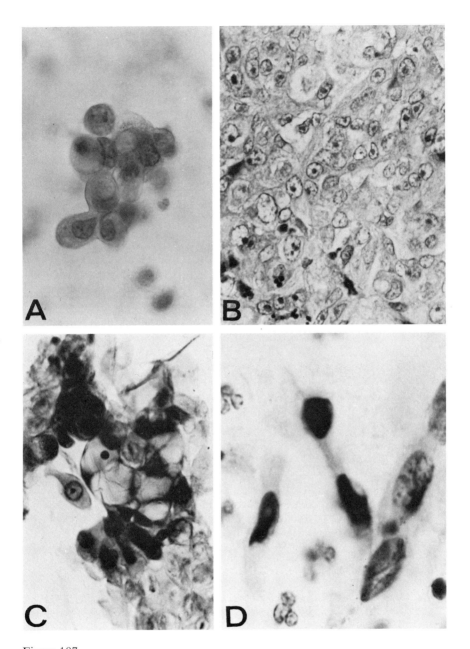

Figure 187
(A) Cluster of poorly differentiated adenocarcinomatous cells in a prostatic massage smear. ($\times$ 450) (B) Histologic section of poorly differentiated adenocarcinoma of the prostate. ($\times$ 450) (C) Cluster of malignant cells in the prostatic smear from a well-differentiated adenocarcinoma. Note the large vacuoles distending the cytoplasm. ($\times$ 450) (D) Atypical cells in a prostatic smear in a patient with a well-differentiated squamous carcinoma. ($\times$ 750)

Metastatic Tumors

Metastatic tumors to the prostate are rare and may exfoliate cells resembling those of the primary neoplasm.

PITFALLS

Some reactive phagocytes with large prominent nuclei may be mistaken for neoplastic cells. Unlike neoplastic cells, the phagocytes do not form cellular sheets, nor do they mold. The cytoplasm is phagocytic with debris-containing vacuoles, and occasionally the nucleus is bean-shaped.

Unknown or unreported estrogen treatment resulting in squamous metaplasia may cause shedding of these cells, which can be mistaken for carcinoma.

The low-grade, well-differentiated adenocarcinoma will shed cells resembling cells from benign prostatic adenoma. Structural differentiation is often impossible in these cases.

The prostatic smear may be contaminated by other tumor cells from neoplasms of adjacent organs, as is often the case in tumors of the bladder, prostate, or urethra.

Large numbers of spermatozoa found in the smears after a seminal vesicle massage may obscure the diagnostic cells.

The seminal vesicle cells, with their large dark nuclei, may be confused with anaplastic cancer cells.

References and Supplementary Reading

Ashton, P. R., and Lamberd, P. Cytodiagnosis of malakoplakia. Report of a case. *Acta Cytol.* 14:92, 1970.

Bamforth, J. The cytological diagnosis of carcinoma of the prostate. *Br. J. Urol.* 30:392–396, 1958.

Berry, A. Evidence of gynecologic bilharziasis in cytological material. *Acta Cytol.* 15:482, 1971.

Bossen, E., et al. The cytologic profile of urine during acute renal allograft rejection. *Acta Cytol.* 14:177, 1970.

Crabbe, J. S. G. Cytology of voided urine with special reference to "benign" papilloma and some of the problems encountered in the preparation of the smears. *Acta Cytol.* 5:233, 1961.

Dorfman, H. D., and Morris, B. Mucin-containing inclusions in multinucleated giant cells and transitional epithelial cells of urine; cytochemical observations on exfoliated cells. *Acta Cytol.* 8:293–301, 1964.

Elwi, A. M., Fam, A., and Ramzy, I. Exfoliative cytology of the bilharzial ulcer and cancer of the urinary bladder. *J. Egypt. Med. Assoc.* 45:235, 1962.

Esposti, P. L. Cytologic diagnosis of prostatic tumors with the aid of transrectal aspiration biopsy. *Acta Cytol.* 10:182, 1966.

Esposti, P., Moberger, G., and Zajicek, J. The cytologic diagnosis of transitional cell tumors of the urinary bladder and its histologic basis. *Acta Cytol.* 14:145, 1970.

Fitzgerald, N. W., and Ludbrook, J. Cytological diagnosis of prostatic cancer. *Br. J. Urol.* 34:326–330, 1962.

Foot, N. C., et al. Exfoliative cytology of urinary sediments. A review of 2,829 cases. *Cancer* 11:127, 1958.

Frank, I. N.   The cytodiagnosis of prostatic cancer. *J.A.M.A.* 203:1698, 1969.

Goldman, E. J., and Samellas, W.   Local extension of carcinoma of the prostate following needle biopsy. *J. Urol.* 84:575–576, 1960.

Hajdu, S. I.   Exfoliative cytology of primary and metastatic Wilm's tumor. *Acta Cytol.* 15:339, 1971.

Hajdu, S. I., et al.   Cytologic diagnosis of renal cell carcinoma with the aid of fat stain. *Acta Cytol.* 15:31, 1971.

Haour, P.   Comparison of radiation cell changes in exfoliated vaginal cells and in exfoliated cells from the urinary tract (urocytogram). *Acta Cytol.* 3:449–450, 1959.

Hazard, J. B., McCormack, L. J., and Belovich, D.   Exfoliative cytology of urine with special reference to neoplasms of urinary tract; preliminary report. *J. Urol.* 78:182–187, 1957.

Johnson, W. D.   Cytopathological correlations in tumors of the urinary bladder. *Cancer* 17:867, 1964.

Kalnins, Z., et al.   Comparison of cytologic findings in patients with transitional cell carcinoma and benign urologic diseases. *Acta Cytol.* 14:243, 1970.

Lieberman, N., Cabaud, G., and Hamm, F. C.   Value of the urine sediment smear for the diagnosis of cancer. *J. Urol.* 89:514, 1963.

Loveless, K.   The effect of radiation upon the cytology of benign and malignant bladder epithelium. *Acta Cytol.* 17:355, 1973.

Macfarlane, E. W. W.   The appearance of multinucleated cells in the urine after purgation. *Acta Cytol.* 10:104, 1966.

Macfarlane, E. W. E., Ceelan, G. H., and Taylor, J. N.   Urine cytology after treatment of bladder tumors. *Acta Cytol.* 8:288, 1964.

Mason, M. K.   The cytological diagnosis of carcinoma of the prostate. *Acta Cytol.* 11:68, 1967.

Masukawa, T., et al.   Herpes genitalis virus isolation from human bladder urine. *Acta Cytol.* 16:416, 1972.

Meisels, A.   Cytology of carcinoma of the kidney. *Acta Cytol.* 7:239, 1963.

Melamed, M. R., Voutsa, N. G., and Grabstald, H.   Natural history and clinical behavior of in situ carcinoma of the human urinary bladder. *Cancer* 17:1533, 1964.

Naib, Z. M.   Exfoliative cytology of renal pelvic lesions. *Cancer* 14:1085, 1961.

Naib, Z. M.   Cytologic diagnosis of cytomegalic inclusion body disease. *Am. J. Dis. Child.* 105:153–159, 1963.

Naib, Z. M., Young, J. D., and Philippidjis, P. J.   Exfoliative cytology of a primary fibrosarcoma of the kidney. *J. Urol.* 90:386, 1963.

Nuovo, V. M.   Tumeur du rein, du bassinet et de l'uretere. *Arch. Anat. Pathol.* 15:182, 1967.

Nuovo, V. M., and Aboulker, P.   Diagnostic cytologique du cancer de la prostate. *Arch. Anat. Pathol.* 15:185, 1967.

O'Morchoe, P. J., and O'Morchoe, C. C. C.   Urinary cytology in reproductive endocrinology. *Am. J. Obstet. Gynecol.* 99:479, 1967.

Powder, J. R., Naib, Z. M., and Younge, J. D.   Cytological examination of the urine sediment as an aid to diagnosis of epithelial neoplasms of the upper urinary tract. *J. Urol.* 84:666, 1960.

Proll, R., Wernett, C., and Mims, M.   Diagnostic cytology in urinary tract malignancy. *Cancer* 29:1084, 1972.

Sano, M. E., and Koprowska, L.   Primary cytologic diagnosis of a malignant renal lymphoma. *Acta Cytol.* 9:194, 1965.

Sicard, A., and Marsan, C. *Atlas de cytologie.* Paris: Editions Varia, 1964. Vol. 2.

Suprun, H., and Bitterman, W. A correlative cytohistologic study on the inter-relationship between exfoliated urinary bladder carcinoma cell types and the staging and grading of these tumors. *Acta Cytol.* 19:265, 1975.

Taylor, J. N., Macfarlane, E. W. E., and Ceelen, G. H. Cytological studies of urine by Millipore filtration technique: Second annual report. *J. Urol.* 90: 113–115, 1963.

Umiker, W. Accuracy of cytologic diagnosis of cancer of the urinary tract. *Acta Cytol.* 8:186, 1964.

Umiker, W., Lapides, J., and Sourenne, R. Exfoliative cytology of papillomas and intra-epithelial carcinomas of the urinary bladder. *Acta Cytol.* 6:255, 1962.

Voutsa, N. G., and Melamed, M. R. Cytology of in situ carcinoma of the human urinary bladder. *Cancer* 16:1307, 1963.

The Breast

15

The cytologic material obtained from the breast may be a spontaneous nipple exudate, a secretion produced by nipple massage, the scraping of an eczematoid lesion of the nipple, or the needle aspiration of a cystic or solid mass. Special precautions should be taken to prevent delay in the fixation of the smear because of the unusually rapid drying capacity of all breast secretions.

ANATOMY AND HISTOLOGY (FIG. 188)

The size and structure of the breasts vary with the age, sex, hormonal status, and heredity of the individual. The areola is the circular, pigmented area of the skin that contains sebaceous glands. In its center, the elevated nipple is covered by wrinkled skin lined with stratified squamous epithelium. It contains 15 to 20 lactiferous ducts lined by a two-layered columnar cell mucosa. Below the nipple, the ducts expand to form an ampulla where milk and other secretions may be stored. Nipple massage will often cause this stored material to be excreted. Below the ampulla, these ducts divide into numerous small terminal tubules, varying in number (maximum during lactation). Each of these terminal tubules emerges from a secretory acinus or lobule lined by one or two layers of secretory columnar cells and surrounded by fibroadipose tissue.

TECHNIQUE FOR COLLECTING SPECIMENS

Nipple Secretion

A nipple discharge can be an exudate from a skin ulceration or a secretion from the excretory ducts. It is physiologic in some instances, as in the newborn (witch's milk), before and after delivery, at puberty, or during the menses. At any other time, it is pathologic and may indicate an inflammation, a chronic glandular cystic hyperplasia, an intraductal papilloma, or a carcinoma. A spontaneous but nonphysiologic nipple discharge is also associated with Chiari-Frommel, Ahumada–del Castillo, and Forbes-Albright syndromes. Oral contraception, especially shortly after discontinuation, may also cause an abnormal but benign discharge.

In collecting the specimen, one should be careful to discard the first several drops of secretion because they will contain mainly degenerate cellu-

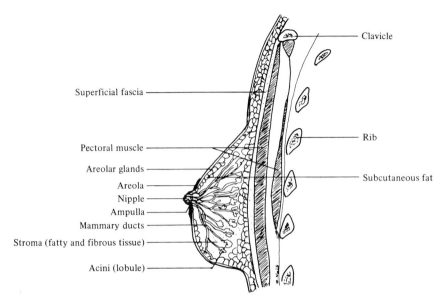

Figure 188
Anatomy of the breast.

lar debris. The surface of the nipple should be carefully wiped to remove anucleated cells and debris, which otherwise will contaminate the smears. This discharge can be serous, sanguinous, milky, cheesy, dark green, or black, acording to its content and the lesions producing it. A bloody or a thin, pink, watery discharge often indicates a neoplastic genesis (14–50%).

Only about 3% of all mammary cancers produce any spontaneous nipple discharge.

Cyst Fluid Aspiration

A 15- to 18-gauge needle attached to a 50 ml syringe with a metallic plunger is recommended to aspirate with good suction the fluid contained in a breast cyst. If scanty, the few droplets are smeared directly on alcohol-moistened slides. If more abundant, the fluid should be immediately mixed with heparin-saline solution for dilution and to prevent coagulation. It may then be processed by passing it through a filter membrane.

Needle Aspiration of a Solid Mass

This is indicated whenever a biopsy is not practical. The mass is immobilized between the fingers and aspirated under negative pressure by a 15- to 19-gauge needle attached to a 50 ml syringe. The needle needs to be moved through the suspected mass in various directions. The retraction of the piston should be discontinued during withdrawal of the needle. The specimen obtained is placed between two previously alcohol-moistened, completely frosted glass slides, pressed, smeared, and immediately fixed.

It has been shown statistically that the danger of tumor cell implantation in the needle track is negligible.

Wound Washing

The incidence of recurrence of a breast carcinoma in the mastectomy scar can be decreased if the operative field is washed with saline solution before the closure of the skin. Because of the usual presence of blood and tissue particles in the washing specimen, the membrane filter technique cannot be used without first lysing the red blood cells. The specimen is mixed with Carnoy's solution, then centrifuged at 3000 rpm for 10 minutes, and the supernatant decanted. The specimen is then either diluted with saline solution and passed through a membrane filter, or the button is smeared directly on several slides.

The presence of loose but viable clusters of malignant cells may indicate the need for irradiation of the operative site.

Imprint Smear of Biopsy Tissue

Imprint smears can be used as an adjunct to frozen section; the freshly cut surface of the suspected biopsy tissue should be pressed to a glass slide and fixed immediately. A benign tumor will exfoliate few cells, whereas the smear from a malignant tumor is usually very cellular. The cells form trabecula-like structures of variable size and shape. Interpreting them is usually relatively easy.

Surface Skin Scraping

If a biopsy is not indicated or possible, smears obtained by an energetic scraping of the surface of a fungating or ulcerative breast lesion may permit diagnosis of a primary Paget's disease or an infiltrating duct carcinoma involving the skin. Furthermore, this technique allows the differentiation without trauma or bleeding of an ulcerated recurrent malignant skin metastasis from the often grossly similar-appearing irradiation necrosis or from keloid scar tissue formation.

CYTOLOGY OF THE NORMAL BREAST

Nonsecretory Duct Cells (Figs. 189A, 190A)

Nonsecretory duct cells are cuboidal cells. They exfoliate in larger numbers during pregnancy and lactation or after energetic breast massage. They are shed singly (20%) or, more often, in tight clusters (80%). They show more cellular molding and form trabeculae of various sizes.

The cytoplasm is usually scanty and uniformly pale pink to blue or transparent. It contains occasional tiny secretory vacuoles compressing the nucleus.

The nucleus is usually single, small (7–8 μ), vesicular, and regular in size with a moderate variation in shape (round, oval, triangular, crescent), especially when present in a tight cellular group.

There is moderate hyperchromatism with coarse but regular, granular chromatin clumping of even distribution. No nucleoli can be seen.

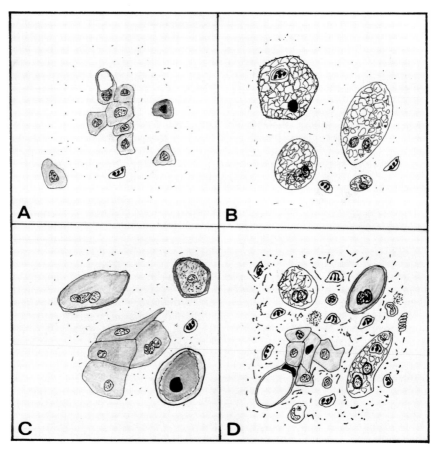

Figure 189
Cytology of benign secretions of the breast. (A) Nonsecretory duct cells. (B) Secretory duct cells. (C) Apocrine metaplasia. (D) Chronic cystic mastitis.

The uniform size of the cells and the regularity of their nuclei differentiate them from malignant cells.

Their presence in a discharge may indicate benign papillary proliferation of ductal epithelium.

Secretory Duct Cells—Foam Cells (Figs. 189B, 190B)

Secretory duct cells are found in larger numbers during pregnancy, lactation, and acute or subacute mastitis. They are usually shed singly or in loose clusters. They are round or oval in shape (98%), with an abundant, vesicular, foamy cytoplasm and fine lipid-containing vacuoles of variable size. They often resemble the cytoplasm of histiocytes. The cytoplasmic membrane is usually sharp and distinct. There is marked variation in their size (15–25 μ), and they stain pale blue to green, or pink to orange with the Papanicolaou stain. The periodic acid–Schiff stain and the fat stain are usually positive.

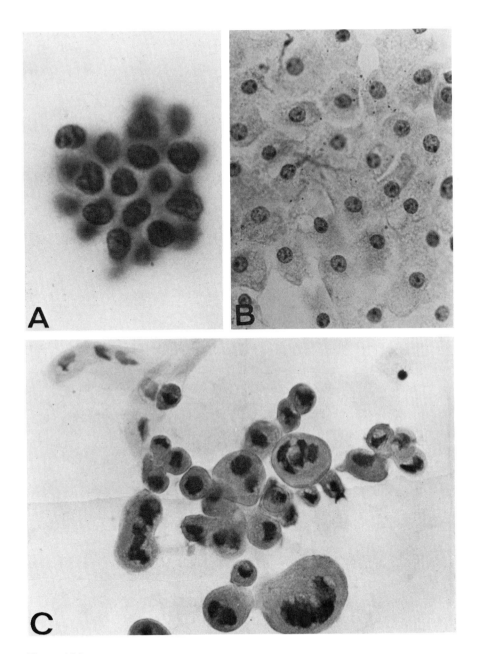

Figure 190
(A) Cluster of benign nonsecretory duct cells in a nipple discharge smear. Note the regularity of the nucleus and the scantiness of the cytoplasm. (× 1300) (B) Foam cells in a nipple discharge smear. Note the abundant vacuolated cytoplasm and the uniform vesicular nuclei. (× 450) (C) Apocrine metaplasia cells in a cyst aspirate. Note the dense cytoplasm, hyperchromatic nuclei, and multinucleated cells. (× 450)

The nuclei are single or multiple (1–5), usually eccentric, but regular in shape, round or oval, with a moderate size variation (7–10 μ).

The chromatin is regular and finely granular with an occasional small, prominent nucleolus.

Nuclear pyknosis may be seen in the degenerate cells, especially when an inflammation is present.

Squamous Cells

Anucleated or normal superficial squamous cells in varying amounts can be seen in the smears originating from the nipple, lactiferous ducts, or areola, or as a contaminant from the hands of the physician. Most of these cells are keratinized and have no specific significance.

Histiocytes (Fig. 191A)

Histiocytes appear irregularly. In some specimens, they are the preponderant cells and are recognized by their typical bean-shaped nuclei and debris- or blood pigment-containing cytoplasm. They can be difficult to differentiate from the foam cells, especially if they have also ingested fatty secretion debris. They are most commonly found in the aspirated cyst fluid.

Inflammatory Cells

Varying numbers of inflammatory cells (polymorphonuclear leukocytes and lymphocytes) are always present. The amount increases prior to and just after delivery. Their cytoplasm often contains phagocytized, small fat droplets stainable with Sudan stain. Small amounts of fresh and old red blood cells that have no pathologic significance may occasionally be found. Because of the frequent coexistence of inflammation and breast carcinomas, an inflammatory smear should be screened for tumor cells with even more care than usual.

CYTOLOGY OF BENIGN LESIONS OF THE BREAST

Acute and Subacute Mastitis

Inflammation occurs more frequently during pregnancy, puerperium, and lactation. It can be primary (trauma) or secondary to a systemic disease (tuberculosis). It can be diffuse or localized and can produce abscess or generalized tenderness. Some inflammations yield an abundance of purulent nipple discharge; in others, secretions must be aspirated. Since cancer can often simulate the symptoms of diffuse mastitis or localized abscess, any inflammatory smear should be carefully screened for the presence of tumor cells.

Large amounts of inflammatory cells, with a predominance of polymorphonuclear cells, are found on a usually thick background composed of cellular debris, microorganisms, and protein precipitates.

Large numbers of naked nuclei from degenerate duct cells are often seen.

Some of the aspirated smears may contain large sheets of normal or reactive duct cells. Their shape, size, nuclei, and chromatin patterns are uniform. The hyperplastic cells have single, round, prominent nucleoli. Degen-

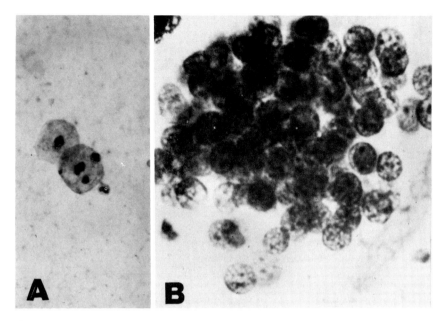

Figure 191
(A) Aspirate from a mass due to a posttraumatic fat necrosis. (× 450) (B)
Aspirate from plasma cell mastitis. Note the cluster of plasmacytes, which need
to be differentiated from small cell carcinoma. (× 1300)

eration of these cells may produce some variation in their size and shape
and condensation of the chromatin toward the edges, which may make them
a pitfall in diagnosis. Small histiocytes and foreign-body giant cells with ingested debris and lipid globules are frequent, especially when posttraumatic
fat necrosis is present (Fig. 191A).

Fresh and old erythrocytes are usually present, especially in the acute
phase. Degenerate squamous or duct cells covered with colonies of microorganisms or contaminant fungi are also found.

Persisting Lactation and Galactocele (Milk Cyst)
The alveoli are usually increased in size and number. Their wall is lined by
a flattened epithelium.

The presence of large numbers of foam cells on a background of abundant protein and lipid deposits is diagnostic. When slightly air dried, the
background of the smear has a honeycomb pattern. Globular fatty droplets
are numerous and may be coated with blue-green granular debris.

The round foam cells with small, round nuclei and cytoplasm containing
lipid droplets may vary enormously in size. Multinucleation is common and
has no significance. Blood and inflammatory cells are usually absent. Clusters of benign nonsecretory duct epithelial cells may also be seen.

413

Apocrine Metaplasia (Figs. 189C, 190C)

Apocrine metaplasia is the result of a chronic irritation or abnormal endocrine stimulation (estrogen). The cells resemble secretory foam duct cells, except for the cytoplasm's being acidophilic in type and amorphous in consistency instead of foamy and basophilic. These cells are mainly found in large numbers, singly or in sheets, in the aspirated material of benign retention cysts.

The abundant cytoplasm will occasionally contain glycogen, staining yellow-brown, mainly in the endoplasm.

The cellular borders are thin with an ectoplasmic rim that is often curled upon itself.

The nuclei are regular, round or oval with slight hyperchromatism. Occasionally, their nucleoli are prominent. They are similar to the small nuclei of the secretory duct cells.

Multinucleation is frequent with variation in size and shape of the nuclei (differing from multinucleated giant phagocytes). The background is often bloody with moderate amounts of debris and protein precipitate. A small number of foam cells may also be present.

Comedomastitis (Duct Ectasia)

This dilatation of the ducts, found mainly in postmenopausal, multiparous women, produces puttylike, cheesy material that can be squeezed from the nipple.

Large amounts of debris, precipitate, and lipid droplets usually obscure the background of the smear.

Numerous lipid-laden macrophages of various sizes are present.

Small and large sheets of uniform, benign-looking duct cells are usually present with a variable degree of degeneration.

Large numbers of inflammatory lymphocytes and plasmocytes are also usually present.

Plasma Cell Mastitis (Fig. 191B)

This occurs mainly in menopausal, multiparous women and is often thought to be the end stage of comedo mastitis. The cloudy nipple discharge contains large amounts of necrotic cells.

Foam cells and sheets of degenerate epithelial duct cells are common.

Numerous lipid-containing macrophages and occasional multinucleated foreign-body giant cells are also seen.

Large amounts of plasmacytes, lymphocytes, eosinophils, and some neutrophils are usually present and are diagnostic.

Fat Necrosis (Fig. 191A)

This degeneration of the adipose tissue may be due to trauma, inflammation, or carcinoma. It may result in cyst formation, hemorrhage, or calcification and fibrosis.

The aspiration of such a mass will contain variable numbers of foam cells and elongated epithelioid cells with dark nuclei and blood pigment-containing or lipid-containing phagocytes.

Large multinucleated foreign-body giant cells with variable intracyto-plasmic inclusions are common.

Variable amounts of inflammatory cells (mainly leukocytes) are present. Cholesterol crystals and calcified necrotic debris may also be seen.

Fibroadenoma (Fig. 192A)

Aspiration biopsy smears of most fibroadenomas are highly cellular. These ovoid and spindle epithelial cells often are seen without their cytoplasm. They are found in tight clusters or large sheets in honeycomb appearance. The size and shape of these benign cells are uniform. Their nuclei are round to oval, regular in size, and slightly hyperchromatic with delicate nuclear membrane and evenly distributed chromatin. They should not be confused with maligant cells shedding from poorly differentiated carcinomas. The presence of clusters of bipolar-shaped naked nuclei is reassuring and should prevent a false-positive cytologic report. The background of the smear is usually clean.

Occasionally, small tubular structures (blood capillaries), interspersed with endothelial cells, may be present; they originate from the usual rich vasculature of this tumor. A few foam cells may also be seen; they have no significance.

Chronic Cystic Mastitis–Fibrocystic Disease (Fig. 189D)

Chronic cystic mastitis is one of the most common causes of spontaneous nipple discharge during the fourth decade of life. The cysts are often the result of chronic endocrine disturbance and inflammation. The secretion can be white, cheesy, or dark green.

Numerous small foam cells and large metaplastic apocrine cells with acidophilic cytoplasm and dark pyknotic nuclei are present in the smears.

The background contains variable amounts of fresh and old red blood cells, cellular debris, and inflammatory cells (polymorphonuclear cells and monocytes).

Sclerosing Adenosis

This benign tumor, considered a subtype of fibrocystic diseases, shows, in the needle-aspirated smear, sheets and clusters of elongated cells that are uniform in size and shape. Their adequate cytoplasm is usually transparent with ill-defined borders and may have dendritic processes. Their single hyperchromatic nuclei are elongated with delicate smooth nuclear membranes. Slight inflammation is seen.

Intraductal Hyperplasia and Papilloma (Figs. 193A, 194A, B)

A localized or diffuse intraductal hyperplasia or papilloma may produce a bloody or lemon-yellow discharge. This discharge should be differentiated from the one resulting from an intraductal carcinoma and from the semi-physiologic serosanguinous nipple discharge occurring during pregnancy, when a mild degree of reversible intraductal hyperplasia often exists. A cytologic diagnosis of a benign papilloma should be made reluctantly during pregnancy or in the first two postpartum months.

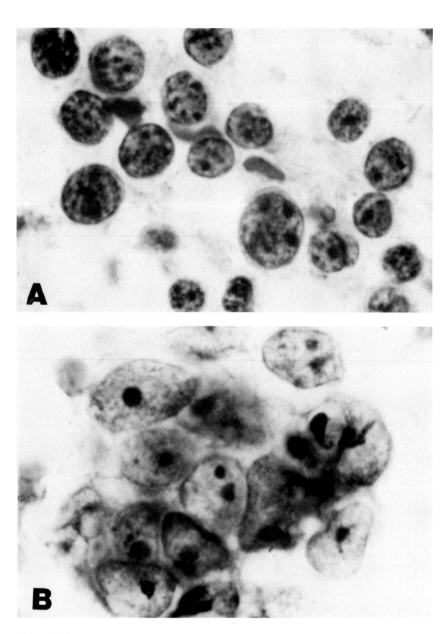

Figure 192
(A) Aspirate from a fibroadenoma. (× 1300) (B) Aspirate from an infiltrating duct carcinoma. Note the absence of nuclear molding in the fibroadenoma. (× 1300)

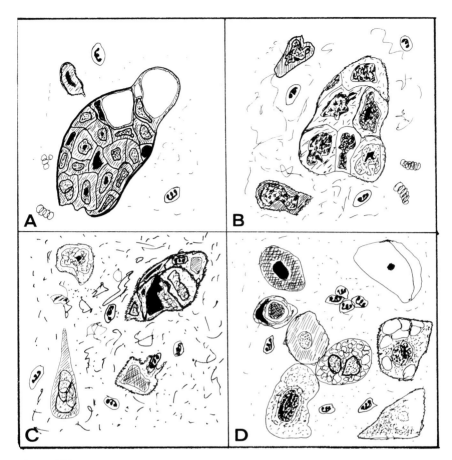

Figure 193
Cytology of a nipple discharge. (A) Intraductal papilloma. (B) Poorly differentiated duct carcinoma. (C) Comedocarcinoma. (D) Paget's disease.

Large numbers of small secretory or passive duct cells are tightly compressed against one another to form papillary structures. The morphologic
analysis needed for the diagnosis is often possible only in the cells located
at the periphery of these clumps.

There is moderate variation in size (10–20 μ) and shape of these cells
with well-defined, sharp boundaries.

The scanty cytoplasm is uniformly basophilic, except when distended by
occasional large vacuoles of the secretory type.

The nuclei are slightly hyperchromatic with marked nuclear molding.
The chromatin is evenly distributed and granular or occasionally pyknotic.

The nucleoli, when present (10%), are prominent and round in shape
with little size variation.

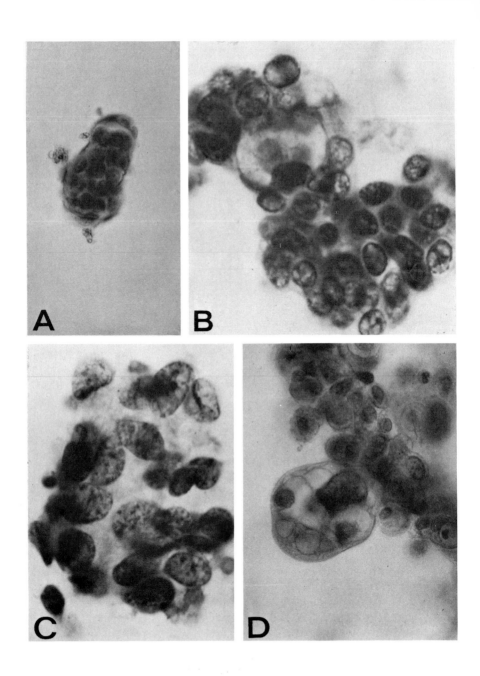

Moderate amounts of cellular debris and red blood cells are found in the background of the smears.

The crowded clusters of irregular dark cells originating from an atypical intraductal papilloma (Fig. 194B) can be very difficult to differentiate from the cells of a poorly differentiated intraductal carcinoma. A conservative cytologic interpretation should usually be rendered, but a biopsy is recommended in all cases.

Apocrine metaplastic cells may also be present. Variable degrees of other atypia may be seen, but in a lesser degree than in cancer.

CYTOLOGY OF MALIGNANT TUMORS OF THE BREAST

Intraductal Adenocarcinoma (Figs. 192B, 193B, 194C)

In intraductal adenocarcinoma, large numbers of sheets, irregular papillary projections, acini, and clusters of single cells may be present, varying in size (12–30 μ) and shape. Their cohesion is poor, and the tumor cells form loose clusters rather than the tight ones seen in benign intraductal papilloma.

The cytoplasm is relatively scanty and transparent with occasional multiple foamy or single large secretory vacuoles distending their often ill-defined cytoplasmic membrane.

In an advanced lesion, numerous single, naked, malignant hyperchromatic, almost pyknotic, nuclei, varying in size and shape, are often seen scattered throughout the smear.

The nuclei of the intact cells, which appear singly or in clusters, are hypertrophic but vesicular. Usually the nuclei of the intact cells are round or oval, molding and overlapping on each other, with marked chromatin clumping or pyknosis. The stripped nuclei are irregular and hyperchromatic.

The chromatin is unequally distributed with angular, dense clumping varying from one cell to the other.

The nuclear membranes are heavy, irregular in thickness and outline, but generally well defined.

The nucleoli, when present (10%), are small and round.

Mitotic figures are seldom seen. A malignancy should be suspected when the atypical cells in the cluster seem to spread in all directions rather than form a cohesive structure as seen in benign papilloma.

Large numbers of phagocytes and occasional multinucleated foreign body

---

Figure 194
(A) Cluster of cells in a nipple discharge from a low-grade intraductal papilloma. Note the tight molding of the cells and the smooth external border. The nuclei are uniform and benign in appearance. ($\times$ 450) (B) Cluster of cells in a nipple discharge smear from a reactive intraductal papilloma. Note the large single vacuole hyperdistending the cytoplasm of one cell and the reactive but uniform nuclei. A biopsy is indicated by such a smear. ($\times$ 750) (C) Cluster of neoplastic cells in a nipple discharge smear from a poorly differentiated infiltrating duct carcinoma of the breast. (D) Cluster of tumor cells in a needle aspirate from a comedocarcinoma. Note the cellular molding, the polymorphism, and the multiple vacuoles in the largest cell. ($\times$ 750)

giant cells are often present (80%). Some contain phagocytized blood pigment.

These malignant features are the same whether the tumor is in the in situ stage or has infiltrated beyond the basement membrane.

Fresh and old blood, cellular debris, and variable amounts of protein deposits obscure the background of the smears.

Comedocarcinoma (Figs. 193C, 194D)

Nipple massage may express wormlike plugs of a mixture of cancer cells and necrotic material. A bloody secretion often indicates secondary infection. In cases of comedocarcinoma variable amounts of hypertrophic cells are found, usually single (70%) or in sheets, papillae, or clusters. There is marked variation in their size (15–40 μ) and shape (round to oval, 40%; irregular, 60%) and amount of cytoplasm present. The degenerate, occasionally multivacuolated cytoplasm stains pale blue with poorly defined borders.

Some of the eccentric nuclei (40%) are coarsely hyperchromatic with jagged clumping; others are large and pale and look edematous. Multinucleation and nuclear pyknosis are common.

Large amounts of cellular debris, fresh and old blood, protein deposits, and mixed inflammatory cells obscure the background of the smears.

Superficial Carcinoma—Paget's Disease of the Nipple (Figs. 193D, 195)

A smear made of the natural serum exudate, as in a touch preparation of the ulcerated Paget's disease, is seldom diagnostic. It will contain mainly anucleated squamous cells, inflammatory cells, serum precipitate, and debris. The cells should be expressed from the ducts or energetically scraped to dislodge the deep diagnostic cells.

The malignant Paget cells may shed singly (60%) or in small clusters (30%) or sheets (10%). They vary in size and shape; they are large (40–80 μ), oval (60%), round (20%), or irregular (20%).

Their cytoplasm is abundant, polyhedral or irregular, foamy, or semikeratinized and stains pale blue to dark pink. The cytoplasmic membrane is smooth, thin, and well defined.

The ovoid nuclei are large, central (60%), and eccentric (40%). They are often multiple (2–3) and have round or ovoid shapes.

Often hyperchromatic, their chromatin is either coarsely granular or irregularly pyknotic. Pseudocannibalism is frequently present.

Large amounts of keratotic cellular debris, phagocytes with foamy cytoplasm, and inflammatory cells (lymphocytes and polymorphonuclear cells) are usually present. Occasional increase of eosinophils has been noted. The finding of Paget's disease without an associated in situ or infiltrating duct carcinoma is extremely rare.

Other Tumors

The majority of the other benign or malignant tumors of the breast do not exfoliate cells through the nipple, but the cells can be studied cytologically

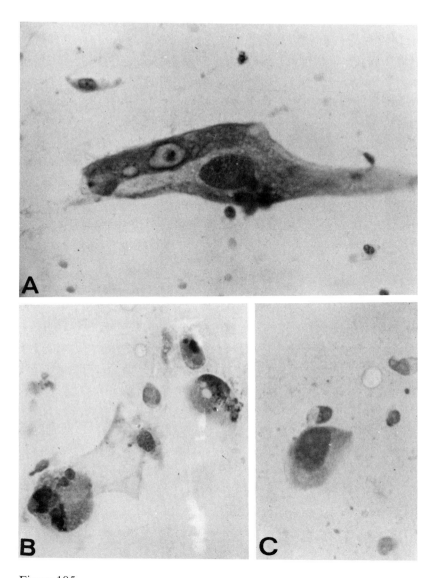

Figure 195
Scraping from Paget's disease of the nipple. (A) Large hyperkeratinized squamous cell with bland nucleus. (× 450) (B) Histiocytes and degenerate binucleated cells in Paget's disease. (× 450) (C) Mononucleated cell in Paget's disease. (× 450)

by needle aspiration. The structures of these cells vary according to the type of neoplasm.

A *colloid carcinoma* is suggested by the presence of sheets or columns and clusters of large tumor cells having compressed crescent-shaped nuclei with prominent nucleoli molded by one or two large cytoplasmic vacuoles (signet ring cells).

An *epidermoid (squamous) carcinoma* is a rare tumor that may originate from the metaplastic epithelium of the duct lining or from the skin covering the nipple. The scraping will produce orange, keratinized malignant cells with abnormal nuclei similar to the ones described for squamous carcinoma of other sites.

The aspiration of a *medullary carcinoma* may produce an abundance of large ovoid or polygonal cells with adequate vesicular, slightly basophilic cytoplasm and round or oval large nuclei with prominent single nucleoli.

Large numbers of lymphocytes may also be present.

PITFALLS AND CAUTIONS
1. The diagnosis should be based only on the features of well-preserved cells. The misleading degenerate or artifactual changes, which are frequently found, should be rejected;
2. Large amounts of blood, leukocytes, foam cells, and protein precipitate may obscure the diagnostic cells;
3. An abundance of epithelial cells does not necessarily indicate a carcinoma; some of the benign papillomas and fibroadenomas exfoliate cells in abundance;
4. An abundance of cytoplasm in an atypical cell does not necessarily rule out cancer;
5. The first few drops of the nipple discharge often contain no diagnostic cells. The fourth or fifth ones should be smeared.

References and Supplementary Reading

Anderson, J. The cytologic examination of smears from the non-lactating breast. *Am. J. Med. Technol.* 32:13, 1966.

Atkins, H., and Wolff, B. Discharge from the nipple. *Br. J. Surg.* 51:602–606, 1964.

Bennington, J. L., Smith, D. C., and Figge, D. C. Detection of cells from extramammary Paget's disease of the vulva in a vaginal smear. *Obstet. Gynecol.* 27:772, 1966.

Cline, J., Frost, K., and Funderburk, W. Nipple discharge. *South. Med. Bull.* 19:26, 1971.

Dutra, F. R. Paget's disease of the breast. Simple method of cytological diagnosis. *J.A.M.A.* 195:185, 1966.

Frost, J. K. Cytology of the Breast. In J. W. Regan (Ed.), *A Manual of Cytotechnology,* National Committee for Careers in Medical Technology, Wichita Falls, Texas, 1962.

Holleb, A. I., and Farrow, J. H. The significance of nipple discharge. *Cancer* 16:182, 1966.

Katayama, K. D., and Masukawa, T.   Ring chromosomes in a breast cancer. *Acta Cytol.* 12:150, 1968.

Kern, W., and Dermer, C.   The cytopathology of hyperplastic and neoplastic mammary duct epithelium. *Acta Cytol.* 16:120, 1972.

Kjellgren, O.   The cytologic diagnosis of cancer of the breast. *Acta Cytol.* 8:216, 1964.

Kline, T. S., and Lash, S. R.   The bleeding nipple of pregnancy and postpartum period. *Acta Cytol.* 8:1336, 1964.

Linsk, J., Kreuzer, G., and Zajicek, J.   Cytologic diagnosis of mammary tumors from aspiration biopsy smears. *Acta Cytol.* 16:130, 1973.

Masukawa, T., Lewison, E., and Frost, J. K.   The cytologic examination of breast secretions. *Acta Cytol.* 10:261, 1966.

Murad, T. M.   Evaluation of the different techniques utilized in diagnosing breast lesions. *Acta Cytol.* 19:499, 1975.

Murad, T. M., and Snyder, M.   The diagnosis of breast lesions from cytologic material. *Acta Cytol.* 17:418, 1973.

Nathan, M.   Diagnostic precoce d'un neoplasm du sein par l'examen histologique de son suintement hemorragique. *Clinique* 60:38–39, 1914.

Papanicolaou, G. N., et al.   Exfoliative cytology of the human mammary gland and its value in the diagnosis of cancer and other diseases of the breast. *Cancer* 11:377–409, 1958.

Ringrose, C. A. D.   The role of cytology in the early detection of breast disease. *Acta Cytol.* 10:373, 1966.

Rosemond, C. P., Maier, W., and Brobyn, T.   Needle aspiration of breast cyst. *Surgery* 128:351, 1969.

Vilaplano, E., and Jimenez, M.   The cytologic diagnosis of breast lesions. *Acta Cytol.* 19:519, 1975.

Webb, A.   The diagnostic cytology of breast carcinoma. *Br. J. Surg.* 57:259, 1970.

Winship, T.   Aspiration biopsy of breast cancers by the pathologist. *Am. J. Clin. Pathol.* 52:438, 1969.

Zajdela, A.   Valeur et interet du diagnostic cytologique dans les tumeurs du sein par ponction. Etude de 600 cas confrontes cytologiquement et histologiquement. *Arch. Anat. Pathol.* 11:85–87, 1963.

Zajicek, J., et al.   Cytologic diagnosis of mammary tumors from aspiration biopsy smears. *Acta Cytol.* 14:370, 1970.

The Eye

16

TECHNIQUE FOR OBTAINING SPECIMENS

The different components of the eye from which cytologic specimens may be obtained are illustrated in Figure 196. The best specimen results from the energetic scraping of the margins of a surface conjunctival or corneal lesion using a metal or plastic spatula after local anesthesia. The scraping should be smeared on one or several alcohol-moistened slides. The alcohol is allowed to evaporate with the slides in a horizontal position; then the slides are immediately placed in a 95% alcohol solution.

The needle aspirate of one or two drops of intraocular fluid from the anterior chamber or of aqueous humor from the vitreous body may be similarly smeared on one or two slides and fixed, or diluted with a small amount of 50% alcohol and passed through a membrane filter with a 5 μ pore opening. If necessary, the anterior chamber can be irrigated with three or four drops of sterile saline solution and this aspirate added to the fluid already obtained.

CYTOLOGY OF THE NORMAL EYE

Nonsecretory Conjunctival Cells (Figs. 197C, 198D)

The conjunctival mucosa is composed mainly of stratified columnar epithelium that varies in texture and thickness according to the location.

The scraping usually produces large numbers of cylindrical basophilic cells, single, or in clusters or sheets. These resemble the previously described basal endocervical cells. They vary moderately in size (15–25 μ in diameter). They are round or pyramidal in shape.

The cytoplasm is semitransparent, adequate to abundant, and stains blue-green with the Papanicolaou stain. The cytoplasmic borders are well defined, sharp, and regular. Occasionally intracytoplasmic, orange-brown melanin pigment granules may be seen.

The single nucleus is eccentric, round to oval (6–9 μ in diameter) and shows a smooth, regular nuclear membrane and moderately coarse granular chromatin, uniformly distributed with an occasional small, red, round, prominent nucleolus.

Conjunctival Goblet Cells (Figs. 197B, 198C)

The presence of conjunctival goblet cells in the specimen depends on the site of the scraping, the age of the patient, and possible disease of the con-

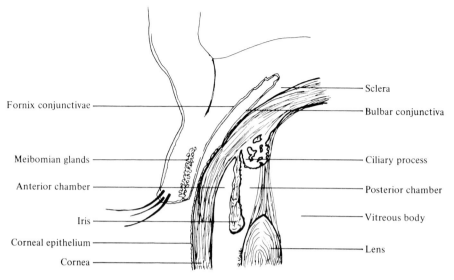

Figure 196
The sagittal section of the anterior portion of the eye.

junctiva. The cells shed singly or in clusters. They are uniformly large (30–40 μ). Their cytoplasm is abundant and often contains one or several large secretory vacuoles compressing the nuclei into crescent shapes. The nucleus is uniform in size (7–10 μ) but varies in shape. The chromatin is finely granular. Minute nucleoli are occasionally seen.

Corneal Cells (Figs. 197D, 198A, B)
The cornea is lined by stratified squamous epithelium. The exfoliated cells are large and flat, resembling vaginal intermediate squamous cells. The cells have an abundant, thin, transparent cytoplasm with no keratinization and central, round or oval, regular vesicular nuclei with fine chromatin granularity. Small nucleoli are occasionally seen.

Ocular Chamber Aspirate (Figs. 197A, 199)
These aspirates normally are very scanty in cells, almost acellular, except for a few mature lymphocytes and histiocytes scattered in the pink-stained, granular protein deposits. Few retinal cells with abundant eosinophilic cytoplasm and central, round, vesicular nuclei may be seen. In cases of retinal detachment, numerous pigmented retinal cells may be found in the aspirate of the fluid accumulated behind the detachment.

Other Cells
On occasion, a few inflammatory cells (lymphocytes or neutrophils) may be present without indicating a disease. Anucleated squamous cells may also be found as the result of contamination from the surrounding skin. The

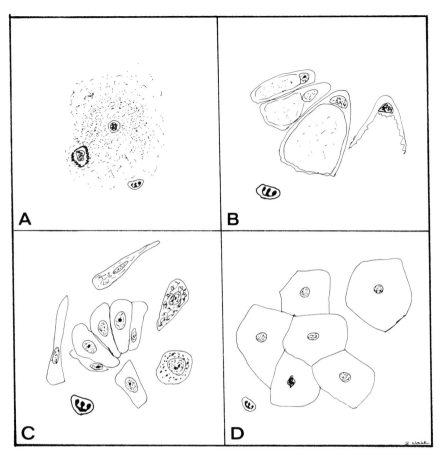

Figure 197
Cytology of the normal eye. (A) Ocular chamber aspirate. (B) Conjunctival goblet cells. (C) Nonsecretory conjunctival cells. (D) Normal corneal cells.

anucleated squamous cells should be differentiated by their transparent, poorly stained cytoplasm from the hyperkeratotic, anucleated, deeply orangeophilic cells shed from conjunctival or corneal keratosis.

INFLAMMATORY DISEASES

Allergic Inflammation (Fig. 200B)
The smears of patients with allergic vernal conjunctivitis usually contain large numbers of intact or fragmented eosinophilic polymorphonuclear cells mixed with variable numbers of lymphocytes, plasmacytes, basophilic and neutrophilic leukocytes. These eosinophils are rarely seen in allergic conjunctivitis due to other sensitizing antigens (pollen, drugs, cosmetics, and so on). An increased number of goblet cells is seen in all allergic conjunctivitis. These hyperactive goblet cells have large hyperchromatic nuclei, prominent nucleoli, and large mucus-containing vacuoles distending their

427

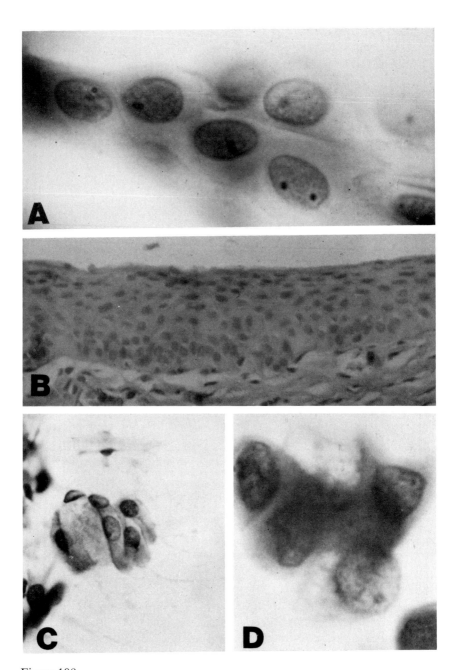

Figure 198
(A, B) Cytology and histology of the normal cornea. (C, D) Normal conjunctival cells.

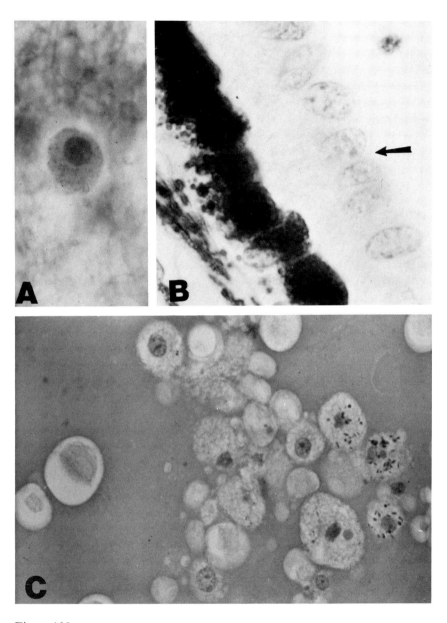

Figure 199
(A) Normal retinal cell in a corpus vitreous aspirate. ($\times$ 450) (B) Histologic
section of normal retinal mucosa showing the origin of the exfoliated cells.
($\times$ 750) (C) Pigmented retinal cells in the aspirate of fluid accumulated behind
a retinal detachment.

429

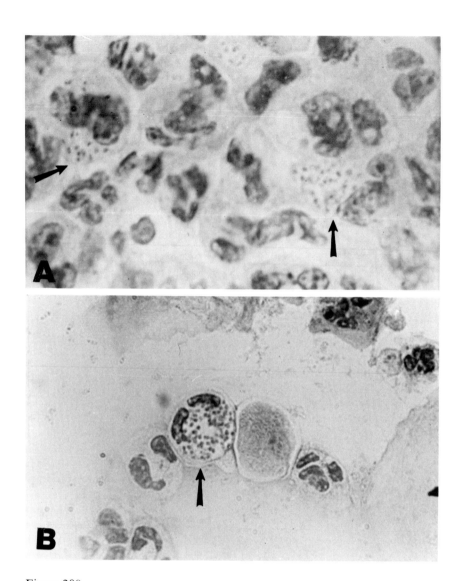

Figure 200
(A) Gonorrheal conjunctivitis. Note the gonococci in the cytoplasm of the neutrophils. (B) Conjunctival scraping in a case of allergic vernal conjunctivitis. Note the eosinophil in the center.

cytoplasm. The other conjunctival or corneal epithelial cells are rare and usually shed singly. They are hypertrophic, often well preserved, and occasionally (10%) show signs of hyperactivity in the hyperchromatism of their nuclei and in the prominence of their nucleoli. The background of the smear generally has a minimum amount of cytoplasmic debris. And compared to the smears of bacterial conjunctivitis, there is relatively little fibrinous exudate in the background.

Bacterial Infection (Fig. 200A)

With the Papanicolaou stain, the presence of pneumococci, gonococci, streptococci, or influenza bacilli can be readily detected in the background of the smear. The conjunctival or corneal epithelial cells are very scanty and show variable degrees of degeneration (vacuolization, eosinophilia, and nuclear pyknosis). The background of the smear is usually obscured by large masses of neutrophils, phagocytes, cellular debris, and fibrinous exudate.

Panophthalmitis

With panophthalmitis the aspiration of the vitreous chamber shows large amounts of clusters of polymorphonuclear leukocytes. Variable amounts of macrophages and foreign-body giant cells are also seen. The responsible microorganisms or fungi are usually readily found.

Mycotic Infection

Mycotic infection is often secondary to a systemic fungal infection or penetrating traumatic injuries or is a complication of prolonged corticosteroid or antibiotic treatment. Usually mycotic infection has a bad prognosis, and the condition often results in the loss of vision. The most common fungus is *Candida* (*Monilia*), which is easily recognized by the round or oval spores (round bodies) or elongated hyphae (long bodies). Other equally catastrophic fungi are the *Actinomyces, Aspergillus,* and *Sporotrichum.* The smears usually show, besides the fungus, large amounts of polymorphonuclear neutrophils, debris, phagocytes, and occasional foreign-body giant cells. The few conjunctival or corneal epithelial cells show marked degenerative changes. Large numbers of epithelial stripped nuclei are common.

Viral Infection

Clinically, viral infections are often difficult to differentiate from other types of ocular inflammatory lesions, but a smear will often pinpoint the exact nature of the infection.

TRACHOMA AND INCLUSION CONJUNCTIVITIS (FIGS. 201A, 202C). The TRIC (trachoma-inclusion conjunctivitis) organism, which is classified with the Bedsonia-Chlamydia agent, is one of the most common causes of acute conjunctivitis in the newborn. It is acquired during the fetus's passage through the mother's infected (inclusion vaginitis) genital region. The infected conjunctival cells are hypertrophic with clusters of basophilic intracytoplasmic inclusions each about 1 to 2 μ in diameter and found inside thin-walled vacuole-like structures located next to the nucleus. These granu-

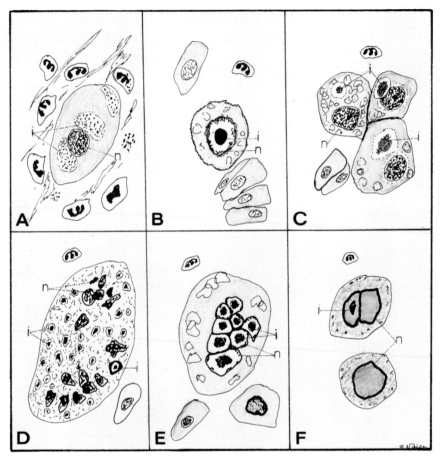

Figure 201
Cytology of various viral eye infections. (A) Inclusion conjunctivitis. (B)
Adenovirus epidemic keratoconjunctivitis. (C) Vaccinia keratitis. (D) Acute
catarrhal measles conjunctivitis. (E) Herpes simplex keratoconjunctivitis. (F)
Herpes zoster keratitis. (i = inclusion; n = nucleus)

lar inclusions are stained reddish brown and are surrounded by faint, microscopic, but individual, halos. In some cases of extensive inflammation, these
inclusions may be seen spilled outside the conjunctival cells in the background of the smear, probably originating from ruptured infected cells.
These chlamydial inclusions should be differentiated from greenish-brown
melanin pigment granules found in the cytoplasm of some conjunctival cells
scraped from dark-skinned patients. They should not be confused with
monococci phagocytized by leukocytes and histiocytes. The smears usually
show large amounts of inflammatory cells (polymorphonuclear cells), thick
mucus, and cellular debris.

ADENOVIRUS INFECTION—ACUTE EPIDEMIC KERATOCONJUNCTIVITIS (FIGS.
201B, 202A). The background of the corneal and conjunctival smears

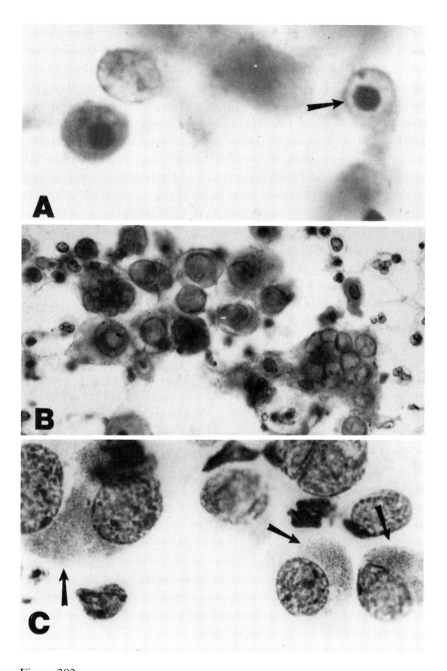

Figure 202
Conjunctival cells with (A) Adenoviral infection. Note the intranuclear inclusion. (× 1300). (B) Herpetic viral changes. Note the bland nuclei and the multinucleation of the cells. (× 450) (C) Inclusion conjunctivitis. Note the intracytoplasmic granular inclusions (*arrows*). (× 1300)

shows only a few lymphocytes and plasma cells with little fibrinous discharge or inflammatory neutrophils. The generally well-preserved infected epithelial cells are few, necessitating a long microscopic search under high-power magnification. The diagnosis is based on the presence of intranuclear inclusions in mononucleated, basal conjunctival or corneal cells. These inclusions are multiple in the early stage of the infection. Then they coalesce and form a single, large, basophilic, central inclusion. Their cytoplasm is not remarkable and shows few degenerative changes, in contrast to the early degeneration of the cells infected with the cytomegalovirus.

VACCINIA VIRAL INFECTIONS (FIG. 201C). The few diagnostic corneal cells present are moderately enlarged with abundant basophilic cytoplasm containing one or several irregularly shaped, large, eosinophilic cytoplasmic inclusions with a well-defined halo. The remainder of the epithelial cells show no specific structural alteration. The background of the smear is usually clean with little inflammatory reaction unless a secondary infection has occurred.

HERPES SIMPLEX AND HERPES ZOSTER VIRAL INFECTIONS (FIGS. 201E, 202B). The scraping from the margin of the ulcerated area produces swollen epithelial cells, usually in sheets. The diagnosis of herpetic infection is based on the occasional multinucleated giant cells with characteristic multiple nuclei molded against one another. These enlarged nuclei have either a bland chromatin pattern or large, eosinophilic intranuclear inclusions surrounded by a prominent halo. The background of the smear is generally free from mucus or debris. Moderate amounts of mixed inflammatory cells are seen.

OTHER INFLAMMATION AND DEGENERATION

Juvenile Intraocular Xanthogranuloma (Fig. 203C)
The anterior chamber aspirate may contain large numbers of oval and elongated mononucleated cells with granular, foamy cytoplasm having indistinct borders. Their size is uniform, and they may occasionally contain phagocytized debris. Their nuclei are small, oval or elongated and central. They are hyperchromatic, almost pyknotic, with fuzzy borders. No nucleoli are seen. A few multinucleated giant cells may be present. Almost no inflammatory cells are seen except for a few lymphocytes and plasma cells. The background is usually moderately cluttered with granular deposits. There is a danger of mistaking these numerous cells with dark nuclei for malignant cells. The regularity of their size and other features is diagnostic.

Xanthelasma (Fig. 203A)
The scraping of these soft, yellow plaques of the rim of the eyelid will produce numerous large histiocytes (foam cells) with abundant, semitransparent cytoplasm containing multiple vacuoles with phagocytized debris and small lipid droplets.

Their eccentric, often bean-shaped, nuclei are usually single and basophilic. Occasional multinucleated giant cells are present.

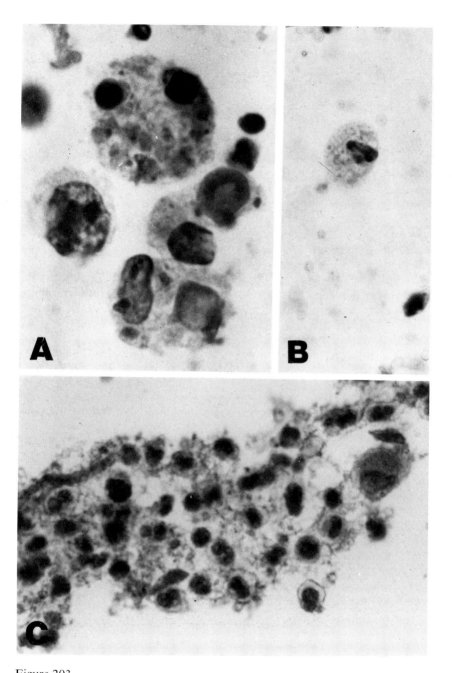

Figure 203
(A) Scraping of a xanthelasma of the eyelid. ($\times$ 1300) (B) Anterior chamber aspirate in a patient with phagolytic glaucoma. Note the phagocytized eosinophilic granules in the cytoplasm of the histiocyte. (C) Anterior chamber aspirate in a patient with juvenile xanthogranuloma. ($\times$ 450)

When the cells are examined with a polarized light, birefringent cholesterol crystals may be present.

Moderate amounts of neutrophils and cellular debris clutter the background.

Some anucleated squamous cells with no remarkable features are usually present.

Glaucoma

The smears from the aspiration of the anterior chamber fluid, in glaucoma, are usually scanty in cells. Large numbers of irregularly shaped and deep blue-stained precipitate of the aqueous fluid forms fiberlike structures that clutter the background of the smear. Among this precipitate, isolated anucleated squamous cells with abundant poorly stained, transparent cytoplasm are usually present. Except for a few monocytes, no other cells are seen in an uncomplicated glaucoma.

Phagolytic Glaucoma (Fig. 203B)

The escape of liquefied lens protein into the anterior and posterior chambers during the extraction of a matured cataract (opacified lens) stimulates an accumulation of macrophages that can lead to the loss of vision. A rapid diagnosis of this occurrence is essential for proper treatment.

The aspirate of the anterior chamber contains numerous vacuolated macrophages with phagocytized pink-to-gray granules (liquefied cortical material). Occasional multinucleated foreign-body giant cells are present. The background of the smear is usually covered with a finely granular eosinophilic precipitate. Except for a few lymphocytes, no other inflammatory cells are present.

BENIGN TUMORS

Papilloma (Fig. 204A, B)

The cytologic diagnosis of papilloma of the conjunctiva and cornea is based on the presence in the direct scraping of the tumor of elongated dark-staining epithelial cells, usually found in sheets. They have abundant, elongated cytoplasm and uniform, oval nuclei with occasional prominent nucleoli and smooth nuclear membranes. These cells are usually well preserved and show a variable degree of keratinization. The absence of goblet cells and of inflammatory background in the smears differentiates these lesions from the grossly similar infected seborrheic warts.

Keratinization (Figs. 205A, 206)

Keratinization of the conjunctival and corneal mucosa may be the result of chronic irritation, vitamin A deficiency, keratoconjunctivitis sicca, or premalignant atypia (Bowen's disease). The pathognomonic enlarged cells of corneal or conjunctival origin have an adequate to abundant, deeply orangeophilic, glassy-appearing cytoplasm. Occasionally keratohyalin granules form concentric perinuclear rings. The central nuclei of the keratinized cells are single or multiple and vary moderately in size and shape. The

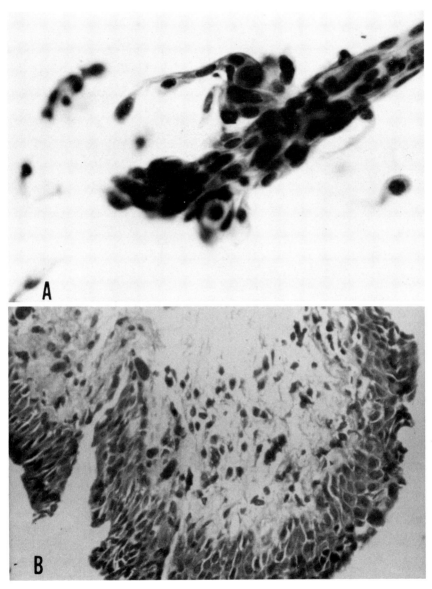

Figure 204
Cytology (A) and histology (B) of a conjunctival papilloma. ($\times$ 450)

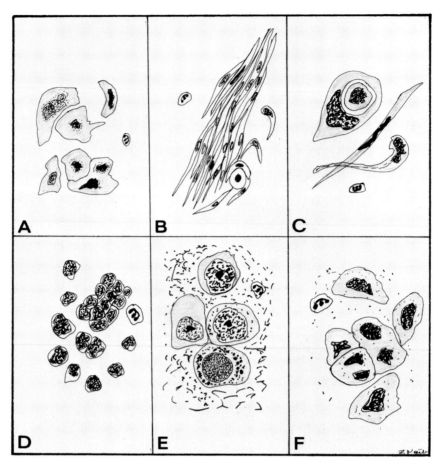

Figure 205
Cytology of various neoplasms of the eye. (A) Squamous dyskeratosis. (B) Conjunctival papilloma. (C) Squamous cell carcinoma. (D) Retinoblastoma. (E) Melanosarcoma. (F) Transitional cell carcinoma involving the orbital content.

chromatin is often very coarsely granular with slight irregularity in distribution. The cells shed singly or in sheets and show marked variation in size and shape, occasionally occurring in tadpole or spindle form. The additional presence of large numbers of conjunctival goblet cells trapped in abundant mucous strands indicates a vitamin A deficiency rather than a premalignant lesion such as Bowen's disease.

These hyperkeratinized cells are sometimes hard to differentiate from similar looking cells exfoliating from benign keratoacanthoma or from malignant invasive squamous carcinoma. In cases of doubt, a biopsy is usually necessary.

Squamous Cell in Situ and Invasive Carcinoma of the Conjunctiva,
Eyelid, or Cornea (Figs. 205C, 206B)

In squamous cell carcinoma of the eye, the exfoliated cells are similar to the deep orange-stained malignant squamous cells previously described for other sites.

Basal Cell Carcinoma

This is the most common malignant neoplasm of the eyelid. The scraping produces numerous sheets of basal cells that are fairly uniform in size (10–15 μ). Their pink cytoplasm is scanty and may contain greenish-brown melanin pigment granules. The single nucleus is round and regular with increased granular chromatin. Nucleoli are rarely seen.

The presence of inflammatory neutrophils is mainly the result of a secondary infection.

Retinoblastoma (Figs. 205D, 206D)

The needle aspirate of the aqueous humor of the vitreous chamber, or of subretinal fluid in the case of retinal detachment, usually contains numerous diagnostic cells. They are found singly, in clusters, and in sheets and, on rare occasions, in typical rosette or tubular formation. A good portion of these cells, having exfoliated some time earlier, show different stages of degenerative changes. The cytoplasm of the well preserved malignant cells is scanty and basophilic. The nuclei are single, hyperchromatic, oval or irregular in shape. The nuclear membranes are irregular in thickness and may show marked indentation (30%). Their chromatin is darkly basophilic, showing coarse clumping with abnormal pointed projections. The nucleoli are rarely seen. Variable amounts of cellular debris, protein, fibrinous mucous material, and degenerative cells can be seen deposited in the background of the aspirated smear. Relatively few inflammatory cells are ever found.

Malignant Melanoma (Figs. 205E, 206C)

With malignant melanoma, the aspirate of the anterior and vitreous chambers in uveal malignant melanoma or of the fluid that often accumulates behind the retinal detachment common in choroidal melanomas may show large numbers of diagnostic cells shed singly or in loose clusters. The usually large malignant cells vary in size (25 $\pm$ 10 μ) and shape, round (30%), oval (30%), or irregular (40%). The cytoplasm is adequate and may contain variable amounts of reddish-brown to black granular melanin pigment. The pleomorphic large, round or oval nuclei are single or multiple, with sharp, irregular nuclear membranes and coarsely granular chromatin with a tendency to condense toward the periphery. Single or multiple prominent nucleoli, often irregular in shape, or generally present. These malignant cells must be differentiated from the benign retinal cells containing melanin pigment that are seen in the aqueous humor in some degenerative retinal diseases and in the subretinal fluid of retinal detachment. In the benign cells,

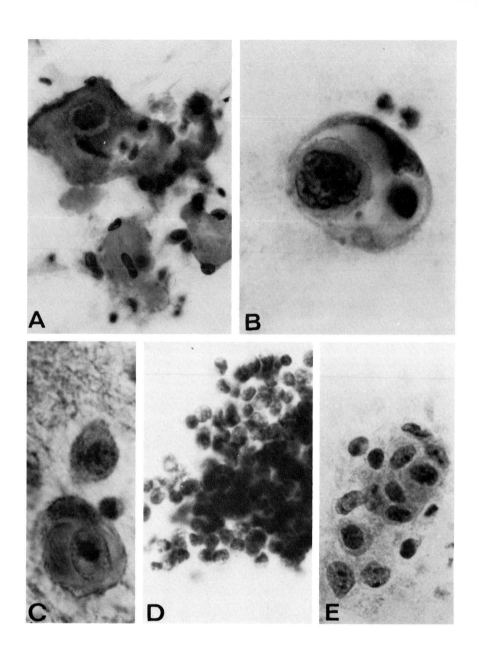

the melanin pigment granules are more regular in shape and uniform in amount from one cell to another. The maligant cells are characterized by their larger size, the increased irregularity of their chromatin, and the prominence of their nucleoli. In hyperpigmented cases of melanosarcoma, it is common to find free melanin pigment granules of varying size and shape cluttering the background of the smear.

Secondary Malignant Neoplasms (Figs. 205F, 206E)
The eye, ciliary body, and orbital content can be invaded by various distant metastases or by the expansion of an adjoining malignant tumor. A needle aspirate of the suspected tumor will usually show diagnostic clusters of neoplastic cells foreign to the eye that vary structurally according to the type of primary neoplasm.

References and Supplementary Reading

Agarwal, L. P., and Malhoutra, R. L.   Conjunctival smear cytology in xerosis. *Ophthalmologica* 130:378–386, 1955.

Azevedo, M. L.   Sex chromatin detection in epithelial cells of the conjunctiva. *Ophthalmologica* 142:425–428, 1961.

Azevedo, M. L.   Citologia ocular. M.D. Thesis, School of Medicine, University of Sao Paulo, 1962.

Azevedo, M. L.   Ocular exfoliative cytology. Basic principles and normal aspects. *Ophthalmol. Iber. Am.* 24:22, 1963.

Bland, J. O. W., and Rubinow, C. F.   The inclusion bodies of vaccinia and their relationship to the elementary bodies studied in cultures of the rabbit's cornea. *J. Pathol. Bacteriol.* 48:381, 1939.

Charlin, C.   Erreurs de diagnostic clinique dans certains tumeurs endoculaires malignes. *Arch. Ophthalmol.* (Paris) 33:103, 1973.

Contifari, N., and Nicolaou, D.   Le Cyto-diagnostie en Ophthalmologie. *Arch. Ophthalmol.* (Paris) 19:360, 1959.

Dawson, C. R., and Schachter, J.   TRIC agent infections of the eye and genital tract. *Am. J. Ophthalmol.* 63:1289, 1967.

Duszynski, R.   Cytology of conjunctival sac. *Am. J. Ophthalmol.* 37:576–578, 1954.

Dykstra, P. C., and Dykstra, B. A.   The cytologic diagnosis of carcinoma and related lesions of the ocular conjunctiva and cornea. *Trans. Am. Acad. Ophthalmol. Otolaryngol.* 73:979, 1969.

Figure 206
(A) Dyskeratotic squamous cells in the scraping of the conjunctiva of a patient with keratoconjunctivitis sicca. ($\times$ 650) (B) Malignant squamous cells in the scraping of an early ulcerated conjunctival squamous cell carcinoma. ($\times$ 1300) (C) Melanocarcinoma cells in an aspirate from the vitreous chamber in a patient with a primary retinal melanoma. ($\times$ 1300) (D) Clusters of neoplastic cells in an aspirate from the vitreous chamber in a boy with retinoblastoma. ($\times$ 750) (E) Cluster of malignant cells in an aspirate from the orbital cavity in a patient with transitional carcinoma of the maxillary sinus invading the orbit. ($\times$ 750)

Gaipa, M. Il Metodo de Papanicolaou vella diagnosi dei tumori dell'occhio e degli annessi. *Boll. Oculist.* 35:491, 1956.

Houser, S. A., Moor, N. A., and Stickle, A. W. Cytology of allergic conjunctivitis. *Arch. Ophthalmol.* 47:728–733, 1952.

Kimura, S., and Thygeson, P. The cytology of external ocular disease. *Am. J. Ophthalmol.* 39:137–144, 1955.

Liolet, S., and Iris, L. Etude analytique de la cytologie conjonctival normale et pathologique. *Arch. Ophthalmol.* (Paris) 23:453-468, 1963.

Mori, N. Morphology of inclusion bodies in trachoma and inclusion bodies in trachomatous conjunctivitis. *Am. J. Ophthalmol.* 45:423–432, 1958.

Naib, Z. M. Cytology of TRIC agent infection of the eye of newborn infants and their mothers' genital tracts. *Acta Cytol.* 14:390, 1970.

Naib, Z. M. Oculogenital infection with TRIC agents. *Clin. Obstet. Gynecol.* 15:976, 1972.

Naib, Z. M., Clepper, A. S., and Elliott, S. R. Exfoliative cytology as an aid in the diagnosis of ophthalmic lesions. *Acta Cytol.* 11:295, 1967.

Norn, M. S. Cytology of the conjunctival fluid. Experimental and clinical studies based on a quantitative pipette method. *Acta Ophthalmol.* (Kbh.) (Suppl. 59), 1960.

Sagiroglu, N., Ozgonul, T., and Muderris, S. Diagnostic intraocular cytology *Acta Cytol.* 19:32, 1975.

Sezer, F. O. N. Cytology of trachoma. *Am. J. Ophthalmol.* 29:1499, 1940.

Thygeson, P. Cytology of conjunctival sac. *Am. J. Ophthalmol.* 34:1709, 1952.

Timsit, E. Le Cyto-diagnostie en ophthalmologie. Note de technique practique. *Algerie Med.* 59:581–583, 1955.

Walter, J. R., and Naylor, B. A membrane filter method used to diagnose intraocular tumor. *J. Pediatr. Ophthalmol.* 5:36, 1968.

The Central Nervous System 17

The Cerebrospinal Fluid

The cytologic interpretation of the exfoliated cells found in spinal fluid is difficult because of the usually small volume of specimen available, the scantiness and distortion of the cells, and the frequent extensive degeneration resulting from the time lapse between the exfoliation of these cells and the collection of the specimen.

The presence of cellular contaminants from other smears (often appearing as the result of a faulty processing technique) may be disastrous because any foreign epithelial cells, benign or malignant, seen in a spinal fluid can be interpreted as an indication of a neoplasm.

TECHNIQUE FOR COLLECTING SPECIMENS

The first few drops from the spinal fluid lumbar tap, if very bloody, may be discarded. Then as much spinal fluid as clinical judgment allows should be collected. If the spinal fluid cannot be processed immediately, or if the centrifugation method is used, the spinal fluid should be mixed with the same amount of 50% ethyl alcohol and saline and refrigerated.

Cell Concentration Techniques

CENTRIFUGATION METHOD. The alcohol-diluted aspirated fluid is centrifuged at 1500 to 2000 rpm for about five minutes and one or two thick smears are made from the sediment on albumin-covered or completely frosted glass slides.

CYTOCENTRIFUGE* OR SEDIMENTATION METHOD. 0.2 to 0.5 milliliters of the nondiluted spinal fluid are placed in small plastic tubes with lateral small holes. These holes are obstructed by an adjacent glass slide covered by a hard filter paper with central small (7 mm) hole matching the size of the centrifuge tube chamber window. The specimens are then centrifuged at 1500 rpm for 10 minutes. The fluid is slowly absorbed by the filter paper, leaving the cells deposited on a small area of the surface of the glass slide.

* Cytocentrifuge, Shandon Scientific Company, Selwicky, Pennsylvania.

443

MEMBRANE FILTER METHOD.   The specimen is passed through a membrane filter with a 5 μ pore opening, and the deposited cells are fixed by the slow passage of 95% alcohol solution through the filter.

Staining Technique

The spinal fluid specimen should be stained by itself; and all the stains, the alcohol, and the xylene solution used should be filtered beforehand. It is very important to eliminate any possibility of cells floating in the staining or fixation fluids attaching themselves to the surface of the smeared slide or membrane filter. The chance of a false-positive interpretation needs to be avoided at all costs. Routine Papanicolaou staining technique is usually used except when a fungal infection is clinically suspected. In these cases a methenamine silver stain or an india ink preparation should be used on a portion of the smeared specimen.

Needle Aspiration

The direct aspiration of a cystic or solid lesion of the central nervous system with a 21-gauge needle attached to a 10 ml syringe may produce large numbers of diagnostic cells. This method is often less traumatic than a biopsy, and it allows an accurate diagnosis if the aspirate is obtained from the lesion proper. The aspirated material is either smeared directly on an alcohol-moistened slide or placed in suspension in saline, which is used to rinse the needle and syringe and then passed through a membrane filter.

Imprint Smear

The freshly cut surface of an excised tissue may be scraped and smeared or rolled over a completely frosted alcohol-moistened slide. This technique can be used as an adjunct to frozen section.

CYTOLOGY OF THE BENIGN CEREBROSPINAL FLUID

The normal spinal fluid is almost acellular, except for a few scattered red blood cells and small shrunken lymphocytes (no more than 5 lymphocytes should be found in 1 ml of fluid). Occasionally, a few histiocytes and leukocytes may be seen without a disease being present.

Ependymal Cells (Fig. 207A)

A few cuboidal ependymal cells originating from the ventricular epithelium are occasionally seen, especially in the presence of degenerative cerebral diseases. These cells are recognized by the regularity of their size (12 $\pm$ 3 μ) and by their transparent, oval, elongated cytoplasm with sharply defined borders. Occasional cilia and multiple small vacuoles in the basophilic cytoplasm may be seen. The nuclei, which are single and central or paracentral, are often degenerate, showing pyknosis or having a pale, bland appearance. The nucleolus is usually small and sharply defined. In cases of chronic irritation (cerebrospinal fluid from patients with hydrocephalus, for example), a marked polymorphism may be seen that sometimes makes these cells difficult to differentiate from tumor cells.

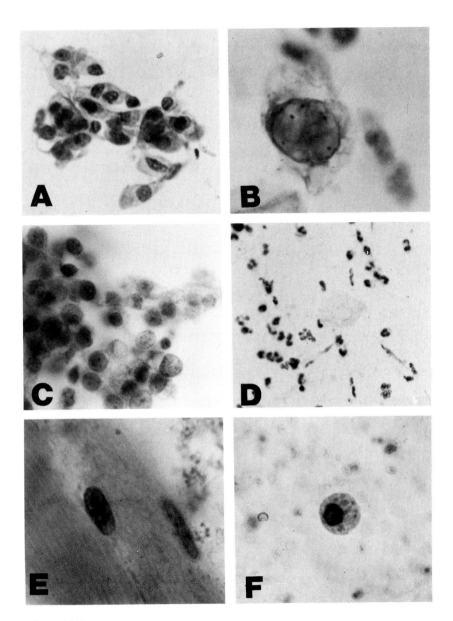

Figure 207
Spinal fluid collected by a ventricular aspiration. (A) Ependymal cells. (× 450)
(B) Pia-arachnoid cells. (× 750) (C) Choroid plexus cells. (× 450) (D) Anucleated squamous cells (contaminant). (× 450) (E) Striated muscle cells (contaminant). (× 450) (F) Hemosiderin-containing histiocytes. (× 450)

Pia-arachnoid Cells (Fig. 207B)

These cells have eccentric, round-to-oval or bean-shaped hyperchromatic nuclei and abundant, foamy cytoplasm with poorly defined borders. The cells are round and usually single and mononucleated. Nucleoli are seldom seen. The cytoplasm is often lysed, and elongated bare nuclei can be seen.

Choroid Plexus Epithelial Cells (Fig. 207C)

These cells, which resemble macrophages, may, in hemorrhagic lesions, contain hemosiderin pigment in their cytoplasm. They are often found in loose papillary clusters. These cells have adequate thick, granular, eosinophilic cytoplasm and no cilia. The dark, ovoid, single nuclei appear pyknotic when degenerate. They may be the cause of false-positive cytologic interpretation.

Squamous Cells (Fig. 207D)

These are usually contaminants from the skin of the patient or the operator during lumbar punctures. One must be aware that some of these keratinized squamous cells may originate from a ruptured craniopharyngioma cyst. In such a case there will also be an abundance of inflammatory cells (mainly lymphocytes).

Other Cells (Fig. 207E)

Occasionally, round, swollen, degenerated normal *astrocytes* may be seen in the spinal fluid of patients with cerebrovascular accident, multiple sclerosis, encephalitis, or subdural hematoma. Their eccentric nuclei are often hyperchromatic or pyknotic, and they should not be mistaken for neoplastic cells.

Fragments of capillary blood vessels, striated muscle, adipose, and fibrous tissue may be carried by the needle puncture from the subcutaneous tissue and musculature. Unless contamination can be ruled out, the presence of any other type of alien cell indicates pathology.

With very few exceptions, the early detection of a primary cerebral neoplasm by cytology is very difficult. The metastatic carcinoma is easier to diagnose.

INFLAMMATORY DISEASE

In *pyogenic infection* (septic meningitis), smears of spinal fluid contain large numbers of inflammatory leukocytes and an abundance of fibrin in deposits in the background. Variable amounts of phagocytes, degenerate ependymal cells, cellular debris, and even colonies of microorganisms may be seen. The amounts of inflammation, cellular debris, and fibrin deposits increase in the case of a brain abscess. Such a smear should be screened very carefully to rule out the possibility of a cancerous meningitis.

In *tuberculous infection* there is an increased number of mature lymphocytes in the smear. They are uniform in size and shape. They are recognized by their coarse, hyperchromatic, round nuclei surrounded by scanty, pale, basophilic cytoplasm. The Langhans' giant cells and the epithelioid histiocytes diagnostic of an active tuberculosis in other sites are almost never seen. The background of the smear shows only a small amount of protein deposit. A special acid-fast stain can be used for the detection of micro-organisms on the surface of the filter membrane.

The predominance of lymphocytes is not specific to tuberculosis; it can also be seen in luetic disease, tabes dorsalis, unruptured abscesses, multiple sclerosis, and cerebral hemorrhage or thrombosis.

In *fungal infection* the diagnosis is made from the presence of various fungi in the forms of spores or hyphae mixed with variable amounts of inflammatory cells. The different fungi infecting the meninges are readily recognized by their structure, as described in Chapter 11. (Meningitis can be due to cryptococcosis, blastomycocis, histoplasmosis, coccidioidomycosis, and actinomycosis.)

The most common fungal infectious disease of the central nervous system is *cryptococcosis* (*torulosis*) (Fig. 208A, B). The causative fungi occur more frequently in debilitated patients with preexisting lymphoma or other advanced malignant diseases. Yeastlike fungi (*Cryptococcus neoformans* or *Torula histolytica*) measuring 5 to 15 μ reproduce by budding. These free, round-to-oval, elliptical, purple-stained yeast cells are easily seen under even low-power magnification and recognized as cryptococci mainly by their thick, semitransparent, refractile, mucinous capsules that create a clear halo around the purple-red–stained spore. Buds are attached to the mother cell by variable-sized but usually narrow necks, which give the mother cell the characteristic teardrop shape. The main danger in not recognizing them is to mistake them for talcum powder–starch crystals, which are often found contaminating the spinal fluid and other cytologic specimens. Occasionally, the capsule may shrink, leaving only a halo around the fungus. If this infection is suspected, an india ink preparation may be helpful. It consists of examining a mixture of a drop of india ink and a drop of spinal fluid sediment smeared on a slide. The gelatinous capsule is seen as a clear or transparent halo around each fungi. Cryptococci are easily isolated on Sabouraud's agar medium.

The background of the smear may contain variable numbers of lymphocytes and mononuclear cells (plasma cells), with occasional phagocytes and foreign-body giant cells. One should be aware that contamination of the filter membrane by opportunistic fungi and various yeasts may occur if the open box of filters is left in a warm, humid place or the staining solution is kept too long and not filtered adequately. The fungus in such a case is usually on both sides of the filter or at a different focusing level from the cells.

In *viral infections,* when the meninges are involved, an increased number of nonspecific mature lymphocytes, plasma cells, and small phagocytes may be seen in the smears. In herpes simplex and herpes zoster encephalitis, occasionally diagnostic giant cells with multiple nuclei containing intranuclear inclusion bodies are found.

In *cytomegalovirus infection,* the inclusion-bearing infected cells are rare and usually very degenerate.

ASTROGLIOSIS

In certain cases of encephalomalacia, necrosis, and softening of the brain substance due to the occlusion of a nutrient artery, large numbers of degenerated glial cells may occasionally be seen in the spinal fluid. The enlarged

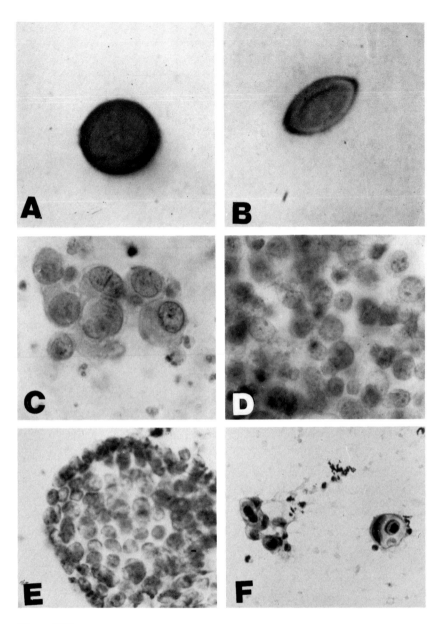

Figure 208
(A, B) Cryptococci in spinal fluid. Note the round and oval shapes. The mucinous capsule, although present, is hard to see. ($\times$ 1300) (C) Cluster of cells in a pinealoma. Note the vesicular nuclei and prominent nucleoli. ($\times$ 750) (D) Aspirate from a suprasellar chromophobic adenoma. ($\times$ 750) (E) Aspiration smear from a medulloblastoma. ($\times$ 750) (F) Squamous cancer cells in the spinal fluid of a patient with metastatic lung carcinoma. ($\times$ 450)

astrocytes are usually single or in loose clusters. Remnants of lacy transparent cytoplasm may be seen still attached to some of the angular nuclei. The majority of these enlarged nuclei are without any cytoplasm.

The ill-defined chromatin is coarsely granular and is condensed toward the nuclear membrane. Nucleoli are fuzzy but prominent in a good portion of these nuclei. The background of the smear is usually cluttered with gray cytoplasmic debris and granular protein deposits. Besides the few leukocytes, plasma cells, and lymphocytes, hemosiderin pigment-containing phagocytes are usually present.

Because of the abundance of these degenerated benign glial cells there is a danger of misinterpreting them as exfoliated tumor cells. The marked degeneration of these cells is diagnostic.

PRIMARY TUMORS
The involvement of the meninges by a primary tumor should be suspected in the presence of abnormal or alien epithelial cells and in an increase in inflammatory cells. The majority of the primary neoplasms are deep lying and do not shed cells into the spinal fluid (false-negative, over 65%). Most of the time no direct communication exists between the tumor and the cerebral or spinal cavities. A positive spinal fluid finding usually indicates invasion of the leptomeninges or the ventricular ependymal membrane. (At autopsy, less than 35% of gliomas involve the leptomeninges.)

Meningioma (Fig. 210A)
A *fibrous meningioma* may be diagnosed by the presence in the spinal fluid of variable numbers of tight groups of cells with marked molding, scanty cytoplasm, and dark, elongated nuclei. The epitheliumlike cells are often arranged in compact whorls or pearl-like formations with occasional hyalinization of the central cells. Their transparent eosinophilic and adequate cytoplasm is often elongated and may resemble glial processes. Their ovoid nuclei are vesicular and uniform in size. No mitotic figures are seen. Occasional psammomalike formations may be seen.

The *malignant meningioma* may produce marked anaplasia of the cells with giant cell formation and numerous abnormal mitoses that are sometimes difficult to differentiate from the metastatic squamous cell carcinomas. These tumors are considered to be of arachnoidal origin. The background of the smear shows the usual increase of mixed inflammatory cells and protein deposit.

Glioblastoma (Astrocytoma) (Fig. 209A, B)
The diagnostic cells indicating a glioma or glioblastoma are found in the lumbar spinal fluid only if the tumor is large and has infiltrated the meninges. Single cells or clusters of cells with marked variation in size (12–25 μ in diameter) are usually present in varying amounts depending on the grading of the tumor. The amount of cytoplasm varies from scanty to abundant. It is usually pale blue to deep blue, according to the degree of cellular degeneration present and has extremely fine vacuoles. The cytoplasmic borders are irregular and indistinct. The shape of the cells is often elongated or

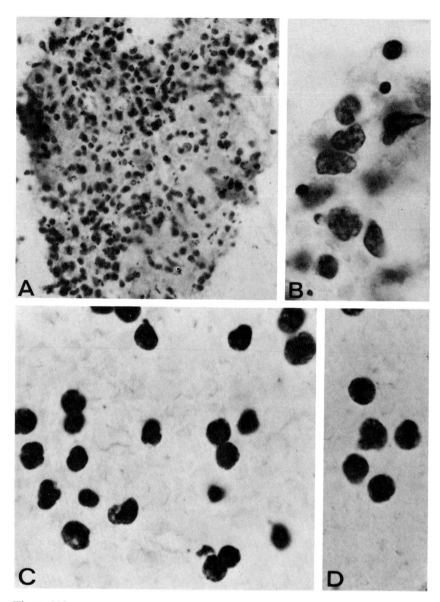

Figure 209
(A) Large clusters of neoplastic cells in the spinal fluid aspirated from the cerebral ventricle of a patient with a glioblastoma. (× 125) (B) The same cells
seen under high-power magnification. Note the cellular molding and irregularity
of the nuclei. (× 1300) (C) Lymphosarcoma cells in the spinal fluid of a patient with an advanced neoplasm. Note the irregularity of size and shape of the
cells. (× 1300) (D) Reticulosarcoma cells in the spinal fluid of a young girl
whose meninges were shown at autopsy to be involved. (× 1300)

triangular. The nuclei, single or multiple, are usually vesicular, enlarged, with variable chromatin clumping (from fine to coarse), and prominent irregular nucleoli, depending on the grading of the malignancy. The background of the smear shows large amounts of cellular debris, inflammatory cells, protein precipitate, and fresh and old blood.

In *glioblastoma multiforme,* multinucleated giant cells may occasionally be present. The glial cells vary in size and shape and may resemble the cells of anaplastic metastatic carcinoma.

Ependymoma and Choroid Plexus Papilloma

Ependymomas occur mainly in children and involve the ventricular passageway. The spinal fluid, if aspirated from one of the ventricles, may be very rich in round, malignant cells showing little polymorphism. The cuboidal cells are shed singly, in loose clusters, in tight papillary projections, or in rosette formation. They have scanty, semitransparent to pale blue, lacy cytoplasm with no glial processes. The often eccentric nuclei are vesicular, hyperchromatic round to oval and vary in size with occasional prominent but small nucleoli.

Mitoses are rarely seen. The cytologic diagnosis of ependymoma is easier when densely cellular papillary projections are present. These papillary structures should be differentiated from the ones seen in cases of choroid plexus papilloma, in which the individual cells show a greater variation in size and shape and in which the nucleoli are often more prominent and multiple.

Medulloblastoma (Fig. 208E)

The cerebrospinal fluid, especially if it has been aspirated directly from the ventricle, may contain large numbers of single neoplastic cells. They are also seen in small loose clusters forming rosettelike structures. The cellular appearance varies from one tumor to another, depending on the grading of the tumor. The cells are generally small (14–20 μ in diameter) and often cuboidal in shape. They resemble the tumor cells seen in pulmonary oat-cell carcinoma and are about the same size as reactive lymphocytes, from which they must be differentiated. Their cytoplasm is very scanty, transparent, vesicular, and poorly defined. The nuclei are dark, centrally located, round or oval, and regular in size. The dense chromatin is coarsely granular in the well-preserved cells and pyknotic and fuzzy in the degenerate ones. Occasional small nucleoli are seen. Some of the cells are binucleated. Few mitoses are seen. An increased number of monocytes and other inflammatory cells is usually present. Large amounts of precipitate may be seen in the background of the smears. In children this tumor grows fairly rapidly, and cytologic study of the ventricular fluid usually is highly successful in permitting early detection.

Sarcoma

Arachnoidal sarcomas may shed large amounts of cells, often seen in tight clusters or sheets. The cells show moderate variation in size. Their eosinophilic cytoplasm varies from scanty to adequate. Their nuclei are often elongated and hyperchromatic, almost pyknotic. No nucleoli are seen.

Chromophobe Adenoma (Fig. 208D)
Large numbers of tumor cells may be seen singly, in clusters, or in small sheets in the spinal fluid after an incomplete resection of the tumor. Their cytoplasm is usually poorly defined and varies in amount. Their round to oval, single nuclei vary in size. The chromatin is finely granular and regularly distributed. Their nucleoli are often prominent. The inflammatory cells are scarce, and only a small amount of protein deposit is seen in the background.

Pinealoma (Fig. 208C)
The cells exfoliating from this rare tumor are usually single or in loose clusters. Polygonal in shape, their vesicular, scanty to adequate cytoplasm is often indistinct. Their round or polyhedral nuclei are large and vesicular with prominent nucleoli. These nucleoli are often multiple and vary in size and shape and constitute one of the main diagnostic features.

Craniopharyngioma
This cystic neoplasm can rupture into the meninges, liberating large numbers of anucleated squamous cornified cells. These cells are often difficult to differentiate from contamination of the specimen by keratinized cells originating from the surface of the skin. The diagnosis is made easier by the marked degeneration of these neoplastic cells and the intense inflammatory reaction (leukocytes) they usually produce.

Lymphoma and Acute Leukemia (Fig. 209C, D)
When the leptomeninges are infiltrated by leukemic or lymphomatous lesions, large numbers of lymphocytes may be seen in the spinal fluid. The majority of these lymphocytes are immature in type and vary in size and shape. They are often seen in clusters. Their nuclei are large, hyperchromatic, and lobular or irregular with occasional prominent nucleoli. Mitoses are occasionally seen.

Secondary Tumors (Figs. 208F, 210B, C)
Secondary tumors (approximately 28% of all intracranial neoplasms) are easier to diagnose than primary ones. Large numbers of single or small sheets of malignant cells can easily be seen, with the structure varying according to the type of the primary neoplasm. The cell size is larger than that of any normal cells seen in the spinal fluid. The malignant cells, often in clusters, show advanced degenerative changes since they have frequently been exfoliated for a long time. Almost all carcinomas or sarcomas have the capacity to metastasize to the brain and the meninges, the most common ones to do so being pulmonary, embryonal, breast, renal, melanomatous, and gastrointestinal carcinomas. The cervical cancer will metastasize to the brain only infrequently. The malignant metastatic squamous cells are usually from poorly differentiated (anaplastic) carcinomas of pulmonary ori-

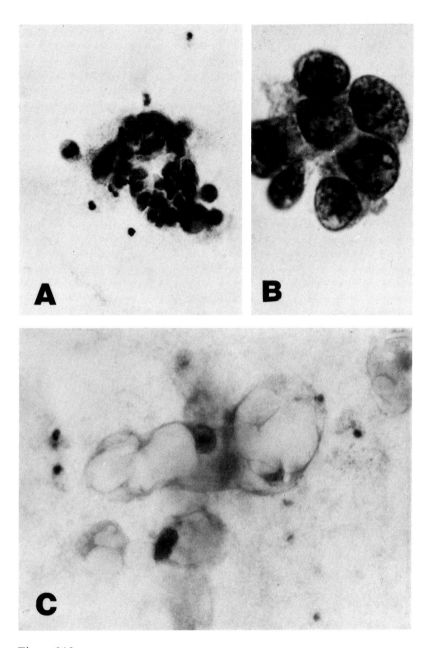

Figure 210
(A) Cluster of neoplastic cells in the spinal fluid of a patient with a meningioma. (× 450) (B) Cluster of neoplastic cells in the spinal fluid of a patient
with meningeal carcinomatosis. (× 750) (C) Hypervacuolated cells in the
spinal fluid of a patient with colonic adenocarcinoma with cerebral metastasis.
Note the large secretory intracytoplasmic vacuole. (× 750)

453

gin. The few well-differentiated carcinomas have abundant orangeophilic cytoplasm. The adenocarcinoma may show hypersecretory cytoplasmic vacuoles and prominent nucleoli. Most of the metastatic carcinomas, especially the ones originating from the stomach, produce a diffuse meningeal carcinomatosis with an abundance of malignant and inflammatory cells. The background of the smears is usually obscured by large amounts of protein deposits, fresh and old red blood cells, and cellular debris. The cytologic interpretation of these alien malignant cells is not difficult. The problems stem from the possibility that some of these cells may be only contaminants from another specimen due to faulty processing technique.

References and Supplementary Reading

Balzereit, F.   Liquor Cytologie in der neurologischen Tumordiagnostik. *Internist* (Berlin) 7:122, 1966.

Baringer, J. R.   A simplified procedure for spinal fluid cytology. *Arch Neurol.* 22:305, 1970.

Billingham, M. E., et al.   The cytodiagnosis of malignant lymphomas and Hodgkin's disease in cerebrospinal fluids. *Acta Cytol.* 19:547, 1975.

Bischoff, A.   Erfahrungen mit der Tumorzelldiagnostik in Liquor Cerebrospinalis. *Acta Neurochir.* (Wein) 9:510, 1961.

Bots, G., Went, L., and Shaberg, A.   Results of a sedimentation technique for cytology of cerebrospinal fluid. *Acta Cytol.* 8:234, 1964.

Burton, J. F.   Tumor cells in cerebrospinal fluid. *Med. Radiogr. Photogr.* 37: 22–23, 1961.

Den Hartog, J.   Cytopathology of the cerebrospinal fluid examined with the sedimentation technique. *J. Neurol. Sci.* 9:155, 1969.

Dufour, H.   Meningite sarcomateuse diffuse avec envahissement de la Moelle et des Racines. Cytologie positive et speciale du liquide cephalorachidien. *Rev. Neurol.* (Paris) 12:104–106, 1904.

El-Batata, M.   Cytology of cerebrospinal fluid in the diagnosis of malignancy. *J. Neurosurg.* 28:317, 1968.

Hashida, Y., and Yunis, E.   Re-examination of encephalitic brains known to contain intranuclear inclusion bodies. *Am. J. Clin. Pathol.* 53:537, 1970.

Jane, J. A., and Bertrand, G.   A cytological method for the diagnosis of tumors affecting the central nervous system. *J. Neuropathol. Exp. Neurol.* 21: 400–409, 1962.

Jane, J. A., and Yashon, D.   *Cytology of Tumors Affecting the Nervous System.* Springfield, Ill.: Thomas, 1969.

Krentz, M. J., and Dyken, P. R.   Cerebrospinal fluid cytomorphology. *Arch. Neurol.* 26:253, 1972.

Lups, S., and Haan, A. M. F.   *The Cerebrospinal Fluid.* New York: American Elsevier, 1954. Pp. 69–71; 261–267.

McCormick, W. F., and Coleman, S. A.   A membrane filter technic for cytology of spinal fluid. *Am. J. Clin. Pathol.* 38:191–197, 1962.

Marks, V., and Marrack, D.   Tumor cells in the cerebrospinal fluid. *J. Neurol. Neurosurg. Psychiatry* 23:194–201, 1960.

Neto, J. B.   Fatal primary amebic meningoencephalitis. *Am. J. Clin. Pathol.* 7:122–130, 1954.

Nies, B. A., Malmgren, R. A., Chu, E. W., Del Vecchio, P. R., Thomas, L. B., and Friereich, E. J.   Cerebrospinal fluid cytology in patients with acute leukemia. *Cancer* 18:1385, 1965.

Oshiro, H.   Clinical study of tumor cells in spinal fluid. *Med. J. Hiroshima Univ.* 14:623, 1966. (Special issue)

Skeel, R. T., Yankee, R. A., and Henderson, E. S.   Meningeal leukemia. Two simple methods for rapid detection of malignant cells in spinal fluid. *J.A.M.A.* 205:155, 1968.

Sornas, R.   Transformation of mononuclear cells in cerebrospinal fluid. *Acta Cytol.* 15:545, 1971.

Spriggs, A. I.   Malignant cells in cerebrospinal fluid. *J. Clin. Pathol.* 7:122–130, 1954.

Wertlake, P. T., and Markovits, B. A.   Cytologic evaluation of cerebrospinal fluid with clinical and histological correlation. *Acta Cytol.* 16:224, 1972.

Wisotzkey, H. M., and Novey, R.   Spinal fluid cytodiagnosis of central nervous system malignancy. *Bull. Univ. Maryland School Med.* 49:40, 1964.

Woodruff, K.   Cerebrospinal fluid cytomorphology using cytocentrifugation. *Am. J. Clin. Pathol.* 60:621, 1973.

The Skin

18

Cytology of Normal Skin

The skin is normally covered with thick layers of keratinized stratified squamous epithelium. The exfoliation of the superficial layers of this epithelium (stratum corneum) produces mainly single, clusters, or sheets of anucleated keratinized squamous cells. A very few may retain their nuclei.

Most of the nucleated cells originate from the deeper stratum granulosum, stratum spinosum, or stratum germinativum and indicate the presence of an erosion or ulceration. The polyhedral abundant cytoplasm of these deep cells often contains keratohyaline granules and shows intercellular bridges. Their round to oval, single nuclei are pyknotic or vesicular with smooth nuclear membranes and regularly distributed finely granular chromatin. No inflammatory cells are normally seen.

INDICATIONS

In patients with a skin lesion in which biopsy is not possible (e.g., patient refusal, inaccessible location) or is not indicated (benign-appearing lesion), cytology of the surface scraping may be of great assistance and can render accurate and specific diagnosis of the nature of the lesion.

In cases where a very rapid diagnosis is needed to begin proper treatment, cytology may often give an early (10 minutes) provisional diagnosis.

Furthermore, some fungal and viral infections are easier to recognize in a smear than in a biopsy.

It has been reported that the biopsy of certain tumors (malignant melanoma) may release neoplastic cells into lymphatic and blood vessels. Gentle scraping for cytologic study decreases the chances of such an occurrence.

TECHNIQUE FOR COLLECTING SPECIMENS

Since the surface of the skin is physiologically dry, most of the superficial lesions should be moistened before scraping. A dripping wet compress left for at least half an hour over a lesion not only makes it softer for easier scraping but also removes most of the loose, degenerated cellular debris and serum crust that otherwise would clutter the smear.

After removal of the crust, the small ulceration produced should be energetically scraped with a sharp curette. With a vesicle or bulla, the dome should be removed and the margins of the ulceration scraped. Some of the nonulcerated subepidermal lesions may need to be aspirated with a needle.

The material collected by scraping or aspiration is spread immediately on an alcohol-moistened, completely frosted slide. The alcohol may be allowed to evaporate before the slide is dropped gently into the fixative.

BACTERIAL INFECTION

In *furunculosis* and *pyoderma* (associated with *Staphylococcus aureus*), in *impetigo* (associated with *Streptococcus pyogenes*), in *erythrasma* (associated with *Corynebacterium*), and in other, less common bacterial skin infections, the scraping of the ulcerated surface contains large numbers of inflammatory polymorphonuclear leukocytes, cellular debris, and different monocytes mixed with variable amounts of fibrinous exudate.

Clusters of pathogenic and nonpathogenic bacteria may be found free in the background of the smears or adherent to the surface of the desquamated squamous cells or ingested in the cytoplasm of leukocytes and phagocytes.

The morphology of these microorganisms is usually distinct. Their recognition and typing often permits a specific diagnosis of the nature of the infection and suggests the type of culture medium to be used for best results.

Large amounts of degenerated epithelial and nonepithelial cellular debris and granular serum protein deposits are usually present, giving evidence of tissue necrosis. Occasional single or sheets of intact squamous cells are present in acute or chronic dermatitis. In the aspiration smears of cutaneous abscesses, no such epithelial cells are usually seen.

Some of these squamous cells have degenerative intracytoplasmic vacuoles, others are enlarged due to hydropic degeneration, and still others show nuclear pyknosis and cytoplasmic loss. These degenerative cells may be the source of an erroneous interpretation of malignancy.

In cases of ulceration and regeneration, connective tissue elements may be found mixed with repair squamous basal and parabasal cells with enlarged nuclei, coarsely granular chromatin, and prominent nucleoli. These regenerating basal cells may appear extremely atypical, especially if viewed with high magnification.

Erythrasma

This mild superficial infection of the body folds and clefts is caused by bacteria of the genus *Corynebacterium*. It is often concomitant with diabetes. The scraping usually shows little inflammatory reaction, with few leukocytes present in the midst of heavy serum protein granular deposits. The diagnosis is based on the presence of the small bacterial rods (1.5 to 2.0 μ in length) and filaments, both found attached to the cytoplasm of keratinized cells and free in the background. Moderate amounts of relatively well preserved single squamous basal cells are usually present. Their vesicular nuclei often have prominent nucleoli, and their cytoplasm may contain degenerative vacuoles.

Impetigo (Fig. 215A)

The bulla characteristic of this streptococcal infection ruptures early and produces an ulceration that is soon covered with a hard yellow crust. In newborns the lesions may be very extensive and have been referred to as pemphigus neonatorum. The smear taken after the removal of the thin crust contains mainly sheets of dense basophilic parabasal cells. The cytoplasm is usually scanty and very irregular in shape (pointed projections). The small, round basophilic nuclei are hyperchromatic with coarse, irregular chromatin granules. Nucleoli are not prominent. Large amounts of inflammatory cells (polymorphonuclear cells) are usually present. Chains of typical streptococci, single or in clusters, are easy to recognize.

Granuloma Inguinale

This venereal disease is caused by bacterium *Donovania granulomatis,* bean-shaped, pink-orange bacilli, also called *Donovan bodies.* The skin and mucosa of the genital and inguinal area are often ulcerated, and the regional lymph nodes enlarged. The extragenital lesions may be located on the lips, cheeks, neck, hands, or chest wall. The scrapings of the margins of these ulcerogranulomatous lesions contain large numbers of leukocytes as well as cellular debris, fibrinous exudate, and granular protein deposits cluttering the background. No lymphocytes are usually seen. The diagnosis is based on the presence of numerous Donovan bodies measuring 1 to 2 μ and found in the cytoplasm of large mononucleated histiocytes.

These bodies are clustered in multiple cystic spaces (2 to 10) with delicate walls often molding against each other in the cytoplasm of the macrophages.

Giemsa-stained bacilli are purple-rose; with the Wright stain they are deep purple; and with silver stain they are black. Because of the reported association of granuloma inguinale with squamous cell carcinoma, the smears from these lesions should be screened carefully for cancer.

FUNGAL INFECTION

More than half of the fungi infecting man limit their activities to the surface of the skin in the keratinized layer. A large number of the so-called deep fungi also infect the skin. The energetic scraping of fungal skin lesions or smearing of the hair follicles may provide a rapid diagnosis of the type of fungal infection. The recognition of the morphology of the fungi is also important because they are often seen as contaminants of various cytologic specimens.

Tinea Pedis (Trichophyton gypsum) (Fig. 211B)

This dermatophyton (superficial fungus) is the cause of the infection known as athlete's foot. The scraped, infected skin of the interdigital areas, where sweating is profuse, contains clumps of pale yellow, short, fat mycelia. Round, budding spores with thick membranes are also found attached to superficial keratinized squamous cells. In chronic infections, the scraping may contain only these spherical spores (6–8 μ in diameter) attached to degenerated anucleated squamous cells. Very few inflammatory cells are seen.

Tinea Capitis (Ringworm) (Fig. 211A)

The scraping of this infection of the scalp, mainly of children after infancy, contains colonies of loose fungi (*Microsporum audouinii* or *Trichophyton*

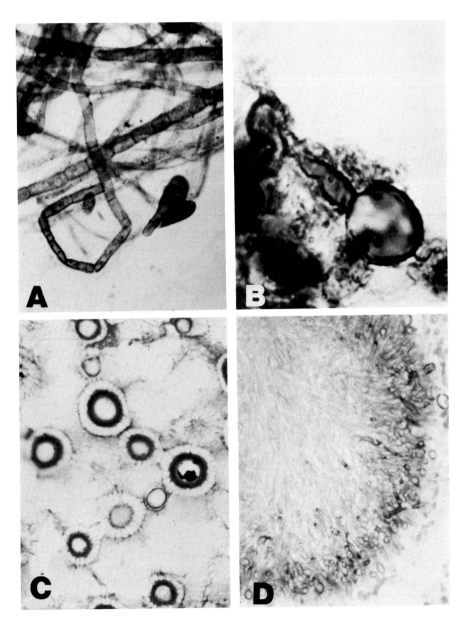

Figure 211
Scraping of superficial fungal lesions of the skin. (A) Tinea capitis. ($\times$ 750) (B) Tinea pedis. ($\times$ 750) (C) Rhinosporidiosis. ($\times$ 750) (D) Mycetoma.

tonsurans) with segmented hyphae and thick-walled multiseptate macroconidia. These spindle-shaped macroconidia have characteristic round, blunt ends and measure 10 to 40 μ in diameter. Moderate amounts of cellular debris and leukocytes are present.

Rhinosporidiosis (Fig. 211C)

The scraping of the polypoid friable nasal mass produced by this fungus generally contains large numbers of round endospores of *Rhinosporidium seeberi* with thick walls and transparent capsules. The spherical spores vary in size and have characteristic, raylike, linear inner structures. Occasional histiocytes and foreign-body giant cells are present. Few inflammatory cells, mainly lymphocytes, are seen.

Tinea Versicolor

The scraping contains numerous round-to-oval blastospores measuring 3 to 8 μ in diameter (*Malassezia furfur*). The hyphae are rare, poorly stained, and rodlike, and vary in size and shape. Their centers appear empty because of the thickness of their walls. Small amounts of fibrin, neutrophils, and necrotic squamous cells are present in the background.

Blastomycosis (North and South American)

The skin lesions usually occur on the exposed parts (face and hands) of the body. They are often clinically first suspected of being a basal-cell carcinoma. The diagnosis is based on the presence of fungi (*Blastomyces dermatitidis*) in the smear obtained from the scraping of a skin ulcer or smearing of the spontaneous drainage from the abscess or fistula that is common in advanced cases. The spherical fungi, measuring 5 to 8 μ have thick walls with double refractile contours and buds with a thick base. Their morphology is similar to those described in sputum (see Chapter 11).

Candidiasis (Candida albicans) (Fig. 212A)

This is a relatively frequent infection of abraded and constantly moist skin and nails of patients with various diseases (e.g., diabetes, malnutrition, obesity). The skin is often destroyed and replaced with a fibrinous membranelike mesh containing fungi and necrotic epithelial cells. The diagnosis is suggested by the presence of yellowish-brown–stained hyphae with regular segmentation and acute angle branching. The presence of single or grapelike clusters of red-to-brown pear-shaped blastospores, 2 to 4 μ in diameter, is diagnostic. Large amounts of fibrinous secretions, granular serum protein deposits, and necrotic epithelial cell debris usually clutter the background of the smear.

Mycetoma (Fig. 211D)

Aspiration of the granulomatous mass in the subcutaneous tissue or the smearing of pus usually draining from multiple sinuses of the infected area may produce large clusters of intermingled, diagnostic broad hyphae of *Allescheria boydii* or *Madurella mycetomi*. Small oval conidia are also usu-

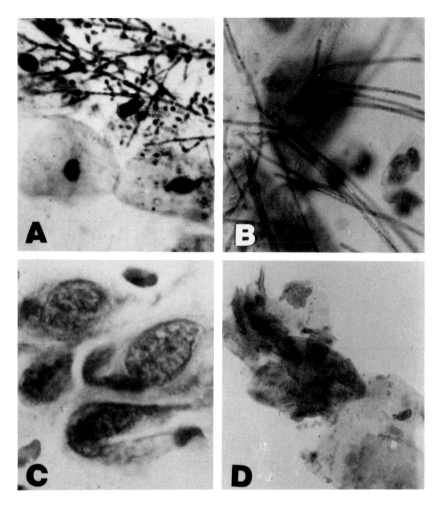

Figure 212
Scraping of (A) skin moniliasis (× 450), (B) nonspecific skin ulceration, with nonpathogenic contaminant fungi, (C) reparative granuloma (× 750), and (D) pityriasis rosea (× 450).

ally present. Large amounts of cellular debris, inflammatory leukocytes, histiocytes, and foreign-body giant cells clutter the smear.

NONSPECIFIC INFLAMMATION
Lichen Planus
This chronic dermatitis of unknown etiology, involving parts of the skin of the extremities and buccal mucosa, appears as papules, ulcers, or bullae that are usually markedly pruritic. A moderate hyperkeratosis, hyperplasia, and an uneven acanthosis with irregular elongation of the rete pegs (sawtooth-shaped) are diagnostic.

The scraping produces nucleated, usually single superficial and intermediate squamous cells with variable amounts of keratin. Degeneration is common especially in single basal cells, where most of the cytoplasm is hypervacuolated or lysed. Leukocytes are often present in these vacuoles. These changes are mostly nonspecific, and accurate cytologic diagnosis of the lesion is difficult. Large numbers of inflammatory lymphocytes and histiocytes are usually present. The background of the smear is often cluttered with debris and granular protein precipitate.

Pityriasis Rosea (Fig. 212D)
The scraping of these round, pinkish (salmon-colored) patches of the skin of the trunk along the line of cleavage produces large numbers of enlarged, edematous, anucleated squamous cells with little keratinization. In clusters some of these cells may retain small, pyknotic nuclei. Mixed inflammation (mainly lymphocytes) is usually present.

Chronic Discoid Lupus Erythematosus (LE)
The scraping of the butterfly lesion over the cheeks is rarely diagnostic. It may contain, in addition to the edematous anucleated squamous cells, fragments of elongated basophilic collagen and elastic fibrils. Monocytic inflammatory cells (lymphocytes, plasmacytes, and macrophages) are usually present. LE cells, leukocytes with their large cytoplasmic diagnostic inclusions, are almost never seen in the skin scraping.

Pemphigus Vulgaris (Fig. 213B)
Bullae of variable size, in various areas of the body are diagnostic of the lesion. The scraping of the raw bleeding surface resulting from the removal of the dome produces large numbers of loose clusters and sheets of round, squamous basal and intermediate cells with adequate brown-to-blue–stained cytoplasm. The round-to-oval central nuclei have coarse regularly distributed chromatin. Their prominent nucleoli vary in size, shape, and number. Binucleation and multinucleation are common. The nuclear membrane is thick but smooth. Degenerate perinuclear halos are often present. Moderate amounts of neutrophils may be seen. No inclusions are present. Because of their active nuclei and enlarged irregular nucleoli, these basal pemphigoid cells may be mistaken for poorly differentiated adenocarcinoma cells.

Bullous Pemphigoid Lesions (Fig. 213C)
The large bulla of this relatively benign disease is a variant of pemphigus vulgaris. The scraping of the base of a ruptured bulla or vesicle contains sheets of degenerate, intermediate type squamous cells with abundant semikeratinized orange-stained cytoplasm and central, oval, vesicular single nuclei. The chromatin pattern is finely granular and regularly distributed. The nuclear membrane is delicate and smooth. The nucleoli are small and seldom prominent. Occasionally groups of reactive parabasal squamous cells with distorted cytoplasm, resembling the repair cells described in Chapter 4, may be seen. No inclusions are seen. During the healing of the

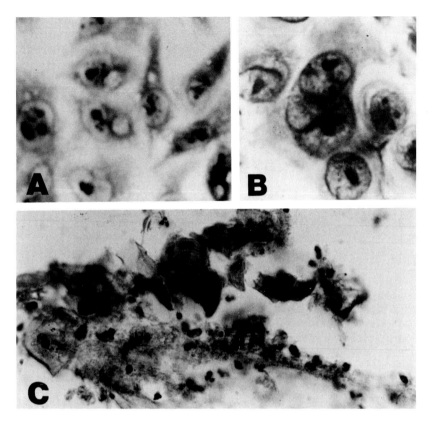

Figure 213
Scraping of (A) pemphigus foliaceus (× 750), (B) pemphigus vulgaris (× 750), and (C) bullous pemphigoid (× 450).

lesion, small sheets of deep basal squamous cells are present. Except for a few neutrophils, the background is relatively clean.

Seborrheic Pemphigus or Pemphigus Foliaceus (Fig. 213A)

Another variant of pemphigus vulgaris, this erythematous and flaccid bulla–producing disease is found on the face and the sweaty folds of the skin of the trunk. The scraping produces large numbers of single and loose clusters of small, round, and spindle basal squamous cells. Their often-elongated adequate to scanty cytoplasm is semikeratinized and orange-stained. Their oval-to-elongated nuclei are often degenerate and show variable degrees of pyknosis. No nucleoli are present. Few leukocytes are present in the massive granular serum protein precipitate usually present in the background.

Pyoderma Gangrenosum (Fig. 214A, B)

The scraping of the large skin ulceration characteristic of this disease contains numerous multinucleated giant cells of variable size and shape. Their

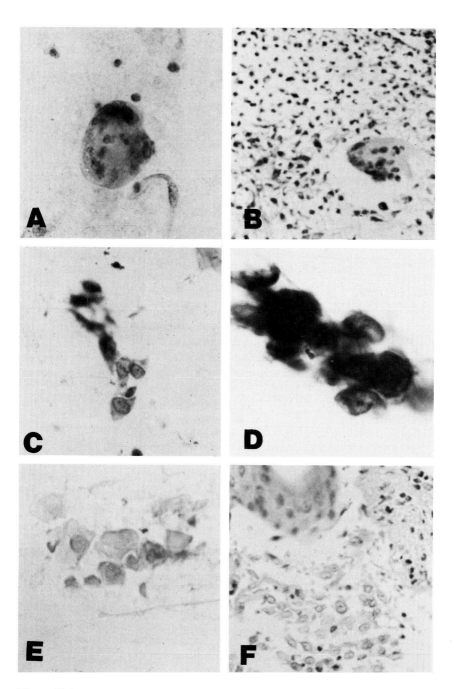

Figure 214
(A) Scraping of pyoderma gangrenosum. (× 450) (B) Section of the scraped lesion. (× 120) (C) Scraping of prurigo nodularis. (× 450) (D) Scraping of prurigo nodularis. (× 1100) (E) Scraping of toxic epidermal necrosis. (× 450) (F) Section of the scraped lesion. (× 120)

cytoplasm is dense with almost no inclusions. The nuclei, which vary in size and number, overlap on each other and occasionally mold.

Scattered elongated epithelioid cells with oval-to-elongated eccentric nuclei are also present. The chromatin is finely granular, and nucleoli are prominent. Their dark-staining cytoplasm is adequate and spindly in shape. Large numbers of eosinophils, neutrophils, and histiocytes are usually present.

Prurigo Nodularis (Fig. 214C, D)

Large numbers of keratinized squamous cells are mixed with clusters of squamous basal cells. These basal cells are spindly or triangular in shape and have adequate cytoplasm with no inclusions. The cytoplasmic membrane is often fuzzy. Their nuclei are usually single, elongated with thick nuclear membranes and fine granular chromatin. Small nucleoli are usually present. The background of these smears is clear of inflammation and contains, in addition to the anucleated keratinized cytoplasmic debris, large amounts of cholesterol crystals with the characteristic broken glass appearance.

Toxic Epidermal Necrosis (Fig. 214E, F)

The scrapings contain, besides the sheets of intact squamous basal cells, large amounts of clusters of anucleated, blue-gray–stained cytoplasmic debris each measuring 10 to 20 μ in diameter. The borders of these anucleated cells are usually sharp. Occasionally a shadow of a nucleus can be seen in the center of some of these cells. Various amounts of histiocytes with lacy cytoplasm containing no inclusions are present. Inflammatory cells (leukocytes) are few. The background is relatively clear except for some granular deposits.

Erythema Multiforme (Fig. 215B)

This acute dermatosis may produce macules, papules, vesicles, or bullae. The scraping produces single cells, clusters, or small sheets of epithelioid parabasal and intermediate cells of variable size. The adequate-to-scanty cytoplasm is usually polychromatophilic and has thick, distinct membranes. The enlarged nuclei are irregular in size and some are bean-shaped. The chromatin is coarsely granular and irregularly distributed. The nucleoli are occasionally prominent. Perinuclear halos and large, eccentric cytoplasmic vacuolization adjacent to the nuclei are common. Pseudocannibalism and aborted pearl-like formations are often present and may become possible pitfalls in diagnosing cancer. The background contains fine granular protein deposits, blood, few leukocytes, and eosinophils.

Foreign-Body Granuloma

The presence of foreign material in the subepidermal tissue may produce a granulomatous reaction and ulceration. The scraping contains variable amounts of mononucleated and multinucleated histiocytes with foamy cytoplasm. Some of these foreign bodies may be seen either free in the background of the smear or phagocytized within the cytoplasm of the histiocytes. Variable amounts of leukocytes are usually present. Special fat stains will show that some of the inclusions in the foamy histiocytes are lipid in nature.

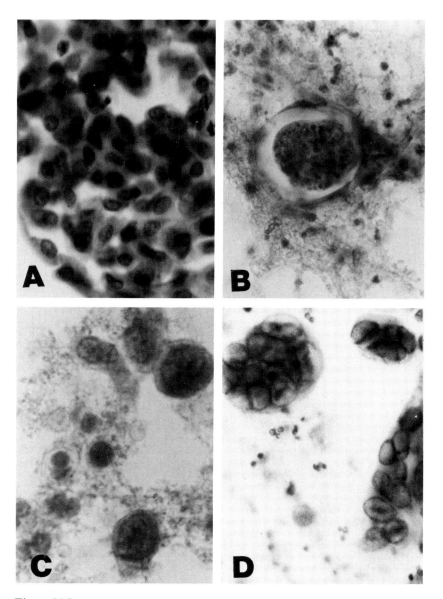

Figure 215
(A) Impetigo. Clusters of reactive dense basal cells. (× 750) (B) Erythema
multiforme. Pearl-like formation. (× 450) (C) Aspirate from a subcutaneous
mass in a patient with Weber-Christian disease. (× 450) (D) Herpes simplex.
(× 450)

Contact (Allergic) Dermatitis

The scraping of the bottom of vesicles and ulcerated areas usually present in contact dermatitis (e.g., poison ivy, drug allergy) is moderately cluttered with debris, granular protein deposits, and inflammatory cells. Large numbers of eosinophils, lymphocytes, and plasma cells are present in the early stage of the inflammation, while polymorphonuclear neutrophils predominate in the late stage, probably because of the common secondary bacterial infection of these lesions. Few isolated intermediate and basal squamous cells are usually present, showing no specific hyperactivity.

Weber-Christian Disease (Fig. 215C)

The smear of the aspirate of a subcutaneous nodule or of the oily discharge from the skin ulceration found in this disease will contain large numbers of lipid-laden macrophages. These cells vary in size, are usually single, and have abundant foamy cytoplasm. Large multinucleated foreign-body giant cells containing large droplets of lipid material are usually present. They differ from the histiocytes of other foreign granulomas by the size of these droplets. Variable amounts of cellular debris, granular precipitate, and inflammatory cells (polymorphonuclear cells) clutter the background of the smear. A special lipid stain of the smear will confirm the diagnosis.

VIRAL INFECTION

Herpes (Simplex, Varicella, Zoster) (Fig. 215D)

This acute viral infection can affect the skin and mucosa in various locations. Herpes type I virus usually affects the nongenital area while type II infects mainly the genitalia.

The scraping of the margins of the denuded base of a vesicle contain typical multinucleated syncytial epithelial cells with bland or inclusion (type A)–containing nuclei. These nuclei have thick regular membranes and are molded against each other. The dense cytoplasm is scanty to abundant, showing variable degrees of degenerative changes. In the scrapings of lesions from dark skinned patients, variable amounts of dark brown melanin pigment may be seen in the cytoplasm of the viral infected cells. Cytology cannot differentiate Herpes type I from type II or from varicella and zoster viral infections.

Vaccinia (Poxvirus) (Fig. 216A)

The scraping produces clusters or sheets of intermediate or basal squamous cells with large poorly stained eosinophilic intracytoplasmic inclusions. In the early stage, the inclusion is small and oval; in the later stage it becomes large, often multiple, and irregular in shape. The nuclei of the infected cells are enlarged but otherwise not remarkable.

Molluscum Contagiosum (Fig. 216B)

The scraping contains mainly free, large, oval, homogeneous, orange-stained cytoplasmic inclusions (molluscum bodies). They are produced by a poxvirus and are liberated from the infected epithelial cells by degenera-

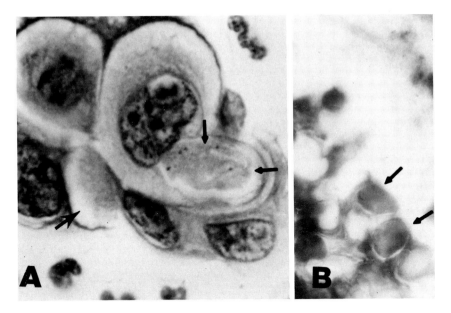

Figure 216
Scraping from lesions due to (A) vaccinia virus (× 1300) and (B) molluscum contagiosum virus (× 450). (Arrows show the cytoplasmic inclusions.)

tion. Occasionally similar inclusions may be seen filling the cytoplasm of intact cells and molding their nuclei in a crescent shape. The main danger is in mistaking these inclusions (molluscum bodies) for artifacts or contaminants (pollen). The background of the smears, aside from some cellular debris, is usually clean with no inflammatory cells.

Measles
Scraping of some of the large skin measles papules and of the Koplik's spots, if done energetically, will produce diagnostic cells. Single and clusters of nonspecific parakeratotic and dyskeratotic squamous cells are usually present, associated with few lymphocytes. Some of these cells show cytoplasmic degenerative vacuoles. The diagnosis of measles virus infection is based on the presence of epidermal syncytial, inclusion-containing giant cells. The number of round nuclei within these giant cells ranges from 4 to more than 100. Numerous granular, eosinophilic inclusions are present in the cytoplasm and nuclei.

BENIGN AND PRECANCEROUS LESIONS

Hyperkeratosis
This lesion is characterized histologically by an increased thickness of the surface (keratin) layer of the skin. It occurs usually as a reaction to chronic irritation. It has very little tendency to become malignant. There is

an increase in the production of keratinized epithelial cells and a decrease in their desquamation. The scraping will produce only numerous large anucleated hyperkeratinized squamous cells. They may be seen singly, in clusters, or in sheets. A few of these cells may still retain small pyknotic nuclei. No inflammatory reaction is usually present.

Seborrheic Keratosis (*Fig. 217A, B*)

The scraping produces many clusters of parabasal squamous cells loaded with intracytoplasmic melanin pigment. In some cells this dark brown pigment completely obscures the features of the cells. In the cells with little pigment the nonkeratinized cytoplasm is usually scanty and surrounds a central single nucleus containing finely granular, regularly distributed chromatin. Nucleoli are usually not present. Although often absent, inflammatory cells (mainly leukocytes) may be seen with a secondary bacterial infection of the lesion.

Keratoacanthoma (*Fig. 217C, D*)

A preponderance of anucleated keratinized squamous epithelial cells is usually present in the smear. Their orange cytoplasm is often semitransparent, indicating that the amount of cytoplasmic keratin is not as great as in hyperkeratosis. Clusters of squamous parabasal cells are also usually present. Their cytoplasm is adequate and often vacuolated. Their single nuclei are regular in size (10 μ) and shape (oval) and contain finely granular, uniformly distributed chromatin with no prominent nucleoli. No inflammatory cells are present in the background.

Sebaceous Adenoma (*Fig. 217E, F*)

The presence of clusters and sheets of columnar cells with vacuolated, eosinophilic cytoplasm is diagnostic of this lesion. The adequate cytoplasm contains no inclusions. Their central single nuclei are large (12–15 μ) and contain coarsely granular chromatin with occasional irregular clumping. The nuclear membrane is delicate and smooth. The small nucleolus is occasionally prominent. The cells stain red with mucicarmine stain. The background of the smear is usually clean.

Nevus (*Pigmented and Nonpigmented*) (*Fig. 218A*)

The smears contain scattered groups of special squamous basal type cells (nevus cells) forming small papillary structures with smooth external borders. Their poorly demarcated, pale blue cytoplasm is often scanty. Variable amounts of finely granular brown-to-black melanin pigment are often present. In some of the cells, the cytoplasm is so densely packed by this pigment that the entire cell is masked. Their round-to-oval single nucleus is uniform in size and shape. The finely granular chromatin is uniformly distributed. The nucleoli are small but often prominent. The background of the smear contains few inflammatory cells (mainly leukocytes) and occasional clusters of free black melanin pigment. The absence of polymorphism of the cells and of giant cells differentiates these smears from the one scraped from malignant melanoma.

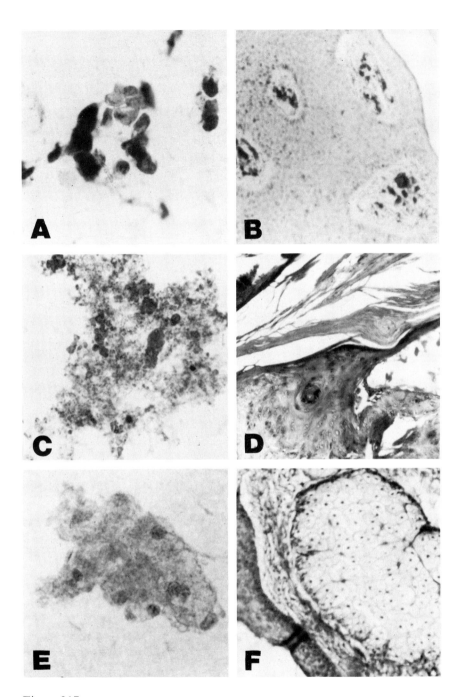

Figure 217
(A) Scraping of seborrheic keratosis. (× 450) (B) Section of the scraped
lesion. (× 120) (C) Scraping of keratoacanthoma. (× 450) (D) Section of the
scraped lesion. (× 120) (E) Scraping of sebaceous adenoma. (× 450) (F) Section of the scraped lesion. (× 120)

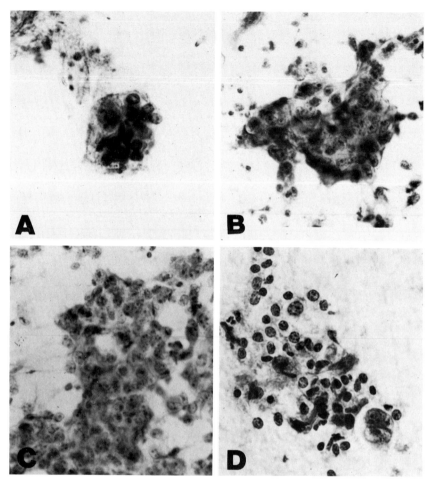

Figure 218
(A) Scraping of nevus. (× 450) (B) Scraping of skin metastasis of a poorly differentiated carcinoma. (× 450) (C) Scraping of skin metastasis of an adeno-carcinoma of the breast. (× 450) (D) Scraping of Hodgkin's lymphoma. (× 450)

MALIGNANT TUMORS

In Situ Squamous Carcinoma

Except for the in situ squamous carcinoma of the vulvar skin (Bowen's disease) the occurrence and morphology of the cancer in other locations on the skin is not well defined. Confusion exists between epithelial in situ cancer and dysplasia. The histologic sections usually show a disorganized proliferation of atypical squamous basal cells and numerous mitoses on top of an intact basement membrane.

The lesion often appears grossly as a benign erythema, but the scraping produces numerous single or clusters of atypical basal and intermediate squamous cells. Varying in size and staining color (deep blue to orange), they usually have bizarre shapes. Their scanty to adequate cytoplasm, some containing keratin, often shows degenerative changes. The amount of nuclear atypia varies and is especially increased when the nucleus is pyknotic. Mitoses are rare. In the vesicular nuclei, the amount of chromatin varies and is often irregularly clumped. Nucleoli are seldom seen. Occasionally multinucleation and pearl-like formations may be present.

The background of the smears is usually free of the necrotic debris commonly seen in invasive carcinoma. Except for variable amounts of monocytes, no inflammatory leukocytes are present.

Basal-Cell Carcinoma (Fig. 219A)
This tumor is most common in exposed areas of the skin (face, scalp, hands) and arises from the deep layers of the skin and appendages. The presence in the scraping of numerous sheets of small basal tumor cells, often in peripheral palisade formation, is diagnostic. Their scanty, sometimes deeply eosinophilic cytoplasm has indistinct borders. Their single nucleus usually shows only moderate variation in size and shape except when differentiated toward the intermediate squamous type, when it shows increased pleomorphism. Their abundant chromatin is granular with an abnormal irregular pattern. The nucleoli are seldom seen and are usually minute. The cytologic diagnosis is based mainly on the small size of these tumor cells and their tendency to exfoliate mainly in sheets. The background of the smears is usually clean. Cellular debris and granular deposits are present when the tumor is ulcerated.

Squamous Cell (Epidermoid) Carcinoma (Fig. 219B)
One of the most common cancers of the exposed area of the skin, squamous cell carcinoma often has an ulcerated necrotic center. The scraping of such ulcerated areas will produce only nondiagnostic degenerate squamous cells, cellular debris, and inflammatory cells. It is important to clean the surface of such a lesion with a dripping wet saline compress before scraping the margins of the ulceration. The degree of differentiation of the tumor varies from nonkeratinized (poorly differentiated) to marked keratinization (well differentiated).

The scraping of a well-differentiated squamous carcinoma produces large numbers of orange-stained polymorphic malignant cells. These cells may be single but are more often seen in sheets and clusters. Some anucleated or ghostlike deep orange-stained cytoplasm is often present. Dark blue-stained plemorphic malignant squamous cells may be seen. The marked variation in size and shape of the enlarged nuclei of these cells and the abnormality of the abundant chromatin clumps are diagnostic. Multinucleation is common. The dense cytoplasm staining deep blue-green to deep pink-orange indicates the squamous origin of these tumor cells. The background of the smear is usually cluttered with necrotic cells, protein granules, and inflammatory cells (leukocytes).

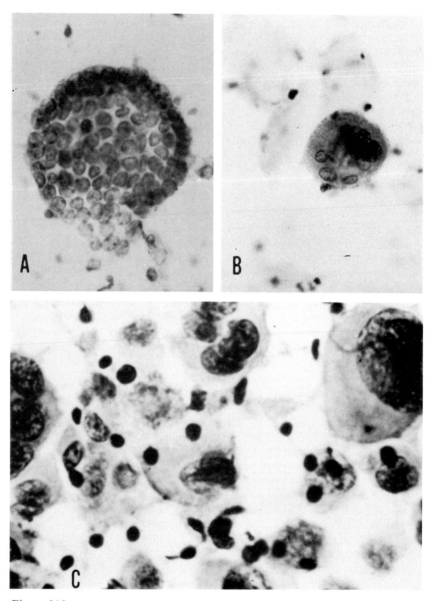

Figure 219
Scrapings from the skin. (A) Basal-cell carcinoma. (× 450) (B) Early squamous cell carcinoma. (× 450) (C) Malignant melanoma. (× 750)

In cases of poorly differentiated, nonkeratinized epidermoid carcinoma, the scraping produces numerous single or clusters of malignant cells. Their basophilic to transparent cytoplasm is scanty and poorly defined. The cells are mononucleated or binucleated. The nuclei are large, polymorphic with thick, wrinkled well-defined nuclear membranes. The increased chromatin is irregularly clumped and distributed. The nucleoli are often prominent and vary in size, shape, and number. Some of the cells are difficult to differentiate from poorly differentiated adenocarcinoma or sarcoma cells. The uniform density of the cytoplasm of some of the squamous cells aids in their recognition.

The background of the smear usually contains less debris and inflammatory cells than in the well-differentiated squamous cells.

Malignant Melanoma (Fig. 219C)

The surface of this tumor, which arises from the dermal-epidermal junction of the skin, is often ulcerating, and the scraping usually produces large numbers of diagnostic cells. Clusters of atypical polyhedral or spindle cells with abnormal cytoplasm and nuclei are diagnostic. The cytoplasm is often adequate to abundant, has sharp borders, and may or may not contain melanin pigment. The nuclei, which may often be multiple, are large and round or irregular in shape, with thick nuclear membranes, abnormally clumped chromatin, and prominent nucleoli. The background of the smear often is cluttered with cellular debris and granular protein deposits. Clusters of free melanin pigment may be present in certain hyperpigmented cases.

Metastatic Neoplasm (Fig. 218 B, C, D)

The small skin metastasis often lies in the dermis and is covered by an intact epidermis. The scraping of the surface of such a lesion will contain no diagnostic cells. A needle aspiration is indicated. In larger lesions the surface of the skin is often ulcerated, and the scraping will produce sheets or clusters of malignant cells resembling the parent tumor cells. These alien cells are usually easy to recognize, and their morphology often indicates the site of the primary cancer. The most common sources of skin metastasis are the cancers of the breast, lung, and gastrointestinal tract, and various lymphomas. A large number of such secondary skin tumors are not true metastases but rather extensions of tumors from an adjacent organ harboring the cancer (lymph node).

References and Supplementary Reading

Berardi, P.   Ultrarapid cytodiagnosis in the practice of ambulatory dermatology. *Arch. Ital. Dermatol. Venereol.* 36:120, 1969.

Blank, H., and Burgoon, C. F.   Abnormal cytology of epithelial cells in pemphigus vulgaris: A diagnostic aid. *J. Invest. Dermatol.* 18:213, 1952.

Blank, H., et al.   Cytologic smears in the diagnosis of herpes simplex, herpes zoster, and varicella, *J.A.M.A.* 146:1410, 1951.

Carvalho, G.   Molluscum contagiosum in a lesion adjacent to the nipple. *Acta Cytol.* 18:532, 1974.

Fortin, R., and Meisels, A.   Rhinosporidiosis. *Acta Cytol.* 18:170, 1974.

Goldman, L., McCabe, M., and Sawyer, F.   The importance of cytology technique for the dermatologist in office practice. *Arch. Dermatol.* 81:359, 1960.

Graham, J. H., Bingul, O., and Burgoon, C. B.   Cytodiagnosis of inflammatory dermatoses. *Arch. Dermatol.* 87:118, 1963.

Graham, J. H., et al.   Papanicolaou smears and frozen sections of selected cutaneous neoplasms. *J.A.M.A.* 178:380, 1961.

Haber, H.   Cytodiagnosis in dermatology. *Br. J. Dermatol.* 66:79, 1954.

Hitch, J. M., Wilson, T. B., and Scoggin, A.   Evaluation of rapid method of cytologic diagnosis in suspected skin cancer. *South. Med. J.* 44:407, 1951.

Medak, H., et al.   The cytology of vesicular conditions affecting the oral mucosa—pemphigus vulgaris. *Acta Cytol.* 14:11, 1970.

Moore, T. D.   A simple technique for the diagnosis of nonlipid histiocytosis. *Pediatrics* 19:438, 1957.

Selbach, C., and Heisel, E.   The cytological approach to skin diseases. *Acta Cytol.* 6:439, 1962.

Steigleder, G. K.   The diagnosis of vesicules and pustules with the aid of smears. *Australas. J. Derm.* 3:11, 1955.

Tzanck, A.   Le cytodiagnostic immediat en dermatologie. *Bull. Soc. Fr. Dermatol. Syph.* 7:68, 1947.

Urbach, F., Burke, E. M., and Traenkle, H. L.   Cytodiagnosis of cutaneous malignancy. *Arch. Dermatol.* 76:343, 1957.

Wilson, G. T.   Cutaneous smears: Diagnostic aid in certain malignant lesions of the skin. *J. Invest. Dermatol.* 22:173, 1954.

Winer, L. H., and Lipschultz, C. E.   Comparative study of histology and cytology in vesiculating eruptions. *Arch Dermatol.* 65:270, 1952.

Woodbourne, A. R., Philport, O. S., and Philport, J. A.   Cytologic studies in skin cancer. *Arch. Dermatol.* 82:992, 1960.

Other Specimens 19

Ear Discharge

The smear of an ear discharge or saline irrigation of the external ear is helpful in the diagnosis of various diseases of the external auditory canal. The smear of a needle aspiration of secretions accumulated in the middle ear may be important in the diagnosis of middle ear diseases and in the choice of the best treatment.

INFLAMMATORY DISEASES

External Otitis (Fig. 220A)

The presence of microorganisms or of various fungi in the midst of anucleated squamous epithelial and mixed inflammatory cells is diagnostic. Large numbers of naked nuclei and degenerate epithelial cells mixed with sheets of reactive squamous cells are common.

Cholesteatoma (Fig. 220B)

The presence of cholesterol crystals (best seen with a polarized light) in the midst of inflammatory and squamous cells in the smears of secretions aspirated directly from the middle ear is diagnostic of cholesteatoma. This direct middle ear needle aspiration is necessary to avoid the cerum or wax normally found in the external auditory canal, which may also contain cholesterol crystals.

Serous Otitis Media

The diagnosis of serous otitis media may be suspected when only a clear, serous, almost acellular fluid is obtained in the aspiration smear of the middle ear. A few lymphocytes and neutrophils are the main cellular components of the smears. A few degenerated intermediate squamous cells and some columnar cells may also be present. In allergic otitis, eosinophils are seldom seen. The main inflammatory cells are lymphocytes, plasmacytes, and small histiocytes.

Catarrhal Otitis Media (Glue Ear Syndrome)

In this chronic infection of the middle ear cleft, which causes a "sudden deafness," mainly in children, the aspiration of the middle ear produces

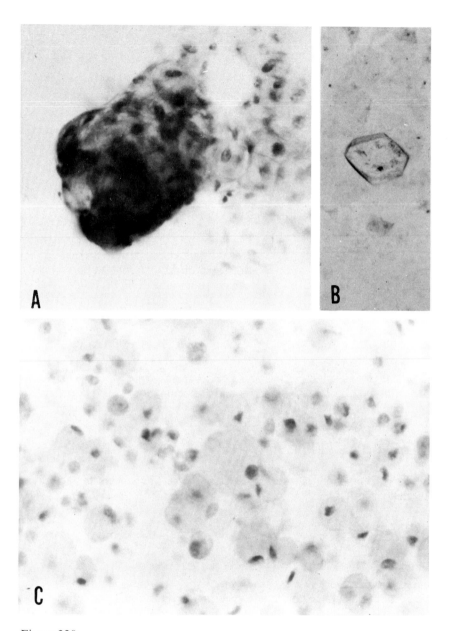

Figure 220
(A) Ear discharge in external otitis. (× 450) (B) Cholesterol crystal in cholesteatoma. (× 450) (C) Aspiration smear from the middle ear in a case of
chronic otitis. (× 450)

variable amounts of thick, glue-like secretions. The smears from these secretions contain large amounts of mucus, some cellular debris, and large numbers of hypervacuolated secretory columnar cells shed singly or in clusters. Variable numbers of lymphocytes and plasma cells are present. Few leukocytes and debris-containing macrophages may also be seen.

Acute Otitis Media and Mastoiditis
The presence of acute inflammation (leukocytes) and large numbers of giant, oval phagocytes in the middle ear or mastoid aspiration smear is indicative of the disease. These often multinucleated phagocytic cells measure 20 to 50 μ in diameter and have abundant foamy or large vacuolated (lipid-containing) cytoplasm. Fat stain (Sudan III) will show the fatty nature of the cytoplasmic inclusions by staining them red.

Chronic Otitis and Mastoiditis (Fig. 220C)
The absence of the giant histiocytes mentioned in the preceding paragraph differentiates a chronic inflammation from an acute one. Variable amounts of degenerate neutrophils and an abundance of squamous epithelial cells with degenerative cytoplasmic vacuolization and nuclear pyknosis are usually present.

Viral Infection
Except for the respiratory syncytial virus, no other respiratory viruses seem to be able to infect the middle ear, in spite of the direct mucosal communication between the middle ear and the pharynx. This is probably due to the presence of some local inhibitors.

NEOPLASMS OF THE EAR
Neoplasms of the ear are rare (0.05% of all neoplasms). The most common cancer of the external and middle ear is the well-differentiated squamous cell cancer. The presence of malignant squamous cells in the scraping or aspiration smear that show classic malignant changes is diagnostic of this rare tumor. The recognition of these cells is easy because of the relative absence of precancerous atypical lesions of the ear.

Basal-cell carcinoma may also involve the external and middle ear (for the features of the diagnostic cells, see page 473).

Rare cases of primary adenocarcinoma of ceruminous glands (ceruminoma) may exfoliate papillae and acini of columnar cells showing few malignant features. Their hyperchromatic nuclei are regular in size and have irregular prominent nucleoli.

Secondary tumors may involve the external and middle ear and the mastoid process (from renal and bronchial carcinoma mainly). The aspiration will contain sheets of diagnostic tumor cells.

Circulating Blood, Bone Marrow, Contact Smears, and Needle Aspiration

CYTOLOGY OF TUMOR CELLS IN CIRCULATING BLOOD
Tumor cells, benign or malignant, may be found in the peripheral circulating blood after manipulation or surgical excision of a neoplasm. The cyto-

logic detection of these neoplastic cells is of no known practical value at this time. The incidence of metastasis does not increase with the discovery of these cells in the bloodstream. Except for investigative purposes, the routine use of this often complicated test is not yet warranted.

To demonstrate these tumor cells the red blood cells are hemolyzed (with distilled water, 50% acetic acid, or streptolysin 0) or separated by sedimentation or centrifugation (flotation method). The resulting mixture of tumor cells and white blood cells is either passed through a filter membrane or centrifuged and smeared.

The neoplastic cells are usually recognized by their large size and their tendency to appear in tight groups of 3 to 4 cells. They are differentiated from ordinary white blood cells by their large nuclei, their marked variation in size and shape, their hyperchromatism and polychromatism, and their enlarged nucleoli. The presence of mitosis and of a high nuclei-cytoplasm ratio is of questionable diagnostic value. The main cells from which they are occasionally difficult to differentiate are:

1. the *metamyelocytes,* which vary in size from 10 to 18 μ. Their cytoplasm is granular, abundant, and unevenly stained, blue to pink. Their nuclei are bean shaped and have irregular chromatin strands;
2. the *megakaryocytes,* which are characterized by their large size (100 μ) and their huge nuclei, which are often lobular to an extent that gives the impression of multinucleation. Their nuclear membrane is thin. Their cytoplasm is usually abundant, pale blue to deep red and uniformly stained. This cytoplasm may contain small vacuoles with granular inclusions. The size of the cells and their hyperchromatism are the main reasons for the confusion.

Some of the megakaryocytes may have scanty cytoplasm with hyperchromatic nuclei that are less lobular than when the cytoplasm is abundant.

CYTOLOGY OF BONE MARROW

Contrary to the cytology of circulating blood, the cytology of bone marrow aspiration is helpful not only in the diagnosis of myelophthisic anemia but also in the detection of metastatic neoplastic cells. This aspiration method can be used in place of a bone biopsy, which must often be decalcified before being embedded and sectioned.

Technique for Collecting Bone Marrow Specimens

The aspirated marrow from the iliac crest or sternum is mixed with a few cubic centimeters of anticoagulant solution and poured into a Petri dish or watch glass, which is placed over a dark surface. The specimen is then examined microscopically through a mobile magnifying glass under oblique lighting, preferably three lights on each side of the examining table. The tissue particles that shine brightly are carefully selected, placed between two completely frosted slides, and smeared. After their fixation, they are stained with routine Papanicolaou stain. The remaining specimen is further diluted with 5 ml of saline solution, passed through a membrane filter with a 5 μ pore opening, and fixed.

Morphology of the Normal and Neoplastic Cells

There is a marked variation in the number of cells, benign or malignant, encountered in the smears. If the bone marrow is hyperplastic or the needle is placed in the middle of a solid metastatic tumor, large numbers of cells are usually present. The background of the smear is usually bloody, but this is seldom a handicap.

The structure of the malignant cells varies according to the type of the primary lesion. When present, they are usually seen with low-power ($\times 10$) magnification. Malignant cells are mainly found in easily identified groups or large sheets containing from 20 to 100 cells each. They can also be found as isolated cells, often degenerate, appearing in the form of stripped nuclei, and their correct interpretation is not easy.

The *isolated tumor cells* have a large, malignant-looking hyperchromatic nucleus and irregular, scanty cytoplasm. They are very difficult to differentiate from some normal cells of the hematopoietic system.

The *atypical hyperplastic reticulum cell* nucleus may also show all the previously described general criteria for malignancy. The nature of the single tumor cells can be suspected by their slightly larger size (25–50 μ), their prominent nucleoli, and the marked irregularity of their chromatin clumping. Similarly, the diagnosis is aided by the presence of large mucus-containing vacuoles or melanin granules in the cytoplasm. The unequivocal diagnosis of a metastasis should be based only on the presence of a sheet of tumor cells, often uniform in their appearance, which no hematopoietic cell can imitate.

The tumor cells should also be differentiated from the occasionally seen *osteoblast*. These cells are also found in clusters. Their size varies from 40 to 60 μ in diameter, and they have an elongated pear-like shape. Their cytoplasm is basophilic and vesicular with irregular borders. The nuclei are eccentric and round or oval with regular filamentous chromatin and no visible nucleoli.

LE cells are found in patients with systemic *lupus erythematosus*. They are neutrophils with large intracytoplasmic inclusions. Their size is slightly larger than that of the normal neutrophil. Round or oval, their cytoplasm is granular and contains a single, large, round, occasionally lobular, homogeneous red inclusion. The nucleus is eccentric, often pyknotic and compressed by the inclusion.

Contact Smears
(Fig. 221A)

This is one application of exfoliative cytology that has been neglected. It can be used instead of, or in addition to, a frozen section. Its main advantages are its simplicity and its sensitivity. Some of the neoplastic cells may be found in larger numbers in a contact smear than in histologic sections because of the decreased mutual adhesiveness of the cancer cells. These contact or surface smears are also very useful in the diagnosis of many skin lesions because they are painless and easily made without disturbing the healing process. They often permit a rapid diagnosis of the majority of the

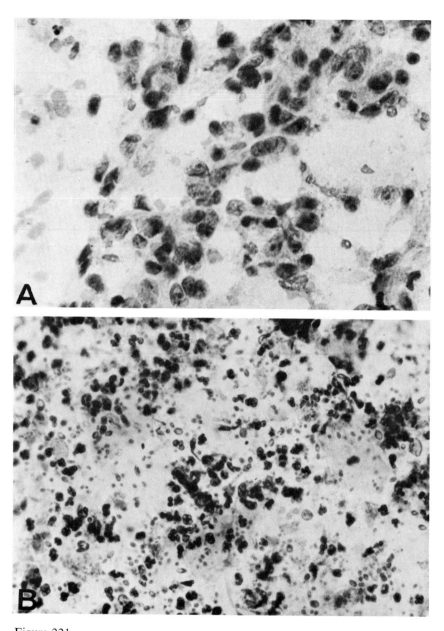

Figure 221
(A) Contact smear of breast adenocarcinoma, showing numerous scattered small tumor cells. (× 450) (B) Smear from aspiration of a lymph node showing the numerous dark, scattered, poorly differentiated adenocarcinoma cells. (× 225)

skin tumors and viral infections. Similarly, repeated contact smears may be taken during the surgical excision of a brain tumor for determination of the exact margins of the neoplasm to permit the removal of the tumor with minimal destruction of normal cerebral tissue. The surface of the freshly sectioned tissue is energetically scraped and the cellular material spread evenly on two or three slides, fixed, and rapidly stained.

SQUAMOUS CELL CARCINOMA

The scraping of the cut surface of a squamous cell carcinoma will generally produce a large number of malignant cells, single or in clusters that are easy to recognize. The cells vary greatly in shape, especially in the well-differentiated carcinomas, and their size varies extensively (10–80 μ). Their cytoplasm is often hyperkeratinized, staining deep orange with the Papanicolaou stain, and has a glassy appearance. Their nuclei are hyperchromatic or pyknotic and show marked variation in structure. Occasionally, tadpole and diagnostic spindle cells are seen. These are anucleated with abundant keratinized cytoplasm. Variation in the shapes of these cells is increased with the differentiation of the carcinoma.

ADENOCARCINOMA

The cells of adenocarcinoma desquamate singly or in sheets. They are often enlarged, and their cytoplasm may contain variable amounts of secretory vacuoles. Their nuclei vary in size and often have prominent irregular nucleoli. At times, these cells are seen in rosette or palisade formation, which helps in diagnosis.

The cells from the surface of a hepatoma, hypernephroma, or thyroid cancer are usually uniform in size. Their cytoplasm is abundant, and their nuclei are eccentric showing a uniform chromatin clumping condensed toward the periphery. The cells from the surface of a prostatic, breast, or pancreatic carcinoma are irregular in size. The scantiness of their cytoplasm varies depending on the differentiation of the neoplasm. The nuclei also frequently show marked irregularity of size and shape. The presence of occasional giant cells may also help in the diagnosis.

SARCOMA

The scraping from a reticulum cell sarcoma shows large numbers of giant cells with single or multiple finely granular or bland nuclei with prominent nucleoli. Their cytoplasm is usually poorly stained. In cases of fibromyosarcoma the cells are often elongated and have pseudopodlike prolongations. Their nuclei, single or multiple, are often enlarged and vesicular with slightly irregular chromatin granularities. In lymphosarcomas and other lymphomatous neoplasms there is usually marked polymorphism of the numerous reactive lymphocytes seen. The cells are shed singly or in tight clusters with marked variation in size and shape and in the prominence of their nucleoli.

LYMPH NODE METASTASIS (FIG. 221B)

The contact smear of a freshly sectioned lymph node will normally show a variable number of immature or mature lymphocytes. A metastatic neoplasm desquamates in large sheets or clusters of tumor cells, which are easily detected, even under low-power magnification.

BREAST CARCINOMA
The contact smear of a breast lesion, if benign, will have only a few isolated cells. Most of them will be seen in clusters or in incomplete acini. No complete glandular structures are seen. In a malignant lesion, the clumps are scarce, and the cells are usually shed singly or in occasional intact glandular structures. The cells also have the features associated with malignancy.

Aspiration Biopsy Smear
(Fig. 222)

Advantages of Aspiration Biopsy Smears
1. Aspiration is usually an office procedure;
2. It may eliminate the need for more complicated diagnostic tests (biopsy, frozen section);
3. The diagnosis may be revealed faster than with a biopsy;
4. Lymph nodes can be explored without removal;
5. For certain diseases (fungal and viral infections and certain tumors) the smear can be easier to interpret than histopathologic sections;
6. It can be repeated, as indicated, with minimal trauma.

Enlarged lymph nodes and subcutaneous or deep, hard-to-reach tumors are the most common site of aspirations.

Material Needed
Silverman needle, 16 to 18 gauge
20 ml syringe
bottle of 95% fixative
6 frosted slides with paper clips

Technique
1. The skin should be sterilized, anesthetized, and nicked with a sharp scalpel;
2. The needle with an obturator is inserted into the lesion;
3. The obturator is removed and replaced with a 20 ml syringe;
4. Repeated suction is applied while the needle is moved in different directions;
5. The needle is withdrawn, and the material collected is smeared on 6 slides and *immediately* dropped into fixative.

Cells can be obtained from deep, inaccessible neoplasms by needle puncture and aspiration. The material obtained should be smeared on one or two slides and immediately fixed. The difference between these cells and the naturally exfoliated ones is that all the cells obtained by aspiration are alive at the time of fixation and do not show physiologic changes of aging and death.

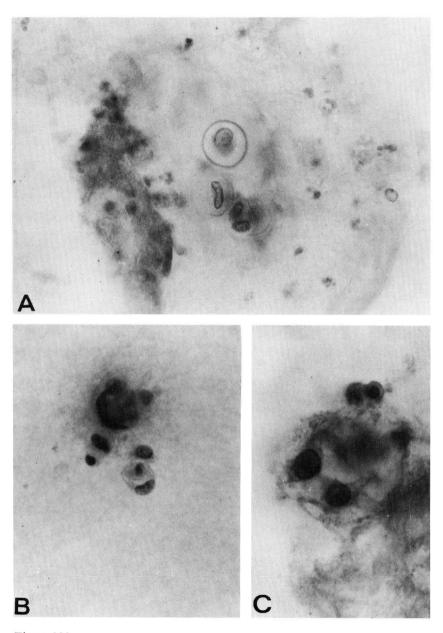

Figure 222
(A) Needle aspirate, knee joint fluid, showing loose cartilaginous cells in acute traumatic arthritis. (× 450) (B) Needle aspirate, mixed salivary gland tumor. (× 450) (C) Needle aspirate, bone lesion metastasized from a prostatic adeno-carcinoma. Note the good preservation of the vacuolated cells. (× 450)

The difficulty of interpretation comes from the presence of unfamiliar mesodermal cells, which normally do not exfoliate. Traumatic structural distortions are also often present. Another difficulty may be an error in aspiration site. For example, during lymph node aspiration the needle may bypass the lymph node and obtain cells from a normal adjoining organ.

Metastatic Cancer

The center of a lymph node with metastasis is often soft and necrotic and easy to aspirate. Metastatic cancer cells in a lymph node may appear better differentiated than the cells of a primary lesion. For example, the nuclei of metastatic squamous cancer cells aspirated from a lymph node will appear less hyperchromatic and anaplastic than the nuclei of malignant cells exfoliated directly from the primary lesion. The cytoplasm is also usually more abundant. Differentiation between various types of carcinomas, although difficult, can be made by cytology. The possible location of the primary cancer can often be narrowed to one or two sites.

Hodgkin's Disease

The aspiration smear of an involved lymph node is highly diagnostic. The recognition of Reed-Sternberg giant cells ($30 \pm 15 \mu$) in the midst of variable numbers of atypical lymphocytes, eosinophils, plasma cells, and other inflammatory cells is usually easy.

The cytoplasm of the Reed-Sternberg cells is adequate and stains pale grayish-blue. It often contains multiple small vacuoles without inclusions. The nuclei are enlarged, often multilobular, and have a vesicular to granular chromatin structure. Binucleation and trinucleation are frequent. The acidophilic nucleoli, often multiple, are prominent; they are usually round but tend to be more irregular with the increase in the degree of anaplasia of the tumor.

Reticulosarcoma

Large numbers of oval cells are scattered throughout the aspiration smear. Measuring $30 \pm 20 \mu$ in diameter each, the malignant reticular cell is more polymorphic than the small-cell anaplastic metastatic carcinoma with which these cells may be confused. Their adequate-to-scanty cytoplasm shows numerous delicate small vacuoles without any inclusions.

Their usually single, oval-shaped nuclei are occasionally multilobular and irregular. The chromatin is densely vesicular. Mitoses are frequent and often abnormal. Small or prominent, single or multiple nucleoli are always present. These nucleoli vary extremely in size and shape from one cell to the next.

Fluorescent Microscopy

Under the action of invisible ultraviolet light ($350-400 \mu$ wavelength), acting as an exciting light, the different cellular components reemit variable lengths of visible light waves, according to their molecular density.

Primary Fluorescence

This refers to natural fluorescence without adjunction of any stain, as found in the skin, fingernails, apocrine gland, adrenal cortex, and testes, all having a faint fluorescent power that is of no practical use in exfoliative cytopathology.

Lipids may have a faint yellow-orange fluorescence, elastic and collagen fibers, a white-blue fluorescence.

Secondary Fluorescence

This is the result of the absorption of fluorescent dyes (fluorochromes) by the different components of a cell that reemit visible light when excited with an ultraviolet light. The most common dyes used for cancer detection are the azo dye, auramine, rhubarb, and acridine orange. Acridine orange (a basic substance), because of its low concentration, slight toxicity, polychromatic properties, and its strong affinity for nuclear acids, is the most common stain used.

The intensity and color of the fluorescence depends on the type of dye used, its concentration, and its pH.

Action of Filters

EXCITER FILTER. These blue filters are placed between the light source and the condenser in order to remove the unnecessary visible light waves (wavelength above 450 μ) and the harmful, very short invisible light waves (wavelength below 350 μ) and allow only the blue and ultraviolet light to pass and fluoresce the cells.

BARRIER OR SUPPRESSION FILTER. These filters (yellow) are placed between the objective and the eye of the observer to remove all the remaining harmful ultraviolet rays and some of the excess of visible light (white light), which decreases the sharpness of the fluorescence.

Fluorescent Stain Technique

Stock solution: 0.10% acridine orange in distilled water, kept in the refrigerator.

Working solution: 10 ml of stock solution mixed with 90 ml of pH 6 phosphate buffer (2 parts of $M/15$ KH_2PO_4 + 5 parts of $M/15$ Na_2HPO_4).

The smears are hydrated by passing them through successive (80%, 70%, 50%) ethyl alcohols, then through distilled water. They are then placed in the working solution for 3 minutes, rinsed for 1 minute in the phosphate buffer solution (pH 6) and differentiated by being placed in $M/10$ calcium chloride for 1 to 2 minutes. The smears are then rinsed in buffer solution, air dried, and eventually mounted in buffer solution.

The fluorescing colors seen by the observer vary from white-yellow to bright orange. The basis for cancer detection is the special bright red fluorescence of the cytoplasmic and nucleolar RNA protein and the bright yellow fluorescence of the DNA protein. These two proteins are generally increased in the malignant cells and consequently give a strong fluorescence.

487

YELLOW FLUORESCENCE. Parabasal, endocervical, endometrial, intermediate cells, squamous metaplasia, and transitional cell nuclei stain dull yellow. Lymphocytic and malignant nuclei stain brilliant yellow.

BRILLIANT RED FLUORESCENCE. Bacteria, *Trichomonas,* viral inclusions, nucleoli, *Candida,* and the cytoplasm of dysplastic and malignant cells stain bright brick red.

DULL RED-ORANGE FLUORESCENCE. Endometrial and endocervical cells, Döderlein's bacillus, foreign-body giant cells, mesothelial cells, trophoblastic cells, parabasal cells, dyskaryotic cells, and bronchial secretory cells appear dull red-orange.

GREEN FLUORESCENCE. Leukocytes, lymphocytes, histiocytes, superficial cells, gastrointestinal cells, ciliated bronchial cells, squamous metaplastic cells, breast cells, prostatic cells, and bronchial basal cells stain a yellow-green of variable intensity.

NONFLUORESCENCE. Red blood cells, fatty tissue cells, and some mucoprotein artifacts remain semitransparent.

ADVANTAGES OF FLUORESCENT STAIN. Fluorescent stain allows rapid screening of negative smears by a relatively untrained technician.
    The staining is simple and rapid.
    An excess of blood is no handicap in screening the smear.
    The process demonstrates the histochemical as well as the structural characteristics of the malignant cells.
    In effusions the fluorescent stain better differentiates the malignant single cell from the reactive mesothelial cell.
    With the use of fluorescent staining, it is easier to detect *Trichomonas, Candida,* and microorganisms (bright red stain).

DISADVANTAGES OF FLUORESCENT STAIN. For the detailed morphologic study of a suspected cell, the smear often must be destained and stained again by the Papanicolaou method, which wastes time. The diagnosis often is not based on the brightness of staining, but rather on the configuration of the cells.
    When the cells are kept too long in the phosphate buffer solution, there will be a loss of detail, including drying and fading of the red fluorescence.
    With fluorescent staining it is more difficult to differentiate the benign cells from the malignant ones if they are found in clusters.
    It is more difficult to determine the type of the detected malignant cell.
    The mere presence of bright luminescence is not always diagnostic of a neoplasm, particularly in an inflammatory smear in which large numbers of leukocytes and regenerating cells are present.
    Degenerate malignant cells may not fluoresce brightly.
    Poorly differentiated cancerous cells are often difficult to recognize.

CONCLUSION. The use of this method for routine cytologic screening is questionable. To be accurate, the diagnosis must be based on minute morphologic details rather than color only, and these details are best seen when the cells are stained with the Papanicolaou technique. Fluorescence microscopy is better used for cytochemistry, antibody studies, or chromosomal banding.

References and Supplementary Reading

CYTOLOGY OF THE EAR

Althaus, S. R., and Ross, J. A. T.   Cerumen gland neoplasia. *Arch. Otolaryngol.* 92:40, 1970.

Bryan, M. P., and Bryan, W.   The use of aural cytology in the diagnosis of various inflammation and malignant tumors. *Acta. Cytol.* 14:411, 1970.

Carr, C. D., and Santuria, B.   Microscopic studies of the secretions of the external auditory canal. *Ann. Otol. Rhinol. Laryngol.* 65:1, 1951.

Friedmann, I.   The pathology of secretory otitis media. *Proc. Roy. Soc. Med.* 56:695, 1963.

Liu, Y. S., et al.   Chronic middle ear effusions. *Arch. Otolaryngol.* 101:279, 1975.

Ojala, L., and Pavla, T.   Macrophages in aural secretions and their clinical significance. *Laryngoscope* 65:670, 1955.

CANCER CELLS IN CIRCULATING BLOOD

Frost, J. K.   Cancer cells in circulating blood: A comparison with cytologic criteria for other sites. *Acta Cytol.* 9:83, 1965.

Frost, J. K., et al.   Cytology filter preparation factors affecting their quality for study of circulating cancer cells in the blood. *Acta Cytol.* 11:363, 1967.

Gazet, J. C.   The detection of viable circulating cancer cells. *Acta Cytol.* 10: 119, 1966.

Goldblatt, S. A., and Nadel, E. M.   Cancer cells in the circulating blood: A critical review II. *Acta Cytol.* 9:6, 1965.

Kierzenbaum, A. L., and Tres, L. L.   Differential diagnosis of megakaryocytes from cancer cells in the peripheral blood. *Acta Cytol.* 8:91, 1964.

McGrew, E.   Criteria for the recognition of malignant cells in circulating blood. *Acta Cytol.* 9:58, 1965.

Nagy, K. P.   A study of normal, atypical and neoplastic cells in the white cell concentrate of the peripheral blood. *Acta Cytol.* 9:61, 1965.

Pruitt, J. C., Powell, R. V., and Prater, T. F. K.   A quantitative comparison of processing techniques for study of circulating malignant cells. *Acta Cytol.* 9:116, 1965.

Rohmsdahl, M. M., et al.   Hematopoietic nucleated cells in the peripheral venous blood of patients with carcinoma. *Cancer* 17:1400, 1964.

Rohmsdahl, M. M., et al.   Cytological changes of megakaryocytes in patients with carcinoma. *Acta Cytol.* 8:343, 1964.

Seal, S. H.   A sieve for the isolation of cancer cells and other large cells from the blood. *Cancer* 17:637, 1964.

Selbach, G. J., Bondar, M., and Klassen, K. P.   The detection of malignant cells in arterial blood. *Acta Cytol.* 8:341, 1964.

CONTACT AND ASPIRATION SMEARS

Bloch, M.   Comparative study of lymph node cytology by puncture and histopathology. *Acta Cytol.* 11:139, 1967.

Bonneau, H., and Varette, I.   Le diagnostic cytologique par ponction a l'aiguille fine des tumeurs des glandes salivarires. *Arch. Anat. Pathol.* 15:156, 1967.

Cardozo, P. L.   The cytologic diagnosis of lymph node punctures. *Acta. Cytol.* 8:194, 1964.

Chromette, G., et al.   Cytologie exfoliatrice des tumeurs malignes humaines (frottis d'apposition). *Bull. Assoc. Fr. Cancer* 51:455, 1964.

Delarue, J., Chromette, G., and Pinaudeau, Y.   Etude cytologique sur produit de ponction des metastases ganglionnaires des tumeurs malignes. *Arch. Anat. Pathol.* (Paris) 11:11–17, 1963.

Duhamel, G.   La biopsie de moelle—Application au diagnostic des metastases cancereuses. *Transfusion* (Paris) 5:179, 1962.

Emerson, C., and Finkel, H. E.   Problem of tumor cell identification in the bone marrow. *Cancer* 19:1527, 1966.

Eneroth, C. M., and Zajicek, J.   Aspiration biopsy of salivary gland tumors III. Morphologic studies on smears and histologic sections from 368 mixed tumors. *Acta Cytol.* 10:440, 1966.

Engzell, U., and Zajicek, J.   Aspiration biopsy of tumors of the neck. *Acta Cytol.* 14:51, 1970.

Sposti, P. L.   Cytologic diagnosis of prostatic tumors with the aid of transrectal aspiration biopsy. *Acta Cytol.* 10:182, 1966.

Franzen, S., Giertz, G., and Zajicek, J.   Cytological diagnosis of prostatic tumors by transrectal aspiration biopsy. A preliminary report. *Br. J. Urol.* 32:193–196, 1960.

Godwin, J. T.   Cytologic diagnosis of aspiration biopsies of solid or cystic tumors. *Acta Cytol.* 8:206–215, 1964.

Hauptmann, E.   The cytologic features of carcinomas as studied by direct smears. *Am. J. Pathol.* 4:1199–1233, 1948.

Johansson, B., and Zajicek, J.   Sampling of cell material from human tumors by aspiration biopsy. *Nature* 200:1333–1334, 1963.

King, E. B., and Russell, W. M.   Needle aspiration biopsy of the lung technique and cytologic morphology. *Acta Cytol.* 11:319, 1967.

McGowan, L., Stein, D., and Miller, W.   Cul-de-sac aspiration for diagnostic cytologic study. *Am. J. Obstet. Gynecol.* 96:413, 1966.

Mavec, P., et al.   Aspiration biopsy of salivary gland tumours. I. Correlation of cytology reports from 652 aspiration biopsies with clinical and histologic findings. *Acta Otolaryngol.* (Stockh.) 58:471–484, 1964.

Pickern, J. W., and Burke, E. M.   Adjuvant cytology to frozen sections. *Acta Cytol.* 7:164, 1963.

Rubenstein, M. A., and Smelin, A.   Superiority of iliac over sternal marrow aspiration in recovery of neoplastic cells. *Arch Intern. Med.* 89:909–913, 1952.

Schoor, L., and Chue, E. W.   Fine needle aspiration in the management of patients with neoplastic disease. *Acta Cytol.* 18:472, 1974.

Stromby, N., and Akerman, M.   Aspiration cytology of the diagnosis of granulomatous liver lesions. *Acta Cytol.* 17:200, 1973.

Stonier, P., and Evans, P.   Carcinoma cells in bone marrow aspirate. *Am. J. Clin. Pathol.* 45:722, 1966.

Slager, U. T., and Reilly, E. B.   Value of examining bone marrow in diagnosing malignancy. *Cancer* 20:1215, 1967.

Ultmann, J. E., Koprowska, I., and Engle, R. L.   A cytological study of lymph node imprints. *Cancer* 11:507–524, 1958.

Zajdela, A., and Rousseau, J.   Etude cytologique des tumeurs ganglionnaires par ponction. *Arch. Anat. Pathol.* (Paris) 15:34, 1967.

FLUORESCENT MICROSCOPY

Atkin, N. B.   The desoxyribonucleic acid content of malignant cells in cervical smears. *Acta Cytol.* 8:68, 1964.

Bastos, A. L., et al.   Preservation of acridine orange fluorescence and meta-chromasia of tumor cells and cell culture exposed to quinacrine. *Acta Cytol.* 12:37, 1968.

Bertalanffy, F. D.   Fluorescence microscopy for the rapid diagnosis of malignant cells by exfoliative cytology. *Mikroskopie* 15:67, 1960.

Bertalanffy, L. Von, Masin, F., and Masin M.   Use of acridine orange fluorescence technique in exfoliative cytology. *Science* 124:1024, 1956.

Bertalanffy, L. Von, Masin, M., and Masin F.   A new and rapid method for diagnosis of vaginal and cervical cancer by fluorescence microscopy. *Cancer* 11:873–887, 1958.

Dart, L. H., Jr., and Turner, T. R.   Fluorescence microscopy in exfoliative cytology. *Lab. Invest.* 8:1513, 1959.

Liu, W.   Fluorescence microscopy in exfoliative cytology. II. *Arch. Pathol.* 71:386, 1961.

Lomas, M. I., Masin, M., and Masin, F.   Appraisal of a fluorescence technic in identification of abnormal cells of the colon. *Dis. Colon Rectum* 6:284, 1963.

Lowhagen, T., Nasiell, M., and Granberg, I.   Acridine orange fluorescence cytology in detection of cervical carcinoma. *Acta Cytol.* 10:194, 1966.

Riva, H. L., and Turner, T. R.   Fluorescence microscopy in exfoliative cytology. *Obstet. Gynecol.* 20:451, 1962.

Seal, S. H.   Silicone flotation: A simple quantitative method for the isolation of free-floating cancer cells from the blood. *Cancer* 12:590–595, 1959.

Stevenson, J.   The accuracy of fluorescence microscopy for the diagnosis of cancer. *Acta Cytol.* 8:224, 1964.

Varga, A.   "Cytological cancer screening," a simple, rapid staining procedure applicable to office use. *Am. J. Obstet. Gynecol.* 15:1, 1960.

Wellman, K. F., McDermott, M. A., and Gray, E. H.   An evaluation of acridine orange fluorescence microscopy in cytology. *Acta Cytol.* 7:111–117, 1963.

Technique

20

Fixation Techniques

METHOD OF CHOICE

For maximum cellular staining the protoplasm must be dehydrated and solidified rapidly with minimum distortion. Because of its penetrating capacity, 95% ethyl alcohol is advocated as the fixative of choice for the various cytologic smears. It gives excellent results. The addition of ether does not improve the appearance of the cells. This fixative should be checked from time to time to make certain that the alcohol concentration remains between 85% and 98%. Small amounts of eosin solution may be added to discourage the misuse of this alcohol. The smear should be left for at least 1 hour and for no more than 3 weeks in this fixative. In the case of a rush specimen, 5 minutes of fixation may be enough for the demonstration of gross nuclear details. After the removal of the slides, the alcohol should be filtered for reuse to eliminate the floating cells that are always present in the used solution.

All the staining solutions should be filtered at least once a day for routine work and immediately after staining a hypercellular specimen, especially certain effusions (nothing appears more malignant than benign reactive mesothelial cells that have settled on a vaginal smear).

The cells must be fixed wet before they air dry. If, by accident, air drying occurs, the smears should not be fixed in alcohol, but rather rehydrated and then fixed while wet. Otherwise the alcohol fixative will preserve forever the air drying artifacts and distortions of the cells, in which the nuclear details and color differentiation are very poor.

ALTERNATE METHODS

THE COATING SPRAY often contains polyethylene glycol and alcohol. These aerosol sprays are commercially available.* They give good results when sprayed on wet smears but are relatively expensive. The use of a cheap water-soluble hair spray found in any drugstore will give similar good results for less money.

* Spray-cyte (Clay-Adams); Cyto-fixer (Lab-Tex); Cyto-Dri Fix (Paragon); Profix (Scientific Products).

This water-soluble coating should be removed from the slides before staining with hematoxylin. One need only wash these slides in tap water for a few minutes before putting them directly in hematoxylin and then stain them routinely.

The staining of squamous cells fixed with a coating spray is comparable to alcohol-fixed cells, but the sprayed columnar cells do not stain as well as the alcohol-fixed ones.

METHYL ALCOHOL with equal parts acetone may be used as a fixative. It has the advantage that it is unaffected by evaporation, but it produces a rather pale cellular staining with poor differentiation. This alternate method of fixation should be employed only when ethyl alcohol is not available.

AIR-DRIED, REHYDRATED SMEARS. Besides simplicity, this method has the convenience of fixing the cell firmly to the slide. The smear may be rehydrated by immersing it in a solution of 50% glycerin and water for 1 to 24 hours, according to the time lapse since its air drying. The squamous cells rehydrate with little damage; the columnar cells do not.

GLYCERIN TECHNIQUE. A few drops of glycerin are placed on the wet slide and covered with a coverslip.

CARNOY'S SOLUTION, a quick-acting fixative, can be used to hemolyze red blood cells. It is made by mixing 60 ml of absolute alcohol with 30 ml of chloroform and 10 ml of glacial acetic acid.

Tap water often needs to be filtered before use because of artifacts (diatoms) that may contaminate the water and assume various shapes, some resembling tumor cells (pitfalls for a false-positive report).

If there is a need to stain smears that have been fixed in acetone and covered with glycerol for fluorescence antigen microscopy, the hematoxylin and eosin stains will penetrate the cells better if the slides are washed with a mild detergent soap solution then progressively dehydrated in alcohol and rehydrated before staining.

Routine Staining

The modified Papanicolaou stain is still the method of choice. The color differentiation of various cellular components used for the diagnosis of malignancy and cell typing is excellent with this stain. The nuclei are stained with Harris hematoxylin, the cytoplasm with an alcoholic polychromatic eosin stain, and the cytoplasmic keratin, when present, with orange G. All stains are available commercially* and ready for use; there is little advantage in preparing them in the laboratory.

HAND STAINING PROCEDURE (MODIFIED PAPANICOLAOU STAIN)
The slides are removed from the fixative (95% ethyl alcohol), then rehydrated by successively transferring them to:

* Clay-Adams, Scientific Products, Paragon, and Ortho Pharmaceutical Co.

80% ethyl alcohol	10 dips
70% ethyl alcohol	10 dips
50% ethyl alcohol	10 dips
distilled or filtered tap water	10 dips
Harris' or Ehrlich's hematoxylin	3 to 6 minutes

The time in hematoxylin may vary, depending on the age of the stain. The slides should be immersed longer in heavily used hematoxylin. Check each time to determine when the stain was made up.

tap water	rinse
sodium acetate buffer (pH 4.2)	1 to 4 quick dips (may vary; the blue smear should start turning red)
tap water	rinse thoroughly
1% saturated aqueous lithium carbonate	2 minutes
running tap water	4 minutes

From this point, in order to prevent any carrying over of excess water from an alcohol of lower percentage to one of higher content, drain the staining rack thoroughly on paper towels without allowing the slides to dry out.

50% ethyl alcohol	10 dips
70% ethyl alcohol	10 dips
80% ethyl alcohol	10 dips
95% ethyl alcohol	10 dips
Orange G (either A, B, or Ortho)	2 minutes
95% ethyl alcohol	5 dips
95% ethyl alcohol	5 dips
EA–65*	2 minutes
95% ethyl alcohol	5 dips
95% ethyl alcohol	5 dips
95% ethyl alcohol	5 dips
Absolute alcohol	10 dips
Absolute alcohol/xylol	2 minutes
Propanol and xylol (1 : 1)	2 minutes
Xylol	5–20 minutes
Xylol	5–20 minutes

If xylol turns milky, the slides have not been thoroughly dehydrated; return to absolute alcohol.

Mount in mounting medium.

---

* EA–65 = Eosin-azure 65, obtained from Scientific Products, Philadelphia, Pennsylvania.

HEMATOXYLIN. Mix 1 part of commercially prepared Harris' hematoxylin with 2 parts distilled water and 4 ml glacial acetic acid. Replenish from this prepared solution.

SODIUM ACETATE BUFFER. A standard 0.5 molar pH 4.2 buffer is prepared by mixing 68 gm of $NaHCO_3 \cdot 3 H_2O$ and 90 ml glacial acetic acid and diluting to 1 liter in volumetric flask.

From the standard buffer a 0.1 molar buffer is prepared for daily use by adding 180 ml 0.5 molar buffer to a graduated cylinder and diluting to 900 ml with distilled water. Prepare fresh buffer daily.

LITHIUM CARBONATE SOLUTION. Keep a saturated solution of $LiCO_3$ on hand. Prepare a 1% solution for daily use: 10 ml saturated solution per 990 ml distilled water. Change when discolored.

EA–36. Mix 3 parts of commercially prepared EA–36 with 1 part of absolute alcohol. Replenish from this prepared solution.

EA–65. Mix 3 parts EA–65 with 1 part absolute alcohol. Replenish from this prepared solution.

AUTOMATIC STAINING PROCEDURE (MODIFIED PAPANICOLAOU STAIN)

The automatic staining apparatus using 12 cups is illustrated in Figure 223. Before immersing in hematoxylin solution, the slides are hydrated by placing them in a series of progressively weaker solutions of alcohol or, if coating spray fixative was used, by rinsing them in running tap water.

In the 12 cup stainer the specimen is immersed in each cup for 2 minutes, for a total of 24 minutes. For the staining of nongenital specimens, EA–65 rather than EA–36 is used and is followed by immersion in propanol, especially if a filter membrane is being stained.

Special Staining

MODIFIED SUDAN STAIN

This stains bright red-orange the extracellular and intracellular lipids in the smear and is especially indicated for sputum of patients suspected of lipid pneumonia. Place the unfixed air-dried smear for 10 to 15 minutes in a saturated solution of Sudan black B in 75% ethyl alcohol.

Transfer and wash in 75% ethanol for 30 seconds.
Wash well with tap water.
Counterstain in Giemsa stain for a few minutes.
Drain, dry, and mount in glycerin jelly.

SHORR'S STAIN FOR HORMONAL EVALUATION

Shorr's stain increases the difference between eosinophilic (cornified) and cyanophilic (noncornified) cells and decreases the transitional hues. It can

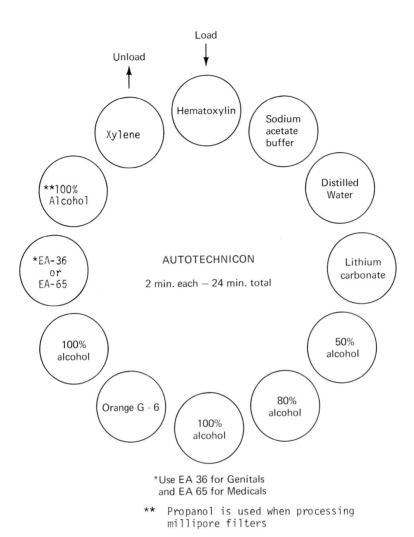

Figure 223
Automatic staining apparatus using 12 cups. The specimen is immersed in each cup for 2 minutes for a total staining time of 24 minutes. EA–36 is used for genital specimens; EA–65 for nongenital specimens. Before staining, hydrate: (1) 50% ethyl alcohol, 10 dips; (2) distilled water, 2 minutes. After staining, dehydrate: (1) ½ 100% ethyl alcohol and ½ xylene, 2 minutes; (2) xylene, 2 minutes. Propanol should be used instead of ethyl alcohol when processing membrane filters.

also be used to stain intracytoplasmic inclusion bodies (bright red). The nuclear features are not as clear as with the Papanicolaou technique.

Composition

50% ethyl alcohol	100 ml
Biebrich scarlet, water-soluble	0.500 gm
orange G	0.250 gm
fast green FCF	0.075 gm
phosphotungstic acid	0.500 gm
phosphomolybdic acid	0.500 gm
glacial acetic acid	1,000 ml

Agitate and mix. Do not use until all the compounds have completely dissolved.

Procedure

1. Place the smear fixed with 95% alcohol in Shorr's staining solution for 1 minute;
2. Wash in 95% alcohol, then in absolute alcohol;
3. Clear in two to three changes of xylene and mount.

The nuclei stain red. The cytoplasm of the squamous superficial cells stains red-orange. The cytoplasm of the intermediate and parabasal cells stains green-blue.

METHENAMINE SILVER NITRATE (MODIFIED GOMORI'S STAIN)
1. Used to stain black fungi in the smears;
2. Place air-dried or 95% alcohol-fixed smears in tap water for 15 minutes;
3. Place in silver methenamine solution made of

silver nitrate, 5% aqueous solution	1 ml
methenamine, 3% aqueous solution, U.S.P. grade	25 ml
borax, 5% solution	2 ml
distilled water	25 ml

and let slide remain in solution for 1 hour in hot oven (56° C to 58° C);
4. Rinse slides thoroughly in distilled water;
5. Place in 0.1% gold chloride solution (10 ml of gold chloride, 1% aqueous solution in 90 ml of distilled water) until the background of the smear becomes clear (about 5 minutes);
6. Rinse slides in distilled water;
7. For 2 minutes, place in a solution of sodium thiosulfate (2 gm) plus distilled water (100 ml) to remove the unreduced silver;
8. Wash in tap water;
9. Dehydrate in 95% ethyl alcohol;
10. Counterstain in EA–36 for 30 seconds;
11. Rinse and clear in alcohol and xylene and mount.

GIEMSA STAIN

Giemsa stain is most useful in the demonstration of cytoplasmic inclusions in ocular and vaginal smears in cases of inclusion conjunctivitis and vaginitis. It can also be used to stain imprint smears. This method often produces: (1) an apparent increase in size of the cells, (2) an apparent increase of the chromatin, and (3) monochromatic stain of the cytoplasm.

The working Giemsa stain is prepared by mixing 1 part Giemsa stock stain with 9 parts of buffer (pH 6.4). This buffer is prepared by mixing:

monobasic potassium phosphate	6.63 gm
anhydrous diabasic sodium phosphate	2.56 gm
distilled water	1 liter

Procedure
1. Giemsa stain for 15 minutes;
2. Rinse in water;
3. Air dry;
4. Rinse in xylene and mount.

MODIFIED WRIGHT-GIEMSA STAIN

This is used mainly to stain air-dried bone marrow and needle aspiration smears.

Procedure
1. Wright stain for 3 minutes;
2. Dilute stain by adding 10 drops of buffer (pH 6.4). The stain should acquire a green sheen; otherwise, add more buffer. Leave stain for an additional 3 minutes. Move or blow gently on the slides to mix solutions;
3. Rinse dye solution in tap water and allow to dry;
4. Cover smear with diluted working Giemsa stain and let stand for 5 minutes;
5. Rinse in tap water and air dry;
6. Dip in xylol, then mount coverglass with any neutral resin.

PERIODIC ACID–SCHIFF (PAS) STAIN

This stain is for the demonstration of fungi, intracytoplasmic glycogen, or mucus. It is used mainly in the recognition of mucous-secreting cells and for typing of adenocarcinomas. In effusions this stain helps to differentiate single reactive mesothelial cells from single adenocarcinoma cells (Chapter 12).

Procedure
1. Rinse the air-dried or alcohol-fixed smear in tap water;
2. Immerse slide for 5 minutes in 0.5% solution of periodic acid made by dissolving 0.5 gm periodic acid crystals in 100 ml of water;
3. Rinse in water;
4. Immerse in Schiff reagent for 15 minutes;
5. Rinse in running tap water for 10 minutes;

6. Stain with Harris' hematoxylin for 5 minutes;
7. Rinse excess stain in tap water;
8. Dip in ammonia water until smears turn blue;
9. Rinse in water;
10. Dehydrate in alcohol and xylene and mount.

RAPID METACHROMATIC STAIN (TOLUIDINE BLUE OR THIONINE)
These basophilic dyes rapidly give satisfactory nuclear details sufficient for the diagnosis of smeared sediment in the majority of effusions and other fluids.

1. With an eyedropper cover the smears with 0.5% alcohol solution of toluidine blue or thionine for 1 minute;
2. The slide is drained, then rinsed gently with water to remove excess stain;
3. The smear is coverslipped with glycerol.

Membrane Filter Technique

1. The 5 μ rectangular, 17 × 42 mm SM Millipore filter,* which acts as a sieve, is used in our laboratory. Other filter membranes may give equally good results;
2. Wet filter with approximately 5 ml of normal saline solution prior to adding specimen to Millipore container;
3. Specimen is poured into Millipore cup, the amount depending on consistency of fluid (example: if the specimen is clear, with little or no blood, use 30 ml of the specimen on each filter);
4. Apply vacuum. If using a portable pump, do not allow the vacuum gauge to exceed 10 mm Hg;
5. When contents of Millipore cup have reached the level of 2 to 5 ml, the cup is rinsed thoroughly with normal saline solution;
6. When the normal saline solution and the specimen have reached the above proportions, 10 ml of 95% ethyl alcohol is added to the cup to fix the material;
7. After the final filter has been placed and before turning on the vacuum in the apparatus, rinse the container in which the fluid was sent to the laboratory in normal saline solution, and pour this into the Millipore cup;
8. The vacuum is turned off when the contents of the cup have reached the approximate amount of 1 ml;
9. Remove the filter with a forceps without allowing it to dry, and place the filter in a dish of 95% ethyl ROH until time for staining. Do not allow the filter to stand for more than 3 hours or it will become brittle;
10. Stain the filter clipped to a clean glass slide as any other nongenital smear.
11. Leave clipped filter for a few minutes in 50% mounting medium and 50% xylol solution;

* Millipore Filter Co., Bedford, Massachusetts.

12. Trim the edges so that no part of the filter will extrude beyond the coverslip, and mount;
13. Seal the edges of the coverslip with red nail polish;
14. Dry the slide overnight in 60° C oven.

References and Supplementary Reading

Anderson, W. A. D., and Gunn, S. A.   The efficacy of a single cytology smear in the detection of cancer of the cervix. *Cancer* 13:92–94, 1963.

Bonime, R. G.   Air-dried smear for cytologic studies. *Obstet. Gynecol.* 73: 783, 1966.

Davis, H. J.   The irrigation smear—Accuracy in detection of cervical cancer. *Acta Cytol.* 6:459–467, 1962.

Hammond, D. O., Seckinger, D. L., and LeDrew, L.   Endometrial cellular study: A new method using membrane filtration. *Acta Cytol.* 11:181, 1967.

Mavec, P.   The quick smear method and the cytologic diagnosis during surgery. *Dag. Mediz. Labor.* (Stuttgart) 20:21, 1967.

Miller, E. M., and Haam, E. von.   Symposium—Advantages and disadvantages of various techniques of obtaining material for routine cytological examinations. *Acta Cytol.* 4:236–237, 1960.

Nieburgs, H. E.   Cytologic techniques for office and clinic. *Postgrad. Med.* 23:309, 1963.

Preece, A.   *A Manual for Histologic Technicians* (2d ed.). Boston: Little, Brown, 1965.

Reagan, J. W., and Lin, F.   An evaluation of the vaginal irrigation technique in the detection of uterine cancer. *Acta Cytol.* 11:374, 1967.

Regan, J. W. (Ed.)   *Manual of Cytotechnology* (2d ed.). National Committee for Careers in Medical Technology, Wichita Falls, Texas, 1963.

Reynaud, A. J., and King, E. B.   A new filter for diagnostic cytology. *Acta Cytol.* 11:289, 1967.

Richart, R. M., and Vaillant, H. W.   Influence of cell collection techniques upon cytological diagnosis. *Cancer* 18:1474, 1965.

Saccomano, C., et al.   Concentration of carcinoma or atypical cells in sputum. *Acta Cytol.* 14:380, 1970.

Shore, E.   Evaluation of clinical applications of vaginal smear method. *J. Mount Sinai Hosp.* (N.Y.) 12:667, 1945.

Wied, G. L.   Importance of the site from which vaginal cytologic smears are taken. *Am. J. Clin. Pathol.* 25:742–750, 1955.

Wilbanks, G. D., et al.   An evaluation of a one slide cervical cytology method for the detection of cervical intraepithelial neoplasia. *Acta Cytol.* 12:257, 1968.

Index

in atrophy, 112
count of, before radiotherapy, 206
deep, 23, 25, 29
dysplasia of, 123–128, 148–149
hyperplasia of, 116
repair cells, 83, 113–115
reserve cell hyperplasia, 119
Basement membrane, in respiratory
tract, 231
Basophils
in respiratory tract, 236, 247
in vaginal smear, 45
Berylliosis, 269
Bicalcium phosphate crystals, urinary,
372
Bilharziasis. *See* Schistosomiasis
Biliary ducts, 354
Biopsy
aspiration, 484–486
surface, 1
Birdseye cells, 133
in cervical carcinoma, 164
Bladder, 365
life span of cells in, 14
squamous cell carcinoma of, and
schistosomiasis, 381, 391
Blastomycosis, 259
skin lesions in, 461
Blood, tumor cells in, 479–480
Bone marrow cytology, 480–481
Bowen's disease
of cervix, 143
of eye, 436, 438
of vulva, 472
Bowman's capsule, 365
Breast, 407–422
acute mastitis, 412–413
adenocarcinoma of, 307
intraductal, 419–420
anatomy and histology of, 407
aspiration of solid mass, 408
carcinoma of, contact smear of, 484
chronic cystic mastitis, 415
collection of specimens from, 407–
409
colloid carcinoma of, 422
comedocarcinoma of, 420
comedo mastitis, 414
cyst fluid aspiration, 408
diagnostic pitfalls in lesions of, 422
epidermoid carcinoma of, 422
fat necrosis of, 414–415
fibroadenoma of, 415
galactocele, 413
histiocytes in, 412
hyperplasia of, intraductal, 415–419
imprint smear of biopsy tissue, 409
inflammatory cells in, 412
lactation persistence, 413

malignant tumors of, 419–422
medullary carcinoma of, 422
metaplasia of, apocrine, 414
nipple secretion, 407–408
nonsecretory duct cells in, 409–410
normal cytology of, 409–412
Paget's disease of nipple, 420
papilloma of, 415–419
plasma cell mastitis, 414
sclerosing adenosis of, 415
secretory duct cells in, 410–412
squamous cells in, 412
surface scraping of, 409
wound washing specimen from, 409
Bronchi, 231
adenoma of, 285
differential features of, 271
aspiration for sputum collection,
233–235
Bronchial cells
benign, clusters of, 283
ciliated, 243–244
degenerated columnar, differential
features of, 283
nonciliated, Papanicolaou, 237, 252
Bronchiectasis, 252
goblet cells in, 244
pulmonary basal cell hyperplasia in,
271
pulmonary squamous metaplasia in,
271
Bronchioles, 231
carcinoma of, 287
terminal, 231
Bronchitis, 252
Curschmann's spirals in, 248
goblet cells in, 244
inflammatory cells in, 247
Brushings, esophageal, 336
Buccal irradiation test, 206
Buccal smears. *See* Oral cavity
Bullae
in pemphigoid lesions, 463
in pemphigus vulgaris, 463
Byssinosis, 265

Calcified debris, in sputum, 248
Calcium bilirubinate crystals, in duodenal drainage, 355
Calcium carbonate crystals, urinary, 374
Calcium oxalate crystals, urinary, 372–
373
Calcium sulfate crystals, urinary, 372
Candida infection
of eye, 431
of oral cavity, 327
pulmonary, 253
skin lesions in, 461
vaginal smear in, 95

505

in coccidioidomycosis, 255
luteal, vaginal smears in, 65
nabothian, 19
ovarian, aspiration of, 191–195. *See also* Ovarian lesions
Cystic duct, 354
Cystine crystals, in urine, 375
Cystitis, 373
  glandular, 386, 389
  radiation, 385–386
Cytogenetics, 215–229
  banding of chromosomes in, 219
  chromosomal nondisjunction, 226
  Down's syndrome, 229
  Edwards syndrome, 229
  hermaphroditism, male, 228
  and identification of chromosome groups, 219
  and karyotype determinations, 216–219
  and karyotyping applications, 215–216
  Klinefelter's syndrome, 228–229
  mosaicism, 226
  normal female, 227
  normal male, 227
  pseudohermaphroditism, female, 227–228
  sex chromosome determination, 220–226
    and abnormalities of Barr body, 222–224
    buccal scraping for, 220–221
    in epithelial cells, 221–222
    factors affecting, 224–226
    in leukocytes, 224
    staining for, 221
  superfemale, 228
  Turner's syndrome, 228
  Y chromosome demonstration, 226
Cytolysis, in vaginal smear, 28, 58, 59, 60, 106
Cytomegalovirus infection
  cerebrospinal fluid in, 447
  respiratory, 239
  of urinary tract, 381–383
  vaginal smear in, 102
Cytopathology, related to histopathology, 1–3
Cytoplasm, 7
  cloudy swelling of, 104
  degenerative changes in, 104–106
  irregular shape of, 105
  keratin in, 106
  in malignant cells, 136–139
    in cervical carcinoma, 150, 151, 158–161
    in endometrial adenocarcinoma, 170–172
  matrix of, 7

organelles in, 7–9
perinuclear halo in, 104–105
polychromasia of, 106
radiation affecting, 201–202
ratio to nucleus
  radiation affecting, 201, 204
  in squamous cells of oral cavity, 322
  in urinary tract carcinoma, 389
vacuolization of, 104

Darier disease, 329–330
Debris, cellular
  in colonic specimen, in ulcerative colitis, 359
  in sputum
    in bronchiectasis, 252
    calcified, 248
Decoy cells, in urine, 374
Degenerate cells
  bronchial columnar, differential features of, 283
  colonic, in ulcerative colitis, 359
  glial cells, 447–449
Degenerative states, cytoplasmic vacuoles in, 211
Dentures
  epithelial changes from, 327
  papillomatosis from, 325
Dermatophytosis, 459
Dermoid cysts of ovary, 193–195
Diabetic coma casts, in urine, 380
Digitalis, affecting vaginal smear, 63
Diploid state, 12
Division of cells, 10–12
DNA
  in mitochondria, 8
  in nuclear chromatin, 9
Döderlein's bacillus, in vaginal smear, 28, 59, 60, 63, 86
  postmenopausal, 86
  in pregnancy, 66, 67
  in squamous basal cell dysplasia, cervical, 128
  and vaginal acidity, 86
Donovan bodies, 459
Down's syndrome, 229
Drugs. *See also* Chemotherapy
  affecting sex chromosome mass determination, 226
Drumsticks, leukocyte, 224
Duodenum, 354–357
  anatomy and histology of, 354
  collection of specimens from, 354
  life span of cells in, 14
  malignant tumors of, 355–357
  normal cytology of, 354–355
Dust cells, in respiratory tract, 245, 247
  life span of, 14

Gomori stain, modified, 498
Gonadal agenesis, 64, 228
Gonadotropic hormones, affecting vaginal smear, 63
Gonorrheal vaginitis, vaginal smear in, 86–88
Goodpasture's syndrome, 264–265
Gouty arthritis, synovial fluid in, 316
Graafian follicles, 19
Granular casts, in urine, 380
Granular cell myoblastoma
    bronchial, 272–273
    of genital tract, 189
Granuloma
    foreign-body, of skin, 466
    inguinale, vaginal smear in, 88
Granulosa-cell tumor of ovary, 195
    endometrial hyperplasia in, 122
    vaginal smear in, 65
Granulosa cells, in follicle cysts of ovary, 191–192
Gynandroblastoma, 195

Halo
    perifungal, in cryptococcosis of central nervous system, 447
    peri-inclusion
        in adenovirus infection of respiratory tract, 238, 239
        in cytomegalic viral infection of respiratory tract, 239
        in herpesvirus infections of eye, 434
        in urine specimens, 381, 383, 385
        in viral infections of oral cavity, 327
    perinuclear, 104–105
        and nuclear hypertrophy, 132
        in prostatic secretion, in prostatitis, 400
        in pulmonary squamous cell carcinoma, 280
        from radiation, 204
        in skin scrapings
            in erythema multiforme, 466
            in pemphigus, 329, 463
        in vaginal smear, 28
            in adenovirus infections, 101
            in condylomata acuminata, 101
            in herpes simplex infection, 98
Hamartoma, pulmonary, 273
Haploid state, 12
Heart-failure cells, in sputum, 245, 251
Helminthosporium, in sputum, 261
Hematuria, 374
Hemophilus vaginalis, in vaginal smear, 86
Hemorrhage, pulmonary, 251

Hemosiderin in sputum
    in histiocytes, 245, 251
    in idiopathic hemosiderosis, 264–265
    in macrophages, 251
    in siderosis, 265
Hemosiderosis, idiopathic, sputum in, 264–265
Henle loop, 365
Hepatic duct, 354
Hepatoma, contact smear of, 483
Hermaphroditism
    female pseudohermaphroditism, 227–228
    male, 228
Herpesvirus infections
    differential diagnosis of, 33
    of esophagus, 338, 342
    of eye, 434
    of respiratory tract, 237
    of skin, 468
    and stomatitis, 327
    of urinary tract, 383
    vaginal smear in, 98–100
Herxheimer's spiral, in tadpole cells, in cervical carcinoma, 161
Hilar cell tumor, 195
Histiocytes
    in breast secretion, 412
    in cerebrospinal fluid, 444
    in colonic specimens, 359
    differential features of, 157, 183, 283
    in effusions, 298
    in oral cavity mucosa, 323
    in respiratory tract, 232, 245–247
        in allergic diseases, 253
        hemosiderin in, 245, 251
    in skin scrapings
        in foreign-body granuloma, 466
        in toxic epidermal necrosis, 466
    in sputum, differential features of, 283
    in urinary tract, 371
    in vaginal smear, 46–49, 85
        in endometrial adenocarcinoma, 170
        in IUD patients, 102
        morphology of, 47–49
        and phagocytosis of superficial squamous cells, 113
        in pregnancy, 66
        radiation affecting, 204
        types of, 46
Histopathology, related to cytopathology, 1–3
Histoplasmosis, pulmonary, 257
Hodgkin's disease
    aspiration smear of, 486
    effusions in, 310
    lungs in, 289

Reed-Sternberg cells in, 289, 310,
    486
  stomach in, 352
Hormonal evaluation, 54–55
  Shorr's stain for, 496–498
  urethral cells in, 55
  vaginal smears in, 22, 55
    abnormal averages in, 64–66
    in erosions, 57–58
    in inflammation, 57, 85
    normal averages in, 58–61
    pitfalls in, 57–58, 121
Hormones
  affecting sex chromosome mass de-
      termination, 225
  affecting urinary tract, 370, 371
  affecting vaginal smear, 27, 61–63
  endocrine cycle in women, 53–63
  estrogen. See Estrogen
  ovarian, 53–54
  therapy with, and diagnosis of endo-
      metrial adenocarcinoma, 173–
      174
  vaginal smear in endocrine disorders,
      65–66
Hyaline casts, urinary, 378
Hyaloplasm, 7
Hydatidiform mole, vaginal smear in,
    75–76
Hyperchromatism
  in cervical carcinoma, 164
  in cervical dysplasia, squamous basal
      cell, 124
  in endocervical ectropion, 116
  in endometrial adenoacanthoma, 176
  in endometrial hyperplasia, 122
  in gastric specimen, in linitis plastica,
      352
  in genital tract metaplasia, squamous,
      120
  in malignant cells, 134
  in pulmonary adenocarcinoma, 286
  in pulmonary squamous cell carci-
      noma, 280
  in urine specimens, 374
  in vaginocervical erosions, 115
Hyperkeratosis of cells
  in condylomata acuminata, 101
  in leukoplakia of genital tract, 120–
      121
  in oral cavity, 325
  in skin, 469–470
Hypernephroma, 393
  contact smear of, 483
Hyperplasia
  adrenal, virilizing, vaginal smear in,
      64
  of breast, intraductal, 415–419
  endometrial, 122

and diagnosis of adenocarcinoma,
    173
  of oral cavity, papillary, 325
  prostatic, 401–402
  pulmonary, basal cell, 271
  reactive cells in. See Reactive cells
  in vaginal smears, 116, 119
Hypertrophy, nuclear, in malignant
    cells, 132
Hyphae
  of *Aspergillus,* 257
  of *Candida albicans,* 253
  of *Geotrichum,* 259
  of *Helminthosporium,* 261
  of *Scopulariopsis,* 261
Hypoparathyroidism, and sex chromo-
    some mass determination, 226
Hypothyroidism, vaginal smear in, 64
Hysterectomy, affecting vaginal smear,
    45, 115

Impetigo, 458–459
Inclusions, cellular, 7, 9
  in conjunctival cells, 431–432
  in malignant cells, 138
  in mesothelial cells, reactive, 301
  in oral cavity epithelium, 322
  in pulmonary giant cell carcinoma,
      283
  in respiratory tract
    in adenovirus infection, 238–239
    in cytomegalic viral infection, 239
    in syncytial viral infection, 239
  in skin scraping
    in molluscum contagiosum, 468–
        469
    in vaccinia, 468
  in urine, 381, 383
  in vaginal smear, 100
Infancy. See Newborn
Infarction, pulmonary, 251
Inflammations
  of central nervous system, 446–449
  of eye, 427–436
  of genital tract, female, 81–108
    hormonal evaluation in, 57
Inflammatory cells
  in breast secretion, 412
  casts in urine, 380
  colonic
    in amebic colitis, 359
    in carcinoma, 362
    in ulcerative colitis, 359
  in eye, 426
  gastric, 345
  with malignant cells, 140–141
  in oral cavity, 322
    in leukoplakia, 325
    in stomatitis, 329

media
  acute, 479
  catarrhal, 477–479
  chronic, 479
  serous, 477
Ovarian lesions
  adenocarcinoma, 307
    poorly differentiated, 195
    pseudomucinous, 195
    serous, 195
  aspiration of cysts, 191–195
  contact smear in, 191
  dermoid cysts, 193–195
  endometrial hyperplasia in, 122
  in endometriosis, 192
  feminizing tumor, 195
    vaginal smear in, 65
  follicle cysts, 191–192
  lutein cysts, 192
    vaginal smear in, 65
  mucinous cysts, 193
  and myxoma peritonei, 193
  psammoma bodies in, 197
  serous cysts, 192–193
    papillary, 193
  vaginal smear in, 28, 64, 65, 189,
    195–197
  virilizing tumors, 195
Ovaries, 19
  hormones of, 53–54
Ovulation, 19, 53

Paget's disease
  of nipple, 420
  of vulva, 166
Pancreatic duct, 354
Panophthalmitis, 431
Papanicolaou cells
  in bacterial diseases of respiratory
    tract, 252
  in parainfluenza viral infection, 237
Papanicolaou stain, modified, 494–496
Papillary adenocarcinoma, endometrial,
    174–176
Papillary hyperplasia, of oral cavity,
    325
Papillary serous cysts, ovarian, 193
Papillary structures in cerebrospinal
    fluid, in ependymomas, 451
Papilloma
  of breast, 415–419
  of choroid plexus, 451
  of eye, 436
  of urinary tract, 386–387
Parabasal cells
  endocervical
    reactive, 165, 173, 183
    reserve, 30
  in oral cavity, 322

in vaginal smears, 23, 25–27, 29
  in atrophy, 111
  in pregnancy, 67
  in squamous metaplasia, 120
Parainfluenza virus infection, 237
Parasites
  in genital tract, 90–95
  intestinal amebiasis, 359
  in oral cavity, 327
  pulmonary, 263
  in urine, 380–381
Parathyroid dysfunction, and sex chro-
    mosome mass determination,
    226
Parietal cells, gastric, 345
Parotid gland, 333
Pearl-like structures
  in effusions, in squamous cell carci-
    noma, 306
  in oral cavity, in squamous cell carci-
    noma, 331
  in respiratory tract
    in pulmonary squamous cell carci-
      noma, 280
    in pulmonary squamous metapla-
      sia, 271
  in skin scrapings, in erythema multi-
    forme, 466
  in vaginal smears, 24
    in cervical carcinoma, invasive, 163
    in endometrial adenoacanthoma,
      176
    in leukoplakia, 121
    in squamous metaplasia, 120
    in squamous superficial cell dyspla-
      sia, cervical, 122
Peg cells, in vaginal smears, 19, 40
Pemphigoid lesions, bullous, 463
Pemphigus, 327–329
  foliaceus, 464
  seborrheic, 464
  vulgaris, 463
Penicillium, in sputum, 257, 261
Peptic ulcer, gastric, 347, 352
Periodic acid-Schiff (PAS) stain, 499–
    500
Peritoneum
  lavage in ovarian lesions, 191
  myxoma peritonei, 193
pH, vaginal
  Döderlein's bacilli affecting, 86
  in inflammation, 83
Phagocytes
  in colonic specimen, in amebic colitis,
    359
  in effusions, differential features of,
    304
  hemosiderin in, in idiopathic hemo-
    siderosis, 265

521

in gastric specimen, in adenocarcinoma, 349
in prostatic secretion, 400
Silicosis, 265–266
Silver nitrate and methenamine stain, 498
Simmonds' disease, vaginal smear in, 65
Sister image of nuclei
    differential features of, 283
    in foreign-body giant cells, 204
Skin, 467–475
    bacterial infections of, 458–459
    benign lesions of, 469–472
    in blastomycosis, 461
    in candidiasis, 461
    carcinoma of
        basal-cell, 473
        in situ, squamous, 472–473
        squamous cell, 473–475
    collection of specimens, 457–458
    in contact dermatitis, 468
    in erythema multiforme, 466
    in erythrasma, 458
    foreign-body granuloma of, 466
    fungal infection of, 459–462
    herpes infections of, 468
    hyperkeratosis of, 469–470
    in impetigo, 458–459
    in keratoacanthoma, 470
    in lichen planus, 462–463
    in lupus erythematosus, 463
    in lymphogranuloma inguinale, 459
    malignant tumors of, 472–475
    in measles, 469
    melanoma of, malignant, 475
    metastasis to, 475
    in molluscum contagiosum, 468–469
    mycetoma of, 461
    necrosis of, toxic epidermal, 466
    nevus of, 470
    normal cytology of, 457–458
    pemphigoid lesions of, bullous, 463
    in pemphigus vulgaris, 463
    in pityriasis rosea, 463
    in prurigo nodularis, 466
    in pyoderma gangrenosum, 464–466
    in rhinosporidiosis, 461
    sebaceous adenoma of, 470
    in seborrheic keratosis, 470
    in seborrheic pemphigus, 464
    tinea infections of, 459–461
    in vaccinia, 468
    viral infections of, 468–469
    in Weber-Christian disease, 468
Slide examination, 4–5
Smears
    aspirated. See Aspiration
    buccal, for sex chromosome determination, 220–221

collection of specimens for. See Collection of specimens
    contact, 481–484
    of effusions, 296
    from gastric washings, 343
    gynecologic, 21–23
    nasopharyngeal, in allergies, 237
    of parotid secretion, 333
    sputum, 232–233
    of urine specimens, 368
    vaginal pool, in ovarian carcinoma, 195–197
Smoking
    and Curschmann's spirals in sputum, 248
    pulmonary squamous metaplasia from, 271
    stomatitis from, 329
Smooth-muscle cells, in respiratory tract, 247
Specimen collections. See Collection of specimens
Spermatozoa
    in prostatic secretion, 399, 404
    in urine, 372
    in vaginal smear, 49
Spider cells, in vaginal smear, 115
Spindle cells
    in genital tract lesions
        in cervical carcinoma, 162
        in leiomyosarcoma, 187
        in melanosarcoma, 187
    in oral cavity
        in reticulum cell sarcoma, 333
        in squamous cell carcinoma, 331
    in pulmonary squamous cell carcinoma, 278, 280
Spinal fluid. See Cerebrospinal fluid
Sporangia, in coccidioidomycosis, 255
Sporotrichosis, eyes in, 431
Sputum
    acellular structures in, 247–251
    amylaceous bodies in, 248, 263
        in alveolar proteinosis, 264
        in pulmonary edema, 251
    calcified debris in, 248
    Charcot-Leyden crystals in, 248
        in asthma, 252, 263
        in Löffler's pneumonia, 252
    collection of specimens, 232–235
        aerosol inhalation for, 233
        bronchial aspiration in, 233–235
        bronchial brushing in, 235
        direct, 233–235
        needle aspiration in, 235
        tracheal aspiration in, 233
    colonies of microorganisms in, 252
    contaminants in, 248, 252, 261–263

postpartum, 69, 83
in pregnancy, 30, 33, 35, 66–76, 83
  and cervical cancer detection, 76
  and glycogen in cells, 60, 67
  intermediate cells in, 66, 67
  navicular cells in, 60, 67
  parabasal cells in, 67
  and prognosis of time of delivery,
    71–73
  trophoblastic cells in, 69
  in tumors of pregnancy, 75–76
progesterone affecting, 27, 61
in prolapse of uterus, 30, 121
in protozoal infections, 90–95
radiation affecting, 25, 46, 63, 115,
  201–213
repair cells in, 83, 113–115
reserve cells in, endocervical, 30
in sarcoma
  botryoides, 188–189
  endometrial, stromal, 183–184
  mixed mesodermal, 184–187
in schistosomiasis, 94–95
in Simmonds' disease, 65
sperm in, 49
squamous epithelium in, stratified,
  23–30. *See also* Squamous
  cells, in vaginal smear
staining defects in, 58
in Stein-Leventhal syndrome, 65
stripped nuclei in, 28, 35–36, 67,
  108
  in cervical squamous basal cell
    dysplasia, 128
  postmenopausal, 112
stromal cells in, 37
superficial cells in, 24, 28–30. *See
  also* Superficial cells, in vaginal smears
talcum granules in, 51
technique for, 21
and terminal bar in cells, 31
in testicular feminization syndrome,
  64
in thyroid dysfunction, 64
toxoplasma in, 91
trauma affecting, 83
in *Trichomonas* infection, 28, 63,
  90–91
in tuberculosis of genital tract, 88
in Turner's syndrome, 64
in uterine anomalies, 64
vermiform bodies in, 42
in viral infections, 97–102
in virilizing tumors, 65, 195
vitamin deficiency affecting, 63
yeast in, 51

Vaginitis
  atrophic, 83–84
  gonorrheal, 86–88
  inclusion, 27, 100
Vater ampulla, 354
  carcinoma of, 357
Vegetable cells, in sputum, 248
Vermiform bodies, in vaginal smear, 42
Viral cells, cytoplasmic vacuoles in, 211
Virilizing adrenal hyperplasia, vaginal
  smear in, 64
Virilizing tumors, vaginal smear in, 65,
  195
Virus infections
  of central nervous system, 447
  ear discharge in, 479
  of eye, 431–434
  of oral cavity, 327
  pulmonary, 252
  of respiratory tract, 237–241
  of skin, 468–469
  vaginal smears in, 97–102
Viscosity test, of synovial fluid, 312
Vitamin deficiency
  eyes in, 436, 438
  genital inflammation in, 85
  vaginal smear in, 63
Vitreous chamber aspirate, 425, 426
Vulva
  Bartholin's gland adenocarcinoma,
    181
  Bowen's disease of, 472
  Paget's disease of, 166
  squamous cell carcinoma of, 165–166
Vulvovaginitis, nonspecific, 86–90

Washings
  bronchial, 233–235
  colonic, 357
  endometrial, 22–23
  esophageal, 335
  gastric, 343
  of mastectomy wound, 409
  peritoneal, in ovarian lesions, 191
Waxy casts, urinary, 378
Weber-Christian disease, 468
Whipple's disease, 349
Wilms' tumor, 394
Witch's milk, 407
Wright-Giemsa stain, modified, 499

Xanthelasma, of eyelid, 434–436
Xanthogranuloma of eye, juvenile, 434

Yeasts
  in urine specimens, 371, 374
  in vaginal smear, 51